AF522209

ENCYCLOPEDIA OF BIOCHEMISTRY

ENCYCLOPEDIA OF BIOCHEMISTRY

Vol. 2

By
Sananda Chatterjee
DAE, DCA, CMCNet
Scientific Consultant
Institute for Natural Sciences
Kolkata
(West Bengal)

DISCOVERY PUBLISHING HOUSE PVT. LTD.
NEW DELHI-110 002

Published by:
Namit Wasan
DISCOVERY PUBLISHING HOUSE PVT. LTD.
4383/4B, Ansari Road, Darya Ganj
New Delhi-110 002 (India)
Phone : +91-11-23279245; 23253475; 43596065
E-mail : discoverybooksindia@gmail.com
discoverypublishinghouse@gmail.com
namitwasan9@gmail.com
web : www.discoverypublishinggroup.com

Reprinted: **2020**
First Edition: **2012**

ISBN: 978-81-8356-876-0 (Set)

Encyclopedia of Biochemistry

Printed at:
Infinity Imaging Systems
Delhi

Preface

It is always an ultimate outlook to give students the best information in a good package. So I took up the job of writing a series of books in chemistry starting from a very introductory level to this advance level of biochemistry.

Biochemistry, to express something about it, in one word, would be difficult for me. It should be ones part of life; the reactions of protein, amino-acids, the DNA RNA—double helix pattern. The nynhydrin reactions are all the wonders of biochemistry—man's anato-biochemical linkages.

Students who want to become doctors or medical practitioners must be a good biochemist first because biochemistry is the theory behind and the medicine is the practical application.

The effects and side effects of the particular medicine, whether it is allergic or non-allergic, to the patients are well understood by the doctors but are taught in biochemistry.

The chemical functions of the different human organs and systems; the kidney, the respiratory system, the blood, the chemical functions of hormones, vitamins, the liver etc are all well covered by this book.

The subject of this book is therefore deeply intended for those upcoming doctors who keep themselves for the real service of the mankind.

The object of this book is not only meant for the doctors but it is the last bangle of that long garland which has started all along from the introduction chapter of The Architecture of Chemistry, The Mans Scientific Attitude.

Biochemistry also comes in use in the field of agriculture, the biochemical configuration of the soil. The biochemical configuration of the plants to be cultivated, the application of the pesticides, and the relation of the pesticides with the human population are well studied under biochemistry.

The Agricultural Scientist, the Pathological Scientist, Forensic Scientist, is definitely inter-joined with each other by a single buffer of the Biochemistry.

This book is dedicated to those promising scientific readers and interested students of biochemistry and advanced chemical science, who wants to make a fruitful attempt in the field of modern scientific world.

The author will be grateful and surely thankful to see the best utilization of this book in any of the field of application.

Sananda Chatterjee

Contents

Secretion—Role of Bile Acids in Fat Digestion and Absorption—Role of Bile Acids in Cholesterol Homeostasis—Enterohepatic Recirculation—Pattern and Control of Bile Secretion—Proteases—Pancreatic Lipase—Amylase—Other Pancreatic Enzymes—Bicarbonate and Water—Transport of lipids—Classification of Lipoproteins—Chemical Structure of Lipoprotein—Metabolisms of Chylomicrons—VLDL, LDL, HDL—Disorders of Lipoprotein Metabolism—Oxidation of Fatty Acids—Beta Oxidation—b-oxidation of unsaturated fatty acids—b-oxidation of odd-numbered chains—Role of Carnitine—Fatty Acid Oxidation Disorders —Ketosis and Ketogenesis.

Introductions

Enzyme comes from the Greek words "Eos " and "Prozyme" meaning in and through life. To define *Enzyme mainly works as a chemical disintegrator and also catalyses many biochemical reactions in the human body*. Enzymes are found in leaving cells, they are comprised of long polypeptide chains with molecular weight from 10,000 to a minion or more. Another important characteristics is its specificity i.e., a given enzyme can catalyze only one particular reaction and not other. Some examples are:

1. Redox (oxyreductase),
2. Transference of special radicals or groups (transferases),
3. Hydrolysis (proteolytic),
4. Removal from or addition to the substrate of specific chemical groups (lyases),
5. Iosmerizations (isomerases),
6. Combination or binding together substrate units (ligases).

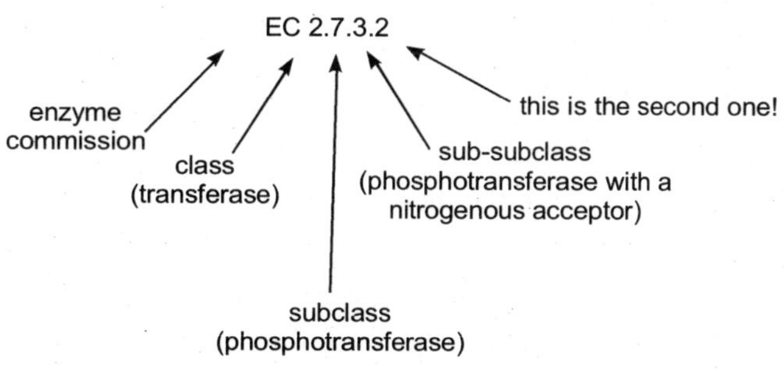

The name of enzyme always end as *-in* or *-ase*

The following list will prove the above statement and also will indicate the definite role played in metabolism.

Enzymes are essential to many biochemical processes especially in the food and beverage, and also in the pharmaceutical industries.

Amylase	Starch hydrolysis
Carboxylase	Decomposes pyruvic acid
Cellulose	Converts cellulose to glucose
Cholinesterase, pepsin, rennin	Inactivates acetyl choline
Chymotrypsin	Hydrolyses Protein
Invertase	Converts sucrose to glucose and fructose
Lipase	Hydrolyses Fat
Maltase	Converts Maltose to glucose
Protease	Hydrolysis of peptide linkage
Ribonuclease	Decomposes RNA
Trypsin	Splits proteins into amino acids
Urease	Decomposes urea to NH_4 and CO_2
zymase	Converts sugar to alcohol and CO_2 (fermentation)

CLASSIFICATIN OF ENZYMES

There are 3000 enzymes known to us but only a few we use in the human metabolism and so we are going to discuss about them in this chapter.

Esterases (EC 3.1.1)

Pancreatic lipase:

This enzyme is present in the pancreas and hydrolyses fat into fatty acids and glycerol, Lipase is found in the gastric juice.

(a) ***Liver Esterase:*** This enzyme is obviously present in the liver, hydrolyses esters into simple alcohols and is an ester – rather than a fat splitting enzyme.

(b) ***Ricinus Lipase:*** This enzyme is found in the seeds of the caster beans and is similar to pancreatic lipase.

(c) ***Chlorophyllase:*** This is an enzyme present in green plants and hydrolyses chlorophyll into phytol and ethyl chlorophyllide.

(d) ***Phosphatases:*** These enzymes are present in various tissues and hydrolyses various esters of phosphoric acid. The phospytase of which they seem to be quite a number are divided into two groups the alkaline phospytase, with an optimum pH of form 8 to 10 found intestinal mucosa, kidneys, liver etc and acid phosphates, with an optimum pH of 4 to 5, present in blood serum etc.

(e) ***Azolesterases:*** These enzymes are named as such because they hydrolyze the nitrogen-alcohol esters (azote-nitrogen). The best known example is cholinesterase, the enzyme which hydrolyses acetylcholine.

Proteinases and Peptidases EC 3.4

(a) ***Pepsin, EC = 3.4.23.1***: This enzyme is present in the stomach and hysdrolyses proteins to protons. Pepsin can also act on synthetic substrates much similar in composition than proteins. The enzymes have been crystallized.

(b) ***Trypsin***: This enzyme is present in pancreas and hydrolyses proteins into peptides and amino acids. Life pepsin, Trypsin can act on substances much simpler chemically.

(c) ***Erepsin***: This substance undoubtedly represents a mixture of enzymes of the peptidase variety.

(d) ***Renisn***: This enzyme is also present in stomach reacts with the casein of the milk and converts it into paracasein, which reacts with calcium to form a milk clot.

(e) ***Papain***: This enzyme is present in the juice of the melon tree and in the plant cells in general. It is similar to the Trypsin into action and has also been isolated in crystallized form.

(f) ***Cathepsin***: This enzyme is found in the various animal cells. There are many types of cathepsin, and they are believed to play important roles in synthesis and hydrolyse within the cell. They all activated by HCN, H_2S and -SH compounds. These activators are for the most part reducing substances. These activators are for the most part reducing substances. The papainases are inactivated by mild oxidation and reactivated by reduction.

(g) ***Ficin:*** This enzyme is found in the milk sap of the fig tree. It digests proteins at about pH 5. It is not activated by -SH compounds.

(h) ***Aminopeptidase:*** This is an enzyme present in the intestine and on polypeptides containing a free amino group.

The essential difference between h and I is the place of rupture in the peptide linkage

(i) ***Carbooxydase:*** This is an enzyme found in the pancreas and acts on carboyl group. It has been crystallized.

(j) ***Dipeptidase:*** This enzyme, present in intestinal juice, this hydrolyses the dipeptides. Since both amino peptides and carboxyl peptides.

Amidiases EC 3.5.1

This acts upon the carbon – nitrogen linkage

(a) ***Urease:*** This is an enzyme which is present in leguminous plants (soy and jack beans) and converts urea into ammonia. It has been isolated in crystalline form.

(*b*) ***Arginase:*** This is an enzyme converts arginine into ornithine and urea.

(*c*) ***Purinase:*** Amidiases, These are the representative of the group of enzymes present in liver. Deaminise purines. Adenase is an example

(*d*) ***Phosphorylases:*** These are enzymes which can decompose polysaccharides as well as bring about the synthesis of the polysaccharides. For example a phosphorylase obtained from muscle converts hexose – 1 phosphate into a polysaccharide. The reaction is a reversible one. An X – Ray diffraction pattern shows it to be similar in structure to starch obtained from plates. In contrast to this when heart or liver phosphorylase is used in place of the one obtained from muscle, the polysaccharide obtained resembles glycogen.

Carbohydrates EC 3.2.1

(*a*) ***Sucrase:*** This enzyme is present in animal and plant tissue and hydrolyses sucrose into glucose and fructose. Maltase and Lactase found with sucrose in the small intestine hydrolyse maltose and lactose respectively.

(*b*) ***Amylase, Alpha amylase or diastase:*** These enzymes are found in plant and animal tissues and hydrolyze starch and glycogen into maltose. The ptyalin found in saliva and the amylase present in pancreatic juice are example.

Oxidase

(*a*) ***Dehyrogenes:*** These are enzymes, found in many tissues (muscles for example) which bring about oxidation by the removal of hydrogen from substances for example succinic dehydrogenase converts succinic acid into fumaric acid.

(*b*) ***Catalase:*** These engyme is found in plant and animal tissue (example liver) and molecular oxygen.

(*c*) ***Peroxydase:*** These enzymes are present in may tissues (instance, the spleen and horse raddish, as they transfer peroxide oxygen to oxydisable substances.

APOENZYME, CO-ENZYME, HOLENZYME AND CO FACTORS

Coenzymes are small organic non-protein molcculcs that carry chemical groups between enzymes.[1] Coenzymes are sometimes referred to as *cosubstrates*. These molecules are substrates for enzymes and do not form a permanent part of the enzymes' structures. This distinguishes coenzymes from prosthetic groups, which are non-protein components that are bound tightly to enzymes - such as iron-sulfur centres, flavin or haem groups. Both coenzymes and prosthetic groups are types of the broader group of cofactors, which are any non-protein molecules (usually organic molecules or metal ions) that are required by an enzyme for its activity.

In metabolism, coenzymes are involved in both group-transfer reactions, for example coenzyme A and adenosine triphosphate (ATP), and redox reactions, such as coenzyme Q_9 and nicotinamide adenine dinucleotide (NAD^+). Coenzymes are consumed and recycled continuously in metabolism, with one set of enzymes adding a chemical group to the coenzyme and another set removing it. For example, enzymes such as ATP synthase continuously phosphorylate adenosine diphosphate (ADP), converting it into ATP, while enzymes such as kinases dephosphorylate the ATP and convert it back to ADP.

Coenzymes molecules are often vitamins or are made from vitamins. Many coenzymes contain the nucleotide adenosine as part of their structures, such as ATP, coenzyme A and NAD^+. This common structure may reflect a common evolutionary origin as part of ribozymes in an ancient RNA world.

Coenzymes as Metabolic Intermediates

The redox reactions of nicotinamide adenine dinucleotide.

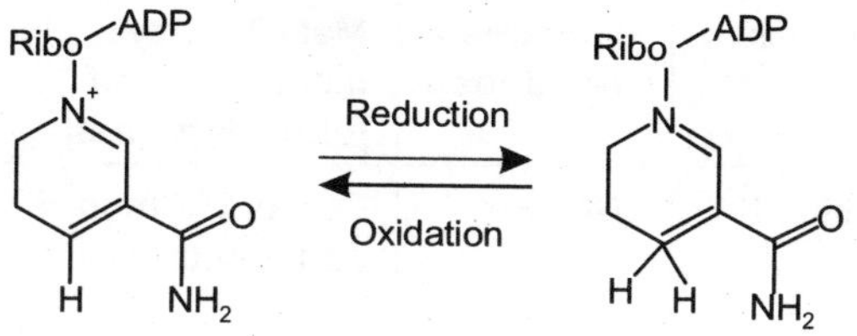

$$NAD^+ + H^+ + 2e^- \longrightarrow NADH$$

Metabolism involves a vast array of chemical reactions, but most fall under a few basic types of reactions that involve the transfer of functional groups. This common chemistry allows cells to use a small set of metabolic intermediates to carry chemical groups between different reactions. These group-transfer intermediates are the coenzymes.

Each class of group-transfer reaction is carried out by a particular coenzyme, which is the substrate for a set of enzymes that produce it, and a set of enzymes that consume it. An example of this are the dehydrogenases that use nicotinamide adenine dinucleotide (NADH) as a cofactor. Here, hundreds of separate types of enzymes remove electrons from their substrates and reduce NAD^+ to NADH. This reduced coenzyme is then a substrate for any of the reductases in the cell that need to reduce their substrates.

Coenzymes are therefore continuously recycled as part of metabolism. As an example, the total quantity of ATP in the human body is about 0.1 mole. This ATP is constantly being broken down into ADP, and then converted back into ATP. Thus, at any given time, the total amount of ATP + ADP remains fairly constant. The energy used by human cells requires the hydrolysis of 100 to 150 moles of ATP daily which is around 50 to 75 kg. Typically, a human will use up their body weight of ATP over the course of the day. This means that each ATP molecule is recycled 1000 to 1500 times daily.

Types

Acting as coenzymes in organisms is the major role of vitamins, although vitamins do have other functions in the body. Coenzymes are also commonly made from nucleotides: such as adenosine triphosphate, the biochemical carrier of phosphate groups, or coenzyme A, the coenzyme that carries acyl groups. Most coenzymes are found in a huge variety of species, and some are universal to all forms of life. An exception to this wide distribution is a group of unique coenzymes that evolved in methanogens, which are restricted to this group of archaea.

Vitamins and Derivatives

Coenzyme	Vitamin	Additional component	Chemical group(s) transferred	Distribution
NAD^+ and $NADP^+$	Niacin (B_3)	ADP	Electrons	Bacteria, archaea and eukaryotes
Coenzyme A	Pantothenic acid (B_5)	ADP	Acetyl group and other acyl groups	Bacteria, archaea and eukaryotes
Tetrahydrofolic acid	Folic acid (B9)	Glutamate residues	Methyl, formyl, methylene and formimino groups	Bacteria, archaea and eukaryotes
Menaquinone	Vitamin K	None	Carbonyl group and electrons	Bacteria, archaea and eukaryotes
Ascorbic acid	Vitamin C	None	Electrons	Bacteria, archaea and eukaryotes
Coenzyme F420	Riboflavin (B_2)	Amino acids	Electrons	Methanogens and some bacteria

Non-vitamins

Coenzyme	Chemical group(s) transferred	Distribution
Adenosine triphosphate	Phosphate group	Bacteria, archaea and eukaryotes
S-Adenosyl methionine	Methyl group	Bacteria, archaea and eukaryotes
3'-Phosphoadenosine-5'-phosphosulfate	Sulfate group	Bacteria, archaea and eukaryotes
Coenzyme Q	Electrons	Bacteria, archaea and eukaryotes
Tetrahydrobiopterin	Oxygen atom and electrons	Bacteria, archaea and eukaryotes
Cytidine triphosphate	Diacylglycerols and lipid head groups	Bacteria, archaea and eukaryotes
Nucleotide sugars	Monosaccharides	Bacteria, archaea and eukaryotes
Glutathione	Electrons	Some bacteria and most eukaryotes
Coenzyme M	Methyl group	Methanogens
Coenzyme B	Electrons	Methanogens
Methanofuran	Formyl group	Methanogens
Tetrahydromethanopterin	Methyl group	Methanogens

Evolution

Coenzymes, such as ATP and NADH, are present in all known forms of life and form a core part of metabolism. Such universal conservation indicates that these molecules evolved very early in the development of living things. At least some of the current set of coenzymes may therefore have been present in the last universal ancestor, which lived about 4 billion years ago. Coenzymes may have been present even earlier in the history of life on Earth. Interestingly, the nucleotide adenosine is present in coenzymes that catalyse many basic metabolic reactions such as methyl, acyl, and phosphoryl group transfer, as well as redox reactions. This ubiquitous chemical scaffold has therefore been proposed to be a remnant of the RNA world, with early ribozymes evolving to bind a restricted set of nucleotides and related compounds. Adenosine-based coenzymes are thought to have acted as interchangeable adaptors that allowed enzymes and ribozymes to bind new coenzymes through small modifications in existing adenosine-binding domains, which had originally evolved to bind a different cofactor.

History

The first coenzyme to be discovered was NAD^+, which was identified by Arthur Harden and William Youndin 1906. They noticed that adding boiled and filtered yeast extract greatly accelerated alcoholic fermentation in unboiled yeast extracts. They called the unidentified factor responsible for this effect a *coferment*. Through a long and difficult purification from yeast extracts, this heat-stable factor was identified as a nucleotide sugar phosphate by Hans von Euler-Chelpin. Other coenzymes were identified throughout the early 20th century, with ATP being isolated in 1929 by Karl Lohmann, and coenzyme A being discovered in 1945 by Fritz Albert Lipmann. The functions of coenzymes were at first mysterious, but in 1936, Otto Heinrich Warburg identified the function of NAD^+ in hydride transfer. This discovery was followed in the early 1940s by the work of Herman Kalckar, who established the link between the oxidation of sugars and the generation of ATP. This confirmed the central role of ATP in energy transfer that had been proposed by Fritz Albert Lipmann in 1941. Later, in 1949, Morris Friedkin and Albert L. Lehninger proved that the coenzyme NAD^+ linked metabolic pathways such as the citric acid cycle and the synthesis of ATP.

KINETICS OF ENZYME REACTION VELOCITY

In most cases, an enzyme converts one chemical (the *substrate*), into another (the *product*). A graph of product concentration vs. time follows three phases as shown in Fig. 3.1.

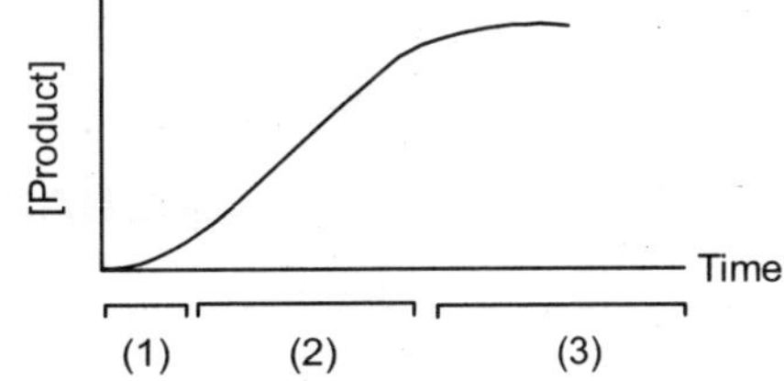

Fig. 3.1 : The Enzyme reaction of enzyme

At very early time points, the rate of product accumulation increases over time. Special techniques are needed to study the early kinetics of enzyme action, since this transient phase usually lasts less than a second (the figure greatly exaggerates the first phase).

For an extended period of time, the product concentration increases linearly with time.

At later times, the substrate is depleted, so the curve starts to level off. Eventually the concentration of product reaches a plateau and doesn't change with time.

It is difficult to fit a curve to a graph of product as a function of time, even if you use a simplified model that ignores the transient phase and assumes that the reaction is irreversible. The model simply cannot be solved to an equation that expresses product concentration as a function of time. To fit these kind of data (called an *enzyme progress curve*) you need to use a program that can fit data to a model defined by differential equations or by an implicit equation. Prism cannot do this.

Rather than fit the enzyme progress curve, most analyses of enzyme kinetics fit the initial velocity of the enzyme reaction as a function of substrate concentration. The velocity of the enzyme reaction is the slope of the linear phase, expressed as amount of product formed per time. If the initial transient phase is very short, you can simply measure product formed at a single time, and define the velocity to be the concentration divided by the time interval. This chapter considers data collected only in the second phase. The terminology describing these phases can be confusing. The second phase is often called the "initial rate", ignoring the short transient phase that precedes it. It is also called "steady state", because the concentration of enzyme-substrate complex doesn't change. However, the concentration of product accumulates, so the system is not truly at steady state until, much later, the concentration of product truly doesn't change over time.

Enzyme Velocity as a Function of Substrate Concentration

If you measure enzyme velocity at many different concentrations of substrate, Fig. 3.2 generally looks like this:

Enzyme velocity as a function of substrate concentration often follows the Michaelis-Menten equation:

$$\text{Velocity} = V = \frac{V_{max}[S]}{[S] + K_M}$$

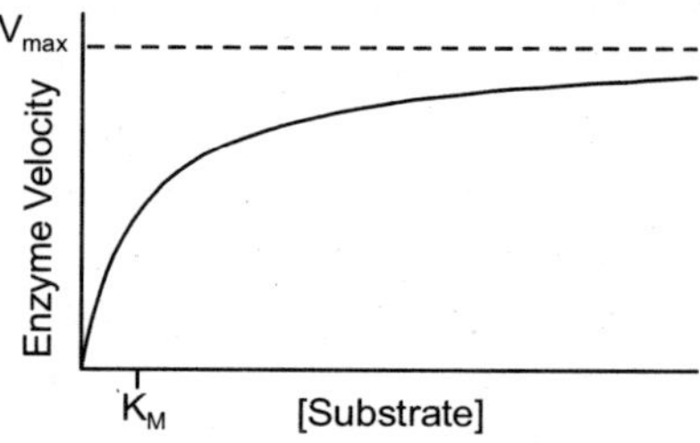

Fig. 3.2 : The Enzyme Velocity

V_{max} is the limiting velocity as substrate concentrations get very large. V_{max} (and V) are expressed in units of product formed per time. If you know the molar concentration of enzyme, you can divide the observed velocity by the concentration of enzyme sites in the assay, and express V_{max} as units of moles of product formed per second per mole of enzyme sites. This is the *turnover number*, the number of molecules of substrate converted to product by one enzyme site per second. In defining enzyme concentration, distinguish the concentration of enzyme molecules and concentration of enzyme sites (if the enzyme is a dimer with two active sites, the molar concentration of sites is twice the molar concentration of enzyme).

MICHAELIS-MENTEN KINETICS

Michaelis–Menten kinetics (occasionally also referred to as **Michaelis-Menten-Henri kinetics**) approximately describes the kinetics of many enzymes. It is named after Leonor Michaelis and Maud Menten. This kinetic model is relevant to situations where very simple kinetics can be assumed, (i.e. there is no intermediate or product inhibition, and there is no allostericity or cooperativity). More complex models exist for the cases where the assumptions of Michaelis-Menten kinetics are not appropriate any more.

The Michaelis-Menten equation relates the initial reaction rate v_0 to the substrate concentration [S]. The corresponding graph is a hyperbolic function; the maximum rate is described as v_{max}.

$$v_0 = \frac{V_{max}[S]}{K_m + [S]}$$

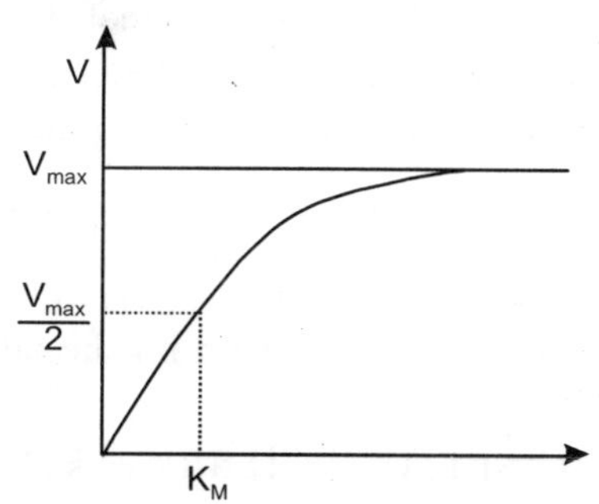

Fig. 3.3 : Michaelis-Menten Plot relating the reaction rate v_0 to the substrate concentration [S].

The Michaelis-Menten equation describes the rates of irreversible reactions. A steady state solution for a chemical equilibrium modeled with Michaelis–Menten kinetics can be obtained with the Goldbeter-Koshland equation.

History

The modern relationship between substrate and enzyme concentration was proposed in 1903 by Victor Henri. A microscopic interpretation was thereafter proposed in 1913 by Leonor Michaelis and Maud Menten, following earlier work by Archibald Vivian Hill. It postulated that enzyme (catalyst) and substrate (reactant) are in fast equilibrium with their complex, which then dissociates to yield product and free enzyme. The current derivation, based on the quasi steady state approximation (that the concentrations of the intermediate complexes remain constant) was proposed by Briggs and Haldane.

Equation

The validity of the following derivation rests on the reaction scheme given below and two key assumptions: that the total enzyme concentration and the concentration of the intermediate complex do not change over time. The most convenient derivation of the Michaelis–Menten equation, described by Briggs and Haldane, is obtained as follows (Note that often the experimental parameter k_{cat} is used but in this simple case it is equal to the kinetic parameter k_2):

The enzymatic reaction is assumed to be irreversible, and the product does not bind to the enzyme.

$$E + S \underset{k\text{-}1}{\overset{k_1}{\rightleftarrows}} ES \xrightarrow{k_2} ES + P \quad (1)$$

The first key assumption in this derivation is the quasi-steady-state assumption (or pseudo-steady-state hypothesis), namely that the concentration of the substrate-bound enzyme ([*ES*]) changes much more slowly than those of the product ([*P*]) and substrate ([*S*]). This allows us to set the rate of change of [*ES*] to zero and also write down the rate of product formation:

$$\frac{dES}{dt} = k_1[E][S] - [ES](k_{-1} + k_2) \overset{!}{=} 0 \quad (2)$$

$$\frac{d[P]}{dt} = k_2\ [ES] \quad (3)$$

The second key assumption is that the total enzyme concentration ([E]) does not change over time, thus we can write the total concentration of enzyme $[E]_0$ as the sum of the free enzyme in solution [E] and that which is bound to the substrate [ES]:

$$[E_0] = [E] + [ES] \overset{!}{=} \text{Constant}$$

Substituting this into equation (2), we obtain an expression for [ES] which in turn we can use in equation (3) to find an expression for the rate of product formation:

$$0 = k_1[S](E_0[ES]) - [ES](k_{-1} + k_2)$$

$$k_1[S][E]_0 = k_1[S][ES] + [ES](k_{-1} + k_2)$$

$$[S][E]_0 = [S][ES] + [ES]\underbrace{\frac{(k_{-1} + k_2)}{k_1}}_{k_M}$$

$$[S][E]_0 = (k_M + [S])[ES]$$

$$[ES] = \frac{[S][E]_0}{k_M + [S]}$$

$$\frac{d[P]}{dt} = v_0 = k_2[ES] = k_2[ES] = k_2[E]_0\frac{[S]}{k_M + [S]}$$

$$v_0 = \frac{v_{max}[S]}{k_M + [S]} \qquad (4)$$

$$\frac{1}{v_0} = \frac{k_M}{v_{max}} \cdot \frac{1}{S} + \frac{1}{v_{max}}$$

(5)

Because the concentration of substrate changes as the reaction takes place, the initial reaction rate v_0 is used to simplify analysis, taking the initial concentration of substrate as [S]. The equation for the reaction rate (4) can also be rewritten in equation (5) which uses the inverse of v_0 and [S]. This makes it easier to determine the constants from measured data (a procedure that results in a Lineweaver–Burk plot or a Hanes–Woolf plot).

Equation (4) results in a so called saturation curve which can be obsorved in the graph on the right. Several interesting cases can be distinguished mathematically and graphically:

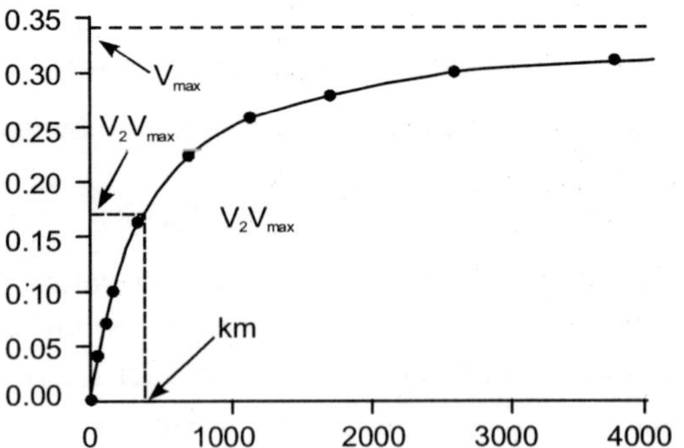

Fig. 3.4 : Saturation curve for an enzyme showing the relation between the concentration of substrate and rate

- If $[S]$ is large compared to K_M then the term $[S]/k_m + [S] \approx$. Therefore, the rate of product formation is

  $$\frac{d[P]}{dt} \gg v_{max} = k_2[E]_0$$

  Thus the product formation rate only depends on the enzyme concentration, the equation resemebles a unimolecular reaction with a corresponding pseudo-first order rate constant k_2. Thus it only matters how fast the [ES] complex turns its bound substrate into product and not how often the enzyme and the substrate meet.

- If $[S] = K_M$ then $[S]/k_m + [S] = [S]/2[S] = \frac{1}{2}$. Therefore, the rate of product formation is

  $$\frac{d[P]}{dt} = 0.5.v_{max} = 0.5k_2[E]_0$$

- If $[S]$ is small compared to K_M then the term $[S]/k_m + [S] \approx [S]/k_m = \frac{1}{2}$ and also very little ES complex is formed, thus $[E]_0 \approx [E]$. Therefore, the rate of product formation is

  $$\frac{d[P]}{dt} \approx v_{max}[S]/k_m \approx \frac{k_2}{k_m}[E][S]$$

Thus the product formation rate depends on the enzyme concentration as well as on the substrate concentration, the equation resembles a bimolecular reaction with a corresponding pseudo-second order rate constant k_2/K_M. This constant is a measure of how efficiently an enzyme converts a substrate into product. The most efficient enzymes reach a k_2/K_M in the range of $10^8 - 10^{10}$ M^{-1} s^{-1} which is the diffusion limit. These enzymes are so efficient they effectively catalyze a reaction each time they encounter a substrate molecule and have thus reached an upper theoretical limit for efficiency, thus these enzymes have often be termed as *perfect enzymes*.

Determination of Constants

To determine the maximum rate of an enzyme mediated reaction, a series of experiments is carried out with varying substrate concentration ([*S*]) and the initial rate of product formation is measured. 'Initial' here is taken to mean that the reaction rate is measured after a relatively short time period, during which complex builds up but the substrate concentration remains approximately constant and the quasi-steady-state assumption will hold. The measurements can then be plotted in a Lineweaver-Burk plot, plotting the inverse of substrate concentration against the inverse of the initial velocity

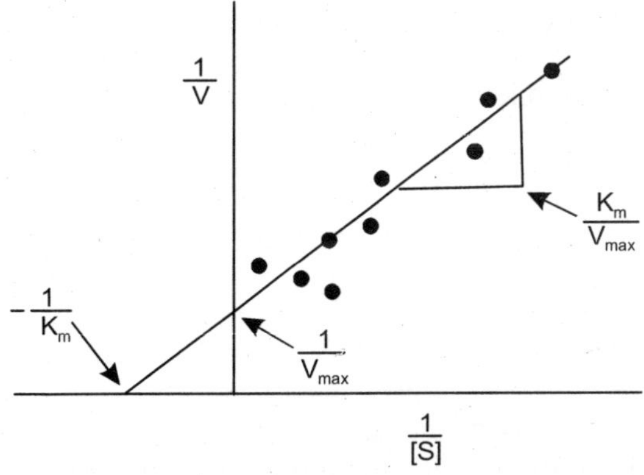

Fig. 3.5 : Lineweaver-Burk plot

$$\frac{1}{v_0} = \frac{k_M + [S]}{v_{max} + [S]} = \frac{k_M}{v_{max}} \cdot \frac{1}{[S]} + \frac{1}{v_{max}}$$

The values of the desired constants K_M and V_{max} can be read directly off the plot. It should be noted that accurate values for K_M and V_{max} can only be determined by non-linear regression of Michaelis-Menten data. The inverse plot while useful for visualization should never be the source of the actual value of the enzyme constant due to large insensitivity to errors inherent in all inverse plots. Should one be forced to derive a value from an inverse plot, the Hanes-Woolf plot is the most accurate.

Michaelis Constant K_M and its Significance

The reaction rate V is the number of reactions per second catalyzed per mole of the enzyme. The reaction rate increases with increasing substrate concentration $[S]$, asymptotically approaching the maximum rate V_{max}. There is therefore no clearly-defined substrate concentration at which the enzyme can be said to be saturated with substrate. A more appropriate measure to characterize an enzyme is the substrate concentration at which the reaction rate reaches half of its maximum value ($V_{max}/2$). This concentration can be shown to be equal to the Michaelis constant (K_M).

For enzymatic reactions which exhibit simple Michaelis–Menten kinetics, the Michaelis constant is defined as (this result comes directly from the derivation of the equation):

$$K_M = \frac{k_{-1} + k_2}{k_1}$$

In the most simple case, when product formation is the rate-limiting step (i.e. when $k_2 << k_{-1}$) the constant will just be equal to the dissociation constant (affinity for substrate) of the enzyme-substrate (ES) complex.

$$K_M \approx \frac{k-1}{k_1} = \frac{[E][S]}{[ES]} = K_d$$

However, often $k_2 >> k_{-1}$, or k_2 and k_{-1} are comparable, in which case there will be significant contributions to K_M in addition to the affinity of the enzyme for the substrate.

Limitations

The first source of limitations for the Michaelis-Menten kinetics is that it is an approximation of the kinetics derived by the law of mass action. In particular, Michaelis-Menten kinetics is based on the quasi-steady state assumption that $[ES]$ does not change,

$$\frac{d\,[ES]}{dt} = 0,$$

which is only approximately true: the rate of change of the complex $[ES]$ is very small but non-zero. The quality of the approximation depends on the timescale separation present in the dynamics on the phase space, which controls the magnitude of the rate of change of $[ES]$. In particular, it has been shown that this timescale separation is measured by a small, positive parameter,

$$\varepsilon = \frac{E_0}{S_0 + K_M},$$

where S_0 is the initial concentration of the substrate and S_0 and K_M have been defined above. The smaller is, the larger the timescale separation present in the system and the more accurate is the quasi-steady state approximation.

The second limitation is that Michaelis-Menten kinetics relies upon the law of mass action which is derived from the assumptions of free (Fickian) diffusion and thermodynamically-driven random collision. However, many biochemical or cellular processes deviate significantly from such conditions. For example, the cytoplasm inside a cell behaves more like a gel than a freely flowable or watery liquid, due to the very high concentration of protein (up to ~400 mg/mL) and other "solutes", which can severely limit molecular movements (e.g., diffusion or collision). This causes macromolecular crowding, which can alter reaction rates and dissociation constants.

For heterogeneous enzymatic reactions, such as those of membrane enzymes, molecular mobility of the enzyme or substrates can also be severely restricted, due to the immobilization or phase-separation of the reactants. For some homogeneous enzymatic reactions, the mobility of the enzyme or substrate may also be limited, such as the case of DNA polymerase where the enzyme moves along a chained substrate, rather than having a three-dimensional freedom. The limitation on molecular mobility (as well as other "non-ideal" conditions) demands modifications on the conventional mass-action laws, and Michaelis–Menten kinetics, to better reflect certain real world situations. Although it has been shown that the law of mass action can be valid in heterogeneous environments.

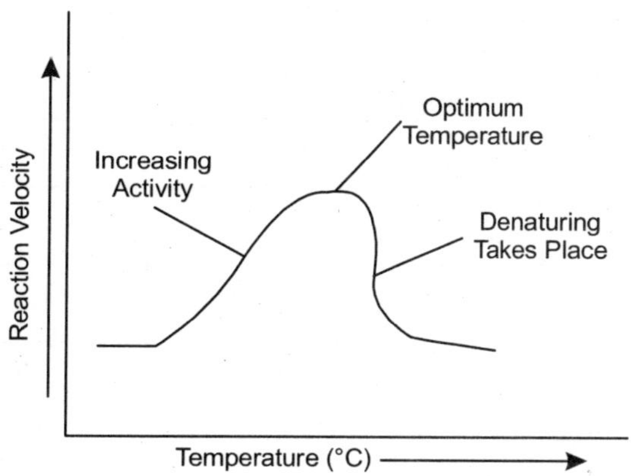

Fig. 3.6 : The effect of temperature on enzyme activity

EFFECT OF TEMPERATURE

Like most chemical reactions, the rate of an enzyme-catalyzed reaction increases as the temperature is raised. A 10°C rise in temperature will increase the activity of most enzymes by 50 to 100%. Variations in reaction temperature as small as 1 or 2 degrees may introduce changes of 10 to 20% in the results. In the case of enzymatic reactions, this is complicated by the fact that many enzymes are adversely affected by high temperatures. As shown in the Figure the reaction rate increases with temperature to a maximum level, then abruptly declines with further increase of temperature. Because most animal enzymes rapidly become denatured at temperatures above 40°C, most enzyme determinations are carried out somewhat below that temperature. Over a period of time, enzymes will be deactivated at even moderate temperatures. Storage of enzymes at 5°C or below is generally the most suitable. Some enzymes lose their activity when frozen.

EFFECT OF PH

Enzymes are affected by changes in pH. The most favorable pH value - the point where the enzyme is most active - is known as the optimum pH. This is graphically illustrated in Fig. 3.7.

Extremely high or low pH values generally result in complete loss of activity for most enzymes. pH is also a factor in the stability of enzymes. As with activity, for each enzyme there is also a region of pH optimal stability. The optimum pH value will vary greatly from one enzyme to another, as Table 3.1 shows :

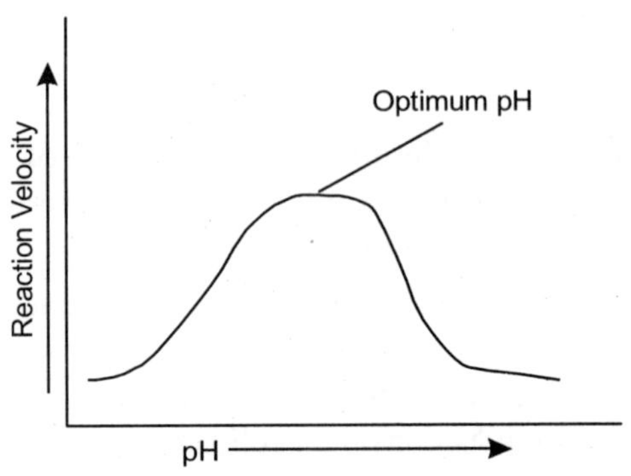

Fig. 3.7 : The Effect of pH on the Enzyme Activity

Table 3.1 : pH for Optimum Activity

Enzyme	pH Optimum
Lipase (pancreas)	8.0
Lipase (stomach)	4.0 - 5.0
Lipase (castor oil)	4.7
Pepsin	1.5 - 1.6
Trypsin	7.8 - 8.7
Urease	7.0
Invertase	4.5
Maltase	6.1 - 6.8
Amylase (pancreas)	6.7 - 7.0
Amylase (malt)	4.6 - 5.2
Catalase	7.0

In addition to temperature and pH there are other factors, such as ionic strength, which can affect the enzymatic reaction. Each of these physical and chemical parameters must be considered and optimized in order for an enzymatic reaction to be accurate and reproducible.

INHIBITORS OF ENZYME ACTION

COMPETITIVE INHIBITORS

A competitive inhibitor is any compound which closely resembles the chemical structure and molecular geometry of the substrate. The inhibitor competes for the same active site as the substrate molecule. The inhibitor may interact with the enzyme at the active site, but no reaction takes place. The inhibitor is "stuck" on the enzyme and prevents any substrate molecules from reacting with the enzyme. However, a competitive inhibition is usually reversible if sufficient substrate molecules are available to ultimately displace the inhibitor. Therefore, the amount of enzyme inhibition depends upon the inhibitor concentration, substrate concentration, and the relative affinities of the inhibitor and substrate for the active site.

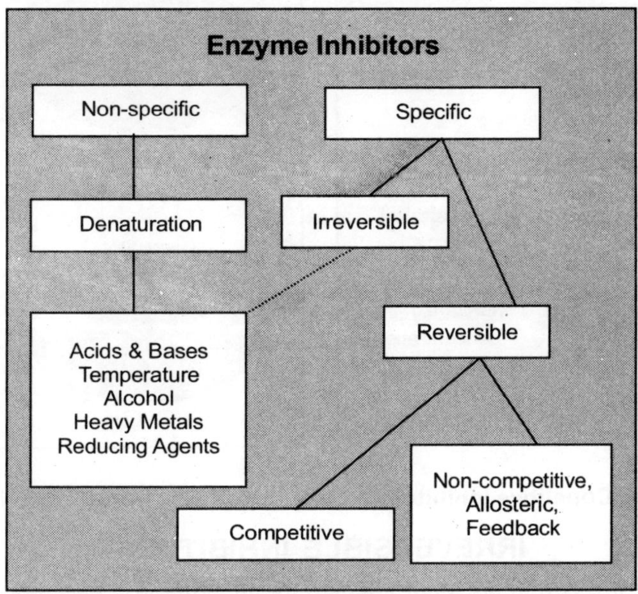

Fig. 3.8 : The relation between Enzyme inhibitors

Example: Ethanol is metabolized in the body by oxidation to acetaldehyde, which is in turn further oxidized to acetic acid by aldehyde oxidase enzymes. Normally, the second reaction is rapid so that acetaldehyde does not accumulate in the body.

A drug, *disulfiram (Antabuse)* inhibits the aldehyde oxidase which causes the accumulation of acetaldehyde with subsequent unpleasant side-effects of nausea and vomiting. This drug is sometimes used to help people overcome the drinking habit.

Methanol poisoning occurs because methanol is oxidized to formaldehyde and formic acid which attack the optic nerve causing blindness. Ethanol is given as an antidote for methanol poisoning because ethanol competitively inhibits the oxidation of methanol. Ethanol is oxidized in preference to methanol and consequently, the oxidation of methanol is slowed down so that the toxic by-products do not have a chance to accumulate.

NON-COMPETITIVE INHIBITORS

A noncompetitive inhibitor is a substance that interacts with the enyzme (Fig. 3.9A), but usually not at the active site. The noncompetitive inhibitor reacts either remote from or very close to the active site. The net effect of a non competitive inhibitor is to change the shape of the enzyme and thus the active site, so that the substrate can no longer interact with the enzyme to give a reaction. Non-competitive inhibitors are usually reversible, but are not influenced by concentrations of the substrate as is the case for a reversible competive inhibitor. (See the Fig. 3.9B)

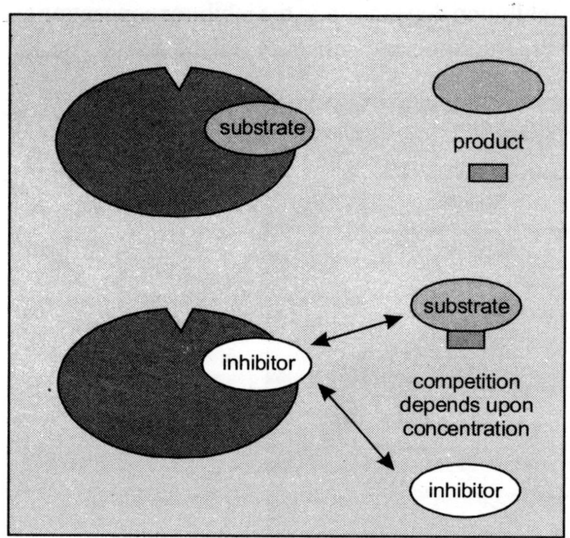

Fig. 3.9A : The Copetitive Inhibitors

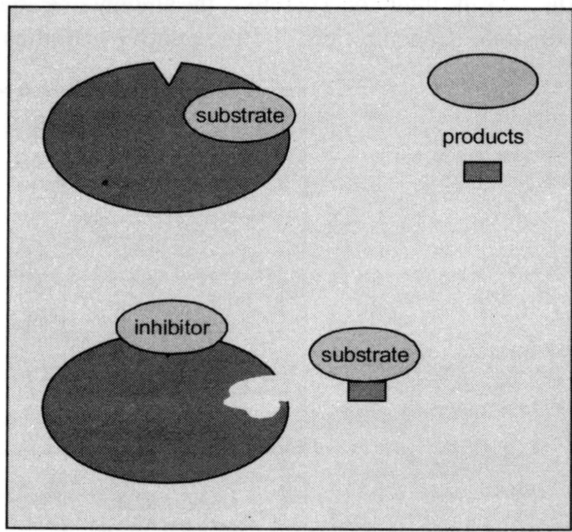

Fig. 3.9B : Non-Competitive Inhibitor

IRREVERSIBLE INHIBITORS

Irreversible Inhibitors form strong covalent bonds with an enzyme. These inhibitors may act at, near, or remote from the active site. Consequently, they may not be displaced by the addition of excess substrate. In any case, the basic structure of the enzyme is modified to the degree that it ceases to work. Since many enzymes contain sulfhydral (-SH), alcohol, or acid groups as part of their active sites, any chemical which can react with them acts as an irreversible inhibitor. Heavy metals such as Ag^+, Hg^{2+}, Pb^{2+} have strong affinities for -SH groups.

Nerve gases such as diisopropylfluorophosphate (DFP) inhibit the active site of acetylcholine esterase by reacting with the hydroxyl group of serine to make an ester. Oxalic and citric acid inhibit blood clotting by forming complexes with calcium ions necessary for the enzyme metal ion activator.

Example: Chymotrpsin is an enzyme which hydrolyzes peptides at the carbonyl side of tyr or phe or trp (i.e. those that have an aromatic side chain. In the graphic on the left, the substrate and the irreversible inhibitor are shown in the active site pocket. In the case of the inhibitor the reaction starts in the same way as with the substrate, but the end result is that the inhibitor is covalently bonded to the histidine-57 in the active site and is not reversible.

SUICIDE INHIBITION

Suicide inhibition, also known as suicide inactivation, is a form of irreversible enzyme inhibition that occurs when an enzyme binds a substrate analogue and forms an irreversible complex with it through a covalent bond during the "normal" catalysis reaction.

Examples: Some clinical examples of suicide inhibitors include:

- Penicillin, which inhibits transpeptidase from building bacterial cell walls.
- Sulbactam, which prohibits penicillin-resistant strains of bacteria from metabolizing penicillin.
- Allopurinol, which inhibits uric acid production by xanthine oxidase in the treatment of gout.
- AZT (zidovudine) and other chain-terminating nucleoside analogues used to inhibit HIV-1 reverse transcriptase in the treatment of HIV/AIDS.
- Eflornithine, one of the drugs used to treat sleeping sickness is a suicide inhibitor of ornithine decarboxylase.

FUNCTIONAL AND NON-FUNCTIONAL ENZYMES

Blood plasma contains many enzymes, which are classified into functional and non-functional plasma enzymes.

Sources of Non-functional Plasma Enzymes

1. Increase in the rate of enzyme synthesis) e.g. bilirubin increases the rate of synthesis of alkaline phosphatase in obstructive liver diseases.
2. Obstruction of normal pathway e.g. obstruction of bile ducts increases alkaline phosphatase.
3. Increased permeability of cell membrane as in tissue hypoxia.

Differences between functional and non-functional plasma enzymes :

	Functional plasma enzymes	Non-functional plasma enzymes
Concentration in plasma	Present in plasma in higher concentrations in comparison to tissues	Normally, present in plasma in very low concentrations in comparison to tissues
Function	Have known functions	No known functions
The substrates	Their substrates are always present in the blood	Their substrates are absent from the blood
Site of synthesis	Liver	Different organs e.g. liver, heart, brain and skeletal muscles
Effect of diseases	Decrease in liver diseases	Different enzymes increase in different organ diseases
Examples	Clotting factors e.g. prothrombin, Lipoprotein lipase and pseudo- choline esterase	ALT, AST, CK, LDH, alkaline phosphatase, acid phosphatase and amylase,

4. Cell damage with the release of its content of enzymes into the blood e.g. myocardial infarction and viral hepatitis.

Medical Importance of Non-functional Plasma Enzymes

Measurement of non-functional plasma enzymes is important for:

1. Diagnosis of diseases as diseases of different organs cause elevation of different plasma enzymes.
2. Prognosis of the disease; we can follow up the effect of treatment by measuring plasma enzymes before and after treatment.

Examples of Medically Important Non-functional Plasma Enzymes

1. Amylase and lipase enzymes increase in diseases of the pancreas as acute pancreatitis.
2. Creatine kinase (CK) enzyme increases in heart, brain and skeletal muscle diseases.
3. Lactate dehydrogenase (LDH) enzyme increases in heart, liver and blood diseases.
4. Alanine transaminase (ALT) enzyme, it is also called serum glutamic pyruvic transaminase (SGPT). It increases in liver and heart diseases.
5. Aspartate transaminase (AST) enzyme, it is also called serum glutamic oxalacetic transaminase (SGOT). It increases in liver and heart diseases.
6. Acid phosphatase enzyme increases in cancer prostate.
7. Alkaline phosphatase enzyme increases in obstructive liver diseases, bone diseases and hyperparathyroidism.

MEASUREMENTS OF ENZYME ACTIVITY AND INTERPRETATION OF UNITS

Amounts of enzymes can either be expressed as molar amounts, as with any other chemical, or measured in terms of activity, in enzyme units.

Enzyme activity = moles of substrate converted per unit time = rate × reaction volume. Enzyme activity is a measure of the quantity of active enzyme present and is thus dependent on conditions, *which should be specified.* The SI unit is the katal, 1 katal = 1 mol s^{-1}, but this is an excessively large unit. A more practical and commonly-used value is 1 enzyme unit (EU) = 1 μmol min^{-1} (μ = micro, $\times 10^{-6}$). 1 U corresponds to 16.67 nanokatals.

The specific activity of an enzyme is another common unit. This is the activity of an enzyme per milligram of total protein (expressed in μmol $min^{-1}mg^{-1}$). Specific activity gives a measurement of the purity of the enzyme.

ISOZYMES

Also known as **isoenzymes** are enzymes that differ in amino acid sequence but catalyze the same chemical reaction. These enzymes usually display different kinetic parameters (i.e. different K_M values), or different regulatory properties. The existence of isozymes permits the fine-tuning of metabolism to meet the particular needs of a given tissue or developmental stage (for example lactate dehydrogenase (LDH)). In biochemistry, isozymes (or isoenzymes) are isoforms (closely related variants) of enzymes. In many cases, they are coded for by homologous genes that have diverged over time. Although, strictly speaking, **allozymes** represent enzymes from different alleles of the same gene, and isozymes represent enzymes from different genes that process or catalyse the same reaction, the two words are usually used interchangeably.

Isozymes were first described by R. L. Hunter and Clement Markert (1957) who defined them as *different variants of the same enzyme having identical functions and present in the same individual*. This definition encompasses:

1. enzyme variants that are the product of different genes and thus represent different loci (described as *isozymes*); and
2. enzymes that are the product of different alleles of the same gene (described as *allozymes*).Isozymes are usually the result of gene duplication, but can also arise from polyploidisation or nucleic acid hybridization.

Over evolutionary time, if the function of the new variant remains *identical* to the original, then it is likely that one or the other will be lost as mutations accumulate, resulting in a pseudogene. However, if the mutations do not immediately prevent the enzyme from functioning, but instead modify either its function, or its pattern of gene expression, then the two variants may both be favoured by natural selection and become specialised to different functions.

For example, they may be expressed at different stages of development or in different tissues. Allozymes may result from point mutations or from insertion-deletion (*indel*) events that affect the DNA coding sequence of the gene. As with any other new mutation, there are three things that may happen to a new allozyme:

1. It is most likely that the new allele will be non-functional—in which case it will probably result in low fitness and be removed from the population by natural selection.
2. Alternatively, if the amino acid residue that is changed is in a relatively unimportant part of the enzyme, for example a long way from the active site then the mutation may be selectively neutral and subject to genetic drift.
3. In rare cases the mutation may result in an enzyme that is more efficient, or one that can catalyse a slightly different chemical reaction, in which case the mutation may cause an increase in fitness, and be favoured by natural selection.

An Example of an Isozyme

An example of an isozyme is glucokinase, a variant of hexokinase which is not inhibited by glucose 6-phosphate. Its different regulatory features and lower affinity for glucose (compared to other hexokinases), allows it to serve different functions in cells of specific organs, such as control of insulin release by the beta cells of the pancreas, or initiation of glycogen synthesis by liver cells. Both of these processes must only occur when glucose is abundant, or problems occur.

Distinguishing Isozymes

Isozymes (and allozymes) are variants of the same enzyme. Unless they are identical in terms of their biochemical properties, for example their substrates and enzyme kinetics, they may be distinguished by a biochemical assay. However, such differences are usually subtle (particularly between *allozymes* which are often neutral variants). This subtlety is to be expected, because two enzymes that differ significantly in their function are unlikely to have been identified as *isozymes*. Whilst isozymes may be almost identical in function, they may differ in other ways. In particular, amino acid substitutions that change the electric charge of the enzyme (such as replacing aspartic acid with glutamic acid) are simple to identify by gel electrophoresis, and this forms the basis for the use of isozymes as molecular markers. To identify isozymes, a crude protein extract is made by grinding animal or plant tissue with an extraction buffer, and the components of extract are separated according to their charge by gel electrophoresis. Historically, this has usually been done using gels made from potato starch, however, acrylamide gels provide better resolution, and cellulose acetate gels are now (as of 2005) the norm. All the proteins from the tissue are present in the gel, so that individual enzymes must be identified using an assay that links their function to a staining reaction. For example, detection can be based on the localised precipitation of soluble indicator dyes such as tetrazolium salts which become insoluble when they are reduced by cofactors such as NAD or NADP, which generated in zones of enzyme activity. This assay method requires that the enzymes are still functional after separation, and provides the greatest challenge to using isozymes as a laboratory technique.

Isozymes and Allozymes as Molecular Markers

Population genetics is essentially a study of the causes and effects of genetic variation within and between populations, and in the past isozymes have been amongst the most widely used Molecular markers for this purpose. Although they have now been largely superseded by more informative DNA-based approaches (such as direct DNA sequencing, single nucleotide polymorphisms and microsatellites), they are still amongst the quickest and cheapest marker systems to develop, and remain (as of 2005) an excellent choice for projects that only need to identify low levels of genetic variation, e.g. quantifying mating systems.

Physical Aspects of Living Matters

ISOTOPES

Isotopes (Greek *esos* = "equal", *tópos* = "site, place") are any of the different types of atoms (nuclides) of the same chemical element, each having a different atomic mass (mass number). Isotopes of an element have nuclei with the same number of protons (the same atomic number) but different numbers of neutrons. Therefore, isotopes have different mass numbers, which give the total number of nucleons, the number of protons plus neutrons.

A nuclide is any particular atomic nucleus with a specific number of an atom Z and mass number A; it is equivalently an atomic nucleus with a specific number of protons and neutrons. Collectively, all the isotopes of all the elements form the set of *nuclides*. The distinction between the terms *isotope* and *nuclide* has somewhat blurred, and they are often used interchangeably. *Isotope* is better used when referring to several different nuclides of the same element; *nuclide* is more generic and is used when referencing only one nucleus or several nuclei of different elements. For example, it is more correct to say that an element such as fluorine consists of one stable nuclide rather than that it has one stable isotope.

Isotopes and nuclides are specified by the name of the particular element, implicitly giving the atomic number, followed by a hyphen and the mass number (e.g. helium-3, carbon-12, carbon-13, iodine-131 and uranium-238). In symbolic form, the number of nucleons is denoted as a superscripted prefix to the chemical symbol (e.g. ^{3}He, ^{12}C, ^{13}C, ^{131}I and ^{238}U).

About 339 nuclides occur naturally on Earth, of which 256 (about 75%) are stable (or, to be careful, have never been observed to decay; this note is necessary because many "stable" isotopes are predicted to be radioactive with very long half-lives). Counting the radioactive nuclides not found in nature that have been created artificially, more than 3100 nuclides are currently known.

History

The term isotope was coined in 1913 by Margaret Todd, a Scottish doctor, during a conversation with Frederick Soddy (to whom she was distantly related by marriage). Soddy, a chemist at Glasgow University, explained that it appeared from his investigations as if several elements occupied each position in the periodic table. Hence Todd suggested the Greek term for "at the same place" as a suitable name. Soddy adopted the term and went on to win the Nobel Prize for Chemistry in 1921 for his work on radioactive substances. Soddy's use of the word isotope was initially with regard to radioactive (unstable) atoms. However, in 1913, as part of his exploration into the composition of canal rays, J. J. Thomson channelled a stream of ionized neon through a magnetic and an electric field and measured its deflection by placing a photographic plate in its path. Thomson observed two patches of light on the photographic plate (see Fig. 4.1), which suggested two different parabolas of deflection. This was the first observation of different stable isotopes for an element. Thomson eventually concluded that some of the atoms in the neon gas were of higher mass than the rest.

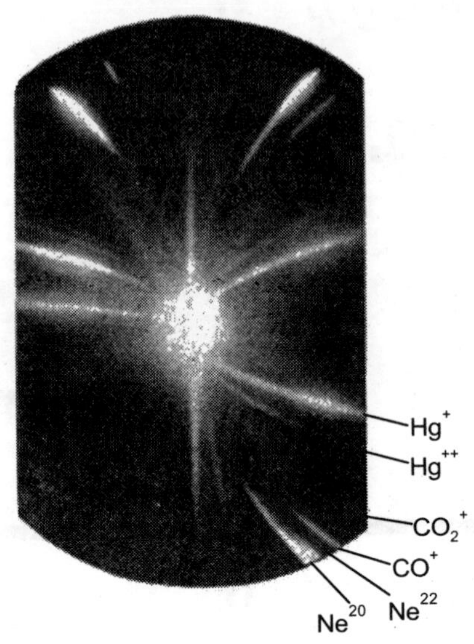

Fig. 4.1 : The bottom right corner of JJ Thomson's photographic plate are markings for the two isotopes of neon: neon-20 and neon-22

Variation in Properties between Isotopes Chemical and Atomic Properties

A neutral atom has the same number of electrons as protons. Thus, different isotopes of a given element all have the same number of protons and electrons and the same electronic structure, and because the chemical behavior of an atom is largely determined by its electronic structure, different isotopes exhibit nearly identical chemical behavior. The main exception to this is the kinetic isotope effect: due to their larger masses, heavier isotopes tend to react somewhat more slowly than lighter isotopes of the same element. This is most pronounced for protium (^1H) vis-à-vis deuterium (^2H), because deuterium has twice the mass of protium. The mass effect between deuterium and the relatively light protium also affects the behavior of their respective chemical bonds, by means of changing the center of gravity (reduced mass) of the atomic systems. However, for heavier elements, the absolute mass of nucleus relative to electrons is far more, and the relative mass difference between isotopes is much less, and thus the mass-difference effects on chemistry are usually negligible.

Similarly, two molecules which differ only in the isotopic nature of their atoms (*isotopologues*) will have identical electronic structure and therefore almost indistinguishable physical and chemical properties (again with deuterium providing the primary exception to this rule). The vibrational modes of a molecule are determined by its shape and by the masses of its constituent atoms. Consequently, isotopologues will have different sets of vibrational modes. Since vibrational modes allow a molecule to absorb photons of corresponding energies, isotopologues have different optical properties in the infrared range.

Nuclear Properties and Stability

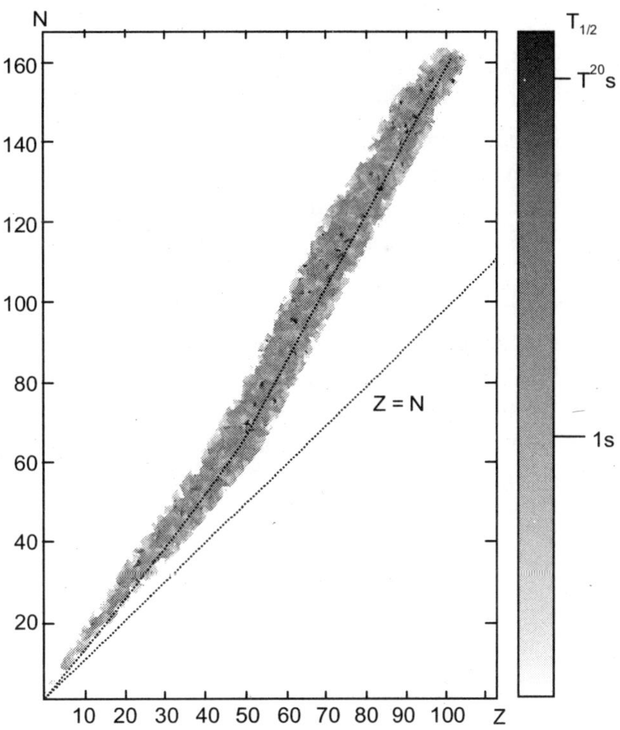

Fig. 4.2 : Isotope half lifes.

Note that the darker more stable isotope region departs from the line of protons Z = neutrons N, as the element number Z becomes larger

Atomic nuclei consist of protons and neutrons bound together by the strong nuclear force. Because protons are positively charged, they repel each other. Neutrons, which are electrically neutral, allow some separation between the positively charged protons, reducing the electrostatic repulsion. Neutrons also stabilize the nucleus because at short ranges they attract each other and protons equally by the strong nuclear force, and this extra binding force also offsets the electrical repulsion between protons. For this reason, one or more neutrons are necessary for two or more protons to be bound into a nucleus. As the number of protons increases, an increasing ratio of neutrons are needed to form a stable nucleus (see Fig. 4.2). For example, although the neutron:proton ratio of ^{3}He is 1:2, the neutron:proton ratio of ^{238}U is greater than 3:2.

Of the 80 elements with a stable isotope, the largest number of stable isotopes observed for any element is ten (for the element tin). Xenon is the only element which has nine stable isotopes. There is no element with exactly eight stable isotopes. See list of elements by nuclear stability for a complete list. Five elements have seven stable isotopes, eight have six stable isotopes, nine have five stable isotopes, nine have four stable isotopes, nine have three stable isotopes, 16 have two stable isotopes (counting Ta-180m as stable), and 26 elements have only a single stable isotope (of these, 19 are so-called mononuclidic elements, having a single primordial stable isotope which dominates and fixes the atomic weight of the natural element to high precision; 3 radioactive mononuclidic elements occur as well). In total, there are 256 nuclides which have not been observed to decay. For the 80 elements which have one or more stable isotopes, the average number of stable isotopes is 256/80 = 3.20 isotopes per element.

Other effects besides the bulk ratio of protons and neutrons affect nuclear stability. For example, the extreme stability of helium-4 due to a double pairing of 2 protons and 2 neutrons prevents *any* nuclides containing five nucleons from existing for long enough to serve as platforms for building up of heavier elements during fusion formation in stars (see triple alpha process). A similar pairing pattern shows in the fact that the 256 known stable nuclides contain only five that have both an odd number of protons *and* an odd number of neutrons (odd-odd nuclei): ^{2}H, ^{6}Li, ^{10}B, ^{14}N, ^{180m}Ta (the last one was predicted to decay but this process was never observed). Also, four long-lived radioactive odd-odd nuclides (^{40}K, ^{50}V, ^{138}La, ^{176}Lu) occur naturally. Most odd-odd nuclides are highly unstable with respect to beta decay, because the decay products are even-even, and are therefore more strongly bound, due to nuclear pairing effects.

Although isotopes exhibit nearly identical electronic and chemical behavior, their nuclear behavior varies dramatically. Adding neutrons to isotopes can vary their nuclear spins and nuclear shapes, causing differences in neutron capture cross-sections and gamma spectroscopy and nuclear magnetic resonance properties.

Occurrence in Nature

Elements are composed of one or more naturally occurring isotopes, which are normally stable. Some elements have unstable (radioactive) isotopes, either because their decay is so slow that a fraction still remains since they were created (examples: uranium, potassium), or because they are continually created through cosmic radiation (tritium, carbon-14) or by decay from an isotope in the first category (radium, radon).

As discussed above, only 80 elements have any stable isotopes, and 26 of these have only one stable isotope. Thus, about two-thirds of stable elements occur naturally on Earth in multiple stable isotopes, with the largest number of stable isotopes for an element being ten, for tin (element number 50). There are about 94 elements found naturally on Earth (up to plutonium, element 94, inclusive), though some are detected only in very tiny amounts, such as plutonium-244. Scientists estimate that the elements which occur naturally on Earth (some only as radioisotopes) occur as 339 isotopes (nuclides) in total. Only 256 of these naturally-occurring isotopes are stable in the sense of never having been observed to decay as of the present time. All the known stable isotopes occur naturally on Earth); the other 85 naturally-occurring isotopes are radioactive, but occur on Earth due to their relatively long half-lives, or else due to other means of ongoing natural production. An additional ~ 2700 radioactive isotopes not found in nature have been created in nuclear reactors and in particle accelerators. Many short-lived isotopes not found naturally on Earth have also been observed by spectroscopic analysis, being naturally created in stars or supernovae. An example is aluminum-26, which is not naturally found on Earth, but which is found in abundance on an astronomical scale.

The tabulated atomic masses of elements are averages that account for the presence of multiple isotopes with different masses. A good example is chlorine, having the composition ^{35}Cl, 75.8%, and ^{37}Cl, 24.2%, giving an atomic mass of 35.5. Values like this confounded scientists before the discovery of isotopes, as most light element atomic masses are close to integer multiples of hydrogen.

According to generally accepted cosmology, only isotopes of hydrogen and helium, and traces of some isotopes of lithium, beryllium and boron were created at the Big Bang, while all other isotopes were synthesized later, in stars and supernovae, and in interactions between energetic particles such as cosmic rays, and previously-produced isotopes. The most common isotope of hydrogen has no neutrons at all. The respective abundances of isotopes on Earth result from the quantities formed by these processes, their spread through the galaxy, and the rates of decay for isotopes that are unstable. After the initial coalescence of the solar system, isotopes were redistributed according to mass, and the isotopic composition of elements varies slightly from planet to planet. This sometimes makes it possible to trace the origin of meteorites.

Atomic Mass of Isotopes

The atomic mass (M_r) of an isotope is determined mainly by its mass number (i.e. number of nucleons in its nucleus). Small corrections are due to the binding energy of the nucleus (see Mass defect), to slightly different masses of neutron and proton, and to the mass of electron shell of the atom. One should take into account that the mass number is an integer dimensionless quantity, whereas the atomic mass is a real number expressed in atomic mass units. But the difference between these quantities is less then 1% in any case.

The atomic masses of naturally occurring isotopes of an element determine the atomic weight of the element. When the element contains N isotopes, the equation below is applied for the atomic weight M:

$$M = M_1 \times \eta_1 + M_2 \times \eta_2 + \ldots + M_N \times \eta N,$$

where $M_1, M_2, \ldots, M_N$ are the atomic masses of each individual isotope, and $\eta_1, \ldots, \eta_N$ are the relative abundances of these isotopes.

Applications of Isotopes

Several applications exist that capitalize on properties of the various isotopes of a given element.

Use of chemical and biological properties:

- Isotope analysis is the determination of isotopic signature, the relative abundances of isotopes of a given element in a particular sample. For biogenic substances in particular, significant variations of isotopes of C, N and O can occur. Analysis of such variations has a wide range of applications, such as the detection of adulteration of food products.[6] The identification of certain meteorites as having originated on Mars is based in part upon the isotopic signature of trace gases contained in them.
- Another common application is isotopic labeling, the use of unusual isotopes as tracers or markers in chemical reactions. Normally, atoms of a given element are indistinguishable from each other. However, by using isotopes of different masses, they can be distinguished by mass spectrometry or infrared spectroscopy (see "Properties"). For example, in 'stable isotope labeling with amino acids in cell culture (SILAC)' stable isotopes are used to quantify proteins. If radioactive isotopes are used, they can be detected by the radiation they emit (this is called *radioisotopic labeling*).
- A technique similar to radioisotopic labelling is radiometric dating: using the known half-life of an unstable element, one can calculate the amount of time that has elapsed since a known level of isotope existed. The most widely known example is radiocarbon dating used to determine the age of carbonaceous materials.
- Isotopic substitution can be used to determine the mechanism of a reaction via the kinetic isotope effect.

Use of nuclear properties

- Several forms of spectroscopy rely on the unique nuclear properties of specific isotopes. For example, nuclear magnetic resonance (NMR) spectroscopy can be used only for isotopes with a nonzero nuclear spin. The most common isotopes used with NMR spectroscopy are ^{1}H, ^{2}D, ^{15}N, ^{13}C, and ^{31}P.

- Mossbauer spectroscopy also relies on the nuclear transitions of specific isotopes, such as ^{57}Fe.
- Radionuclides also have important uses. Nuclear power and nuclear weapons development require relatively large quantities of specific isotopes. The process of isotope separation represents a significant technological challenge, but more so with heavy elements such as uranium or plutonium, than with lighter elements such as hydrogen, lithium, carbon, nitrogen, and oxygen. The lighter elements are commonly separated by gas diffusion of their compounds such as CO and NO. Uranium isotopes have been separated in bulk by gas diffusion, gas centrifugation, laser ionization separation, and by a type of production mass spectroscopy.

RADIO-ISOTOPES

In most cases, elements like to have an equal number of protons and neutrons because this makes them the most stable. Stable atoms have a binding energy that is strong enough to hold the protons and neutrons together. Even if an atom has an additional neutron or two it may remain stable. However, an additional neutron or two may upset the binding energy and cause the atom to become unstable. In an unstable atom, the nucleus changes by giving off a neutron to get back to a balanced state. As the unstable nucleus changes, it gives off radiation and is said to be radioactive. Radioactive isotopes are often called *radioisotopes.*

IONIZING RADIATION

Ionizing radiation consists of subatomic particles or electromagnetic waves that are energetic enough to detach electrons from atoms or molecules, ionizing them. The occurrence of ionization depends on the energy of the impinging individual particles or waves, and not on their number. An intense flood of particles or waves will not cause ionization if these particles or waves do not carry enough energy to be ionizing. Roughly speaking, particles or photons with energies above a few electron volts (eV) are ionizing.

Examples of ionizing particles are energetic alpha particles, beta particles, and neutrons. The ability of electromagnetic waves (photons) to ionize an atom or molecule depends on their wavelength. Radiation on the short wavelength end of the electromagnetic spectrum - ultraviolet, X-rays, and gamma rays - is ionizing.

Ionizing radiation comes from radioactive materials, X-ray tubes, particle accelerators, and is present in the environment. It has many practical uses in medicine, research, construction, and other areas, but presents a health hazard if used improperly. Exposure to radiation causes microscopic damage to living tissue, resulting in skin burns and radiation sickness at high doses and cancer, tumors and genetic damage at low doses.

Types of Radiation

Alpha (α) radiation consists of Helium 4 (^{4}He) nuclei and is stopped by a sheet of paper. Beta (β) radiation, consisting of electrons, is halted by an aluminium plate. Gamma (γ) radiation, consisting of energetic photons, is eventually absorbed as it penetrates a dense material. Neutron (n) radiation consists of free neutrons which are blocked using light elements, like hydrogen, which slow and/or capture them.

Various types of ionizing radiation may be produced by radioactive decay, nuclear fission and nuclear fusion, extremely hot objects via blackbody radiation, and by particle accelerators.

In order for a particle to be ionizing, it must both have a high enough energy and interact with the atoms of a target. Photons interact strongly with charged particles, so photons of sufficiently high energy also are ionizing. The energy at which this begins to happen with photons (light) is in the ultraviolet region of the electromagnetic spectrum; sunburn is one of the effects of ionization. Charged particles such as electrons, positrons, and alpha particles also interact strongly with electrons of an atom or molecule. Neutrons, on the other hand, do not interact strongly with electrons, and so they cannot directly cause ionization by this mechanism. However, fast neutrons will interact with the protons in hydrogen (in the manner of a billiard ball hitting another, sending it away with all of the first ball's energy of motion), and this mechanism produces proton radiation (fast protons). These protons are ionizing because of their strong interaction with electrons in matter. A neutron can also interact with an atomic nucleus, depending on the nucleus and the neutron's velocity; these reactions happen with fast neutrons and slow neutrons, depending on the situation. Neutron interactions in this manner often produce radioactive nuclei, which produce ionizing radiation when they decay, they then can produce chain reactions in the mass that is decaying, sometimes causing a larger effect of ionization.

In the picture at left, gamma rays are represented by wavy lines, charged particles and neutrons by straight lines. The little circles show where ionization processes occur.

An ionization event normally produces a positive atomic ion and an electron. High-energy beta particles may produce bremsstrahlung when passing through matter, or secondary electrons (δ-electrons); both can ionize in turn.

Unlike alpha or beta particles (see particle radiation), gamma rays do not ionize all along their path, but rather interact with matter in one of three ways: the photoelectric effect, the Compton effect, and pair production. By way of example, the figure shows Compton effect: two Compton scatterings that happen sequentially. In every scattering event, the gamma ray transfers energy to an electron, and it continues on its path in a different direction and with reduced energy.

In the same figure, the neutron collides with a proton of the target material, and then becomes a fast recoil proton that ionizes in turn. At the end of its path, the neutron is captured by a nucleus in an (n,γ)-reaction that leads to a neutron capture photon.

The negatively-charged electrons and positively charged ions created by ionizing radiation may cause damage in living tissue. If the dose is sufficient, the effect may be seen almost immediately, in the form of radiation poisoning. Lower doses may cause cancer or other long-term problems. The effect of the very low doses encountered in normal circumstances (from both natural and artificial sources, like cosmic rays, medical X-rays and nuclear power plants) is a subject of current debate. A 2005 report released by the National Research Council indicated that the overall cancer risk associated with background sources of radiation was relatively low.

Radioactive materials usually release alpha particles, which are the nuclei of helium, beta particles, which are quickly moving electrons or positrons, or gamma rays. Alpha and beta particles can often be stopped by a piece of paper or a sheet of aluminium, respectively. They cause most damage when they are emitted inside the human body. Gamma rays are less ionizing than either alpha or beta particles, and protection against gammas requires thicker shielding. The damage they produce is similar to that caused by X-rays, and include burns and also cancer, through mutations. Human biology resists germline mutation by either correcting the changes in the DNA or inducing apoptosis in the mutated cell.

Non-ionizing radiation is thought to be essentially harmless below the levels that cause heating. Ionizing radiation is dangerous in direct exposure, although the degree of danger is a subject of debate. Humans and animals can also be exposed to ionizing radiation internally: if radioactive isotopes are present in the environment, they may be taken into the body. For example, radioactive iodine is treated as normal iodine by the body and used by the thyroid; its accumulation there often leads to thyroid cancer. Some radioactive elements also bioaccumulate.

The units used to measure ionizing radiation are rather complex. The ionizing effects of radiation are measured by units of exposure:

- The coulomb per kilogram (C/kg) is the SI unit of ionizing radiation exposure, and measures the amount of radiation required to create 1 coulomb of charge of each polarity in 1 kilogram of matter.

Table 4.1 : Weighting factors W_R for equivalent dose

Radiation	Energy	wR
X-rays, gamma rays, electrons, positrons, muons	1	
neutrons	< 10 keV	5
	10 keV - 100 keV	10
	100 keV - 2 MeV	20
	2 MeV - 20 MeV	10
	> 20 MeV	5
protons	> 2 MeV	2
alpha particles, fission fragments, heavy nuclei		20

- The Roentgen (R) is an older traditional unit that is almost out of use, which represented the amount of radiation required to liberate 1 esu of charge of each polarity in 1 cubic centimeter of dry air. 1 Roentgen = 2.58×10^{-4} C/kg

However, the amount of damage done to matter (especially living tissue) by ionizing radiation is more closely related to the amount of energy deposited rather than the charge. This is called the absorbed dose.

Interaction of ionizing Radiation with Matter

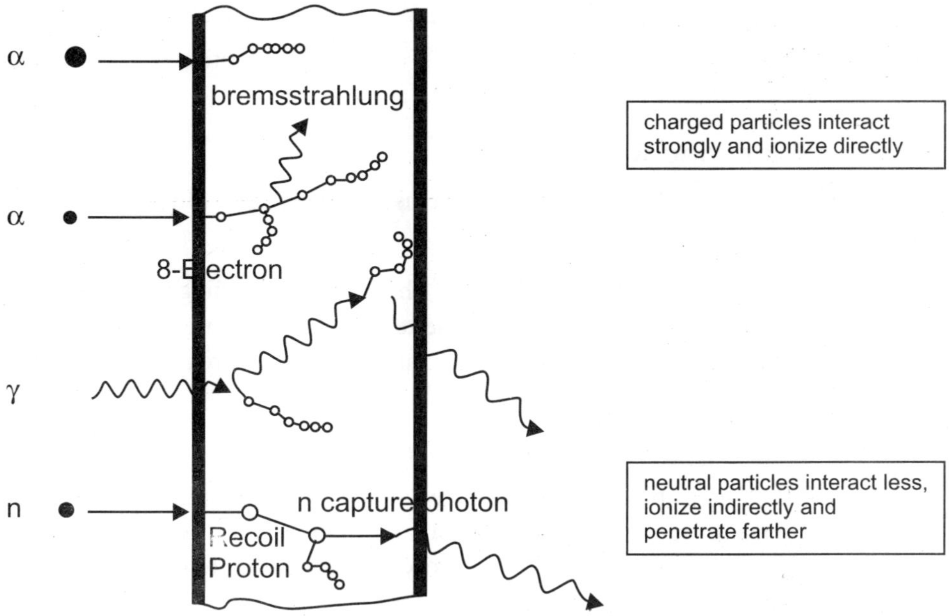
α
bremsstrahlung
charged particles interact strongly and ionize directly
α
8-Electron
γ
n
n capture photon
Recoil Proton
neutral particles interact less, ionize indirectly and penetrate farther

Fig. 4.3 : The Ionization of radiation

- The gray (Gy), with units J/kg, is the SI unit of absorbed dose, which represents the amount of radiation required to deposit 1 joule of energy in 1 kilogram of any kind of matter.
- The rad (Roentgen absorbed dose), is the corresponding traditional unit which is 0.01 J deposited per kg. 100 rad = 1 Gy.

  Equal doses of different types or energies of radiation cause different amounts of damage to living tissue. For example, 1 Gy of alpha radiation causes about 20 times as much damage as 1 Gy of X-rays. Therefore the equivalent dose was defined to give an approximate measure of the biological effect of radiation. It is calculated by multiplying the absorbed dose by a weighting factor W_R which is different for each type of radiation (Table 4.1).
- The sievert (Sv) is the SI unit of equivalent dose. Although it has the same units as grays, J/kg, it measures something different. It is the dose of any type of radiation in Gy that has the same biological effect on a human as 1 Gy of X-rays or gamma radiation.
- The rem (Roentgen equivalent man) is the traditional unit of equivalent dose. 1 sievert = 100 rem. Because the rem is a relatively large unit, typical equivalent dose is measured in millirem (mrem), 10^{-3} rem, or in microsievert (μSv), 10^{-6} Sv. 1 mrem = 10 μSv.

For comparison, the 'background' dose of natural radiation received by a US citizen is around 3 mSv (300 mrem) per year. The lethal dose of radiation for a human is around 4 - 5 Sv (400 - 500 rem).

Uses

Ionizing radiation has many uses, such as to kill cancerous cells. However, although ionizing radiation has many applications, overuse can be hazardous to human health. For example, at one time, assistants in shoue shops used X-rays to check a child's shoue size, but this practice was halted when it was discovered that ionizing radiation was dangerous.

Technical Uses of Ionizing Radiation

Since ionizing radiations can penetrate matter, they are used for a variety of measuring methods.

Radiography by means of gamma or X rays

This is a method used in industrial production. The piece to be radiographed is placed between the source and a photographic film in a cassette. After a certain exposure time, the film is developed and it shows internal defects of the material if there are any.

Gauges

Gauges use the exponential absorption law of gamma rays

- Level indicators: Source and detector are placed at opposite sides of a container, indicating the presence or absence of material in the horizontal radiation path. Beta or gamma sources are used, depending on the thickness and the density of the material to be measured. The method is used for containers of liquids or of grainy substances
- Thickness gauges: if the material is of constant density, the signal measured by the radiation detector depends on the thickness of the material. This is useful for continuous production, like of paper, rubber, etc.

Applications using ionization of gases by radiation

- To avoid the build-up of static electricity in production of paper, plastics, synthetic textiles, etc., a ribbon-shaped source of the alpha emitter ^{241}Am can be placed close to the material at the end of the production line. The source ionizes the air to remove electric charges on the material.
- Smoke detector: Two ionisation chambers are placed next to each other. Both contain a small source of ^{241}Am that gives rise to a small constant current. One is closed and serves for comparison, the other is open to ambient air; it has a gridded electrode. When smoke enters the open chamber, the current is disrupted as the smoke particles attach to the charged ions and restore them to a neutral electrical state. This reduces the current in the open chamber. When the current drops below a certain threshold, the alarm is triggered.
- Radioactive tracers for industry: Since radioactive isotopes behave, chemically, mostly like the inactive element, the behavior of a certain chemical substance can be followed by *tracing* the radioactivity. Examples:
  - Adding a gamma tracer to a gas or liquid in a closed system makes it possible to find a hole in a tube.

– Adding a tracer to the surface of the component of a motor makes it possible to measure wear by measuring the activity of the lubricating oil.

Potential electricity generation through nanomaterials

Using layers of carbon nanotubes interlaced with gold and lithium hydride, has been shown to produce a current when the gold particles are hit by radiation, releasing electrons which can travel through the carbon nanotubes to the lithium hydride, and then to electrodes in order to generate electricity.

Biological and Medical Applications of Ionizing Radiation

In biology, radiation is mainly used for sterilization, and enhancing mutations. For example, mutations may be induced by radiation to produce new or improved species. A very promising field is the sterile insect technique, where male insects are sterilized and liberated in the chosen field, so that they have no descendants, and the population is reduced.

Radiation is also useful in sterilizing medical hardware or food. The advantage for medical hardware is that the object may be sealed in plastic before sterilization. For food, there are strict regulations to prevent the occurrence of induced radioactivity. The growth of a seedling may be enhanced by radiation, but excessive radiation will hinder growth.

Electrons, X-rays, gamma rays or atomic ions may be used in radiation therapy to treat malignant tumors (cancer). Furthermore, just like in industrial application, X-rays can also be used in radiography to create images of hard-to-image objects, such as inside one's body. Tracer methods are used in nuclear medicine in a way analogous to the technical uses mentioned above.

RADIOIMMUNOASSAY

Radioimmunoassay (RIA) is a scientific method used to test antigens (for example, hormone levels in the blood) without the need to use a bioassay. It was developed by Rosalyn Yalow and Solomon Aaron Berson in the 1950s. In 1977, Rosalyn Sussman Yalow received the Nobel Prize in Medicine for the development of the RIA for insulin, the precise measurement of minute amounts of such a hormone was considered a breakthrough in endocrinology.

Although the RIA technique is extremely sensitive and extremely specific, it requires a sophisticated apparatus and is costly. It also requires special precautions, since radioactive substances are used. Therefore, today it has been largely supplanted by the ELISA method, where the antigen-antibody reaction is measured using colorometric signals instead of a radioactive signal.

To perform a radioimmunoassay, a known quantity of an antigen is made radioactive, frequently by labeling it with gamma-radioactive isotopes of iodine attached to tyrosine. This radiolabeled antigen is then mixed with a known amount of antibody for that antigen, and as a result, the two chemically bind to one another. Then, a sample of serum from a patient containing an unknown quantity of that same antigen is added. This causes the unlabeled (or "cold") antigen from the serum to compete with the radiolabeled antigen for antibody binding sites.

As the concentration of "cold" antigen is increased, more of it binds to the antibody, displacing the radiolabeled variant, and reducing the ratio of antibody-bound radiolabeled antigen to free radiolabeled antigen. The bound antigens are then separated from the unbound ones, and the radioactivity of the free antigen remaining in the supernatant is measured. A binding curve can then be plotted, and the exact amount of antigen in the patient's serum can be determined. With this technique, separating bound from unbound antigen is crucial. Initially, the method of separation employed was the use of a second "anti-antibody" directed against the first for precipitation and centrifugation.

General Procedure for Performing a RIA Analysis

- Mix sample containing drug with fixed quantity of labeled drug and antibody
- Allow to equilibrate - incubate
- Separate drug bound to antibody from unbound drug:
  - Charcoal adsorption of antibody (and bound drug)
  - Antibody - antibody binding precipitates bound drug
  - Antibody bonded to container
- Measure radioactivity associated with bound labeled drug:
  - low drug concentration means more bound radioactivity and higher measurement
  - high drug concentration means less bound radioactivity and lower measurement
- Determine standard curve:
  - Non-linear plot of radioactivity versus concentration
  - Logit-log concentration plot is linear

A Blank and Three Standard Samples

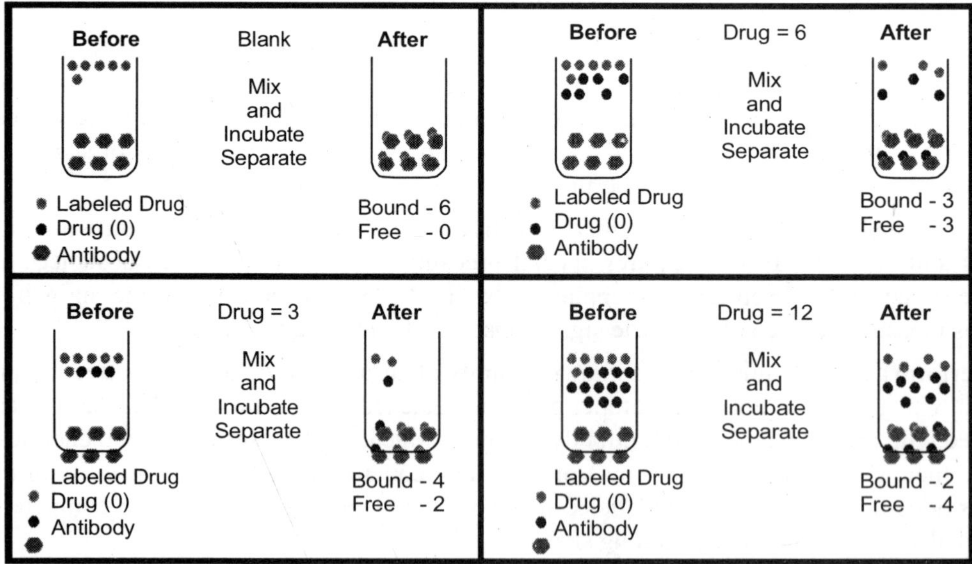

Fig. 4.4 : RIA before and after Incubation—Blank and Three Standard Samples

Table 4.2 : Total Bound and Free Drug Concentrations

Total [Drug]	Bound [Drug]	Free [Drug]
0	6	0
3	4	2
6	3	3
12	2	4

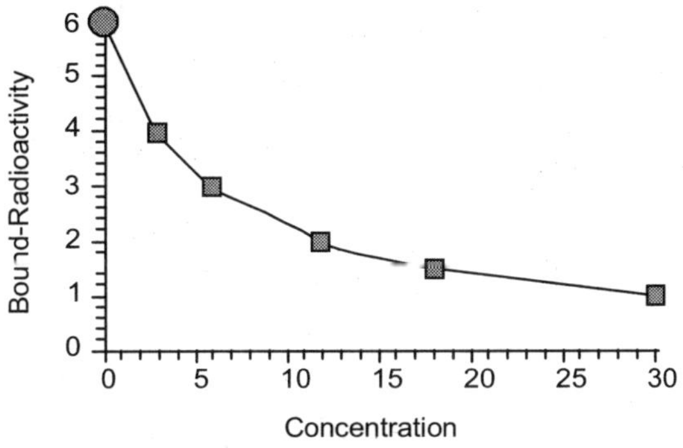

Fig. 4.5 : Plot of Bound versus Total Drug Concentration

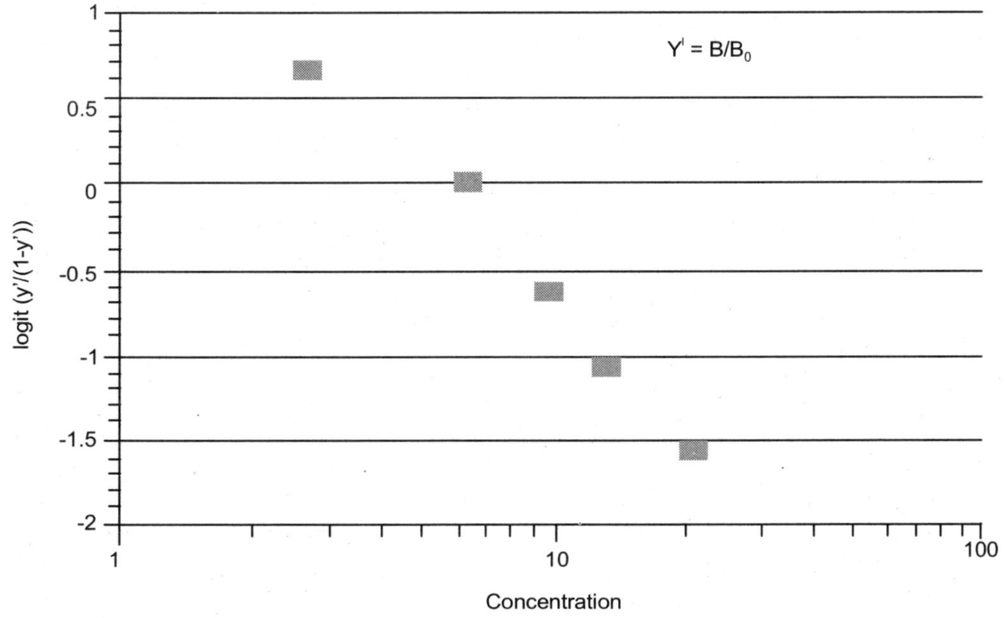

Fig. 4.6 : Logit versus Log Total C Plot

COLLOID AND CRYSTALLOID

A body that in solution can pass through a semipermeable membrane, as distinguished from a colloid, which cannot do so.

OSMOTIC PRESSURE

Generaly if two solutions of the same substance but of different concentrations be placed in contact with each other there will be a tendency on the one part of the solutions to equalize concentrations by movements of the solute from the more concentrated solution to the dilute one by diffiusion amd movement of solvent from the dilute solution to the concentrated one. This tendency proves the fact that the free energy of the solute in the concentrated solution is greater than that in the dilute one and hence there will be a tendency for equalization of concentration by movement of growth. The free energy of the solvent is greater in the dilute solution and hence the solvent will also flow in the opposite direction. If, however, we do not allow the solute by the help of a semi-permeable membrane which allows the solvent molecules to pass through but stops movement of the solute (like animal bladder or copper ferricyanide paper membrane) the solvent will flow from the dilute solution to the more concentrated one and will tend to dilute it. This phenomenon is osmosis. The reason for this phenomenon of osmosis can be explained by the help of Le Chatelier's Principle because the solution (under strain) will tend to go back to the state of pure solvent in its attempt undergo the effect of the stain.

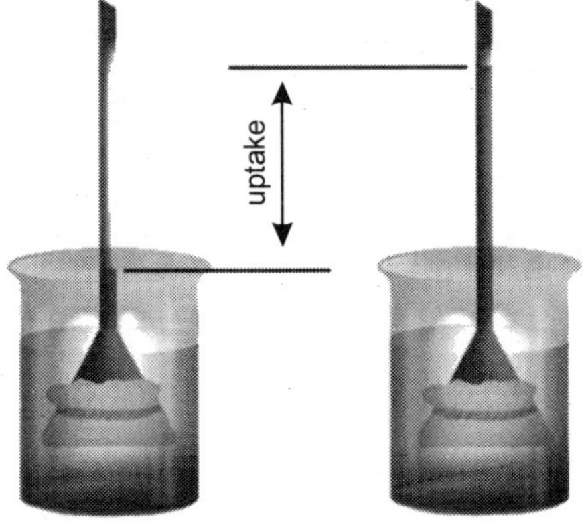

Fig. 4.7 : The Experiment of Osmosis

This osmosis experiment can be demonstrated simply by closing the broad end of a thistle funnel by an animal bladder or a parchment paper (see Figs.). This is then filled up with a sugar solution and dipped in the beaker containing plain water, such that the level of the solution inside the funnel is equal to the water level inside the beaker, and outside the funnel. After about 4 to 5 hours the solution level inside the funnel will be seen raised up indicating the entrance of the water from the beaker.

After 24 hours the level will not rise any further but halt at some height. This height will exert a pressure which is equal and opposite to what is called the osmotic pressure of the solution. The tendency on the part of the solution to get itself diluted it attracts the solvent towards itself. This

Fig. 4.8 : The Lab set up for Osmosis Experiment

attractive force per unit area is the *osmotic pressure.* This being balanced by the hydrostatic pressure set up in the funnel. This hydrostatic pressure is therefore the measure of the osmotic pressure in the funnel.

COLLOIDAL OSMOTIC PRESSURE

Oncotic pressure, or colloid osmotic pressure, is a form of osmotic pressure exerted by proteins in blood plasma that usually tends to pull water into the circulatory system.

Throughout the body, dissolved compounds have an osmotic pressure. Because large plasma proteins cannot easily cross through the capillary walls, their effect on the osmotic pressure of the capillary interiors will, to some extent, balance out the tendency for fluid to leak out of the capillaries. In other words, the oncotic pressure tends to pull fluid into the capillaries. In conditions where plasma proteins are reduced, e.g. from being lost in the urine (proteinuria) or from malnutrition, the result of low oncotic pressure can be excess fluid buildup in the tissues (Oedema).

Oncotic pressure is represented by the symbol ð in the Starling equation and elsewhere.

Used in Intravenous Therapy

Clinically, there are two types of fluids that are used for intravenous drips; crystalloids and colloids. Crystalloids are aqueous solutions of mineral salts or other water-soluble molecules. Colloids contain larger insoluble molecules, such as gelatin; blood itself is a colloid.

Colloids preserve a high *colloid osmotic pressure* in the blood, while, on the other hand, this parameter is decreased by crystalloids due to hemodilution. However, there is still controversy to the actual difference in efficacy by this difference. Another difference is that crystalloids generally are much cheaper than colloids.

Osmotic pressure and oncotic pressure can be *measured* by suitable instruments. They can also be *calculated* for an ideal solution by appropriate substitutions in the van't Hoff equation

Osmotic pressure = $n \times (c/M) \times RT$

where:

- n is the number of particles into which the substance dissociates (n = 1 for plasma proteins)
- c is the concentration in G/l
- M is the MW of the molecules
- c/M is thus the molar concentration of the substance
- R is the universal gas constant
- T is the absolute temperature (K)

As an example, if values are substituted in this equation for a typical plasma sample:

- T = 310K (ie temp of 37C)
- R = 0.082
- n = 1 (for plasma proteins as they do not dissociate)

and

- Multiplying by 0.001 to convert from Osmoles to mOsmoles
- Multiplying by 760 to convert the result from atmospheres to mmHg
- Multiplying by 280 to convert the osmotic pressure per mOsm/kg to a value for plasma with an osmolality of 280 mOsm/kg

then:

Total plasma osmotic pressure = $1 \times 0.082 \times 310 \times 0.001 \times 760 \times 280 = 5409$ mmHg

For a plasma osmolality of 280 mOsm/kg at 37C, total osmotic pressure is about 5409mmHg (i.e., about 7.1 atmospheres!)

Each mOsm/kg of solute contributes about 19.32mmHg to the osmotic pressure

Now consider the case of plasma proteins alone and calculate the colloid osmotic (oncotic) pressure.

Using typical values for concentration & MW of the plasma proteins, the protein concentration is about 0.9 mOsmol/kg which predicts an oncotic pressure of 17.3 mmHg (ie 19.32×0.9). Measurement in an oncometer shows the actual plasma oncotic pressure is about 25 mmHg which is equivalent to a plasma protein concentration of 1.3 mmol/kg.

ACID BASES AND PH

For thousands of years people have known that vinegar, lemon juice and many other foods taste sour. However, it was not until a few hundred years ago that it was discovered why these things taste sour - because they are all acids. The term acid, in fact, comes from the Latin term *acere*, which means "sour".

In the seventeenth century, the Irish writer and amateur chemist Robert Boyle first labeled substances as either acids or bases (he called bases alkalies) according to the following characteristics:

- **Acids** taste sour, are corrosive to metals, change litmus (a dye extracted from lichens) red, and become less acidic when mixed with bases.
- **Bases** feel slippery, change litmus blue, and become less basic when mixed with acids.

While Boyle and others tried to explain why acids and bases behave the way they do, the first reasonable definition of acids and bases would not be proposed until 200 years later.

In the late 1800s, the Swedish scientist Svante Arrhenius proposed that water can dissolve many compounds by separating them into their individual ions. Arrhenius suggested that **acids** are compounds that contain hydrogen and can dissolve in water to release hydrogen ions into solution. For example, hydrochloric acid (HCl) dissolves in water as follows:

$$\mathrm{HCl} \xrightarrow{\mathrm{H_2O}} \mathrm{H^+}(aq) + \mathrm{Cl^-}(aq)$$

Arrhenius defined bases as substances that dissolve in water to release hydroxide ions (OH-) into solution. For example, a typical base according to the Arrhenius definition is sodium hydroxide (NaOH):

$$\mathrm{NaOH} \xrightarrow{\mathrm{H_2O}} \mathrm{Na^+}(aq) + \mathrm{OH^-}(aq)$$

The Arrhenius definition of acids and bases explains a number of things. Arrhenius's theory explains why all acids have similar properties to each other (and, conversely, why all bases are similar): because all acids release H^+ into solution (and all bases release OH^-). The Arrhenius definition also explains Boyle's observation that acids and bases counteract each other. This idea, that a base can make an acid weaker, and vice versa, is called neutralization.

Neutralization: As you can see from the equations, acids release H^+ into solution and bases release OH^-. If we were to mix an acid and base together, the H^+ ion would combine with the OH^- ion to make the molecule H_2O, or plain water:

$$H^+ (aq) + OH^-(aq) \rightarrow H_2O$$

The neutralization reaction of an acid with a base will always produce water and a salt, as shown below:

Acid		Base		Water		Salt
HCl	+	NaOH	→	H_2O	+	NaCl
HBr	+	KOH	→	H_2O	+	KBr

Though Arrhenius helped explain the fundamentals of acid/base chemistry, unfortunately his theories have limits. For example, the Arrhenius definition does not explain why some substances, such as common baking soda ($NaHCO_3$), can act like a base even though they do not contain hydroxide ions.

In 1923, the Danish scientist Johannes Brønsted and the Englishman Thomas Lowry published independent yet similar papers that refined Arrhenius' theory. In Brønsted's words, ". . . acids and bases are substances that are capable of splitting off or taking up hydrogen ions, respectively." The Brønsted-Lowry definition broadened the Arrhenius concept of acids and bases.

The Brønsted-Lowry definition of acids is very similar to the Arrhenius definition, any substance that can donate a hydrogen ion is an acid (under the Brønsted definition, acids are often referred to as **proton** donors because an H^+ ion, hydrogen minus its electron, is simply a proton).

The Brønsted definition of **bases** is, however, quite different from the Arrhenius definition. The Brønsted base is defined as any substance that can accept a hydrogen ion. In essence, a base is the opposite of an acid. NaOH and KOH, as we saw above, would still be considered bases because they can accept an H^+ from an acid to form water. However, the Brønsted-Lowry definition also explains why substances that do not contain OH^- can act like bases. Baking soda ($NaHCO_3$), for example, acts like a base by accepting a hydrogen ion from an acid as illustrated below:

Acid		Base				Salt
HCl	+	$NaHCO_3$	→	H_2CO_3	+	NaCl

pH

Under the Brønsted-Lowry definition, both acids and bases are related to the concentration of hydrogen ions present. Acids increase the concentration of hydrogen ions, while bases decrease the concentration of hydrogen ions (by accepting them). The acidity or basicity of something therefore can be measured by its hydrogen ion concentration.

In 1909, the Danish biochemist Soren Sorensen invented the pH scale for measuring acidity. The pH scale is described by the formula:

pH = -log [H^+]-Note: concentration is commonly abbreviated by using square brackets, thus [H^+] = hydrogen ion concentration. When measuring pH, [H^+] is in units of moles of H^+ per liter of solution.

For example, a solution with [H^+] = 1 x 10^{-7} moles/liter has a pH equal to 7 (a simpler way to think about pH is that it equals the exponent on the H^+ concentration, ignoring the minus sign). The pH scale ranges from 0 to 14. Substances with a pH between 0 and less than 7 are acids (pH and [H^+] are inversely related - lower pH means higher [H^+]). Substances with a pH greater than 7 and up to 14 are bases (higher pH means lower [H^+]). Right in the middle, at pH = 7, are neutral substances, for example, pure water. The relationship between [H^+] and pH is shown in the table below alongside some common examples of acids and bases in everyday life.

HENDERSON-HASSELBACH EQUATION

In chemistry, the **Henderson–Hasselbalch** (often misspelled as ***Henderson–Hasselbach***) **equation** describes the derivation of pH as a measure of acidity (using pK_a, the acid dissociation constant) in biological and chemical systems. The equation is also useful for estimating the pH of a buffer solution and finding the equilibrium pH in acid-base reactions (it is widely used to calculate isoelectric point of the proteins).

Two equivalent forms of the equation are

$$pH = pK_a + \log\frac{[A]^-}{[HA]}$$

and

$$pH = pK_a + \log\left(\frac{[base]}{[acid]}\right).$$

Here, pK_a is $-\log(K_a)$ where K_a is the acid dissociation constant, that is:

for the non-specific Bronsted acid-base reaction:

In these equations, A^- denotes the ionic form of the relevant acid. Bracketed quantities such as [base] and [acid] denote the molar concentration of the quantity enclosed.

In analogy to the above equations, the following equation is valid:

Where BH^+ denotes the salt of the corresponding base B.

BUFFER

A buffer solution is an aqueous solution consisting of a mixture of a weak acid and its conjugate base or a weak base and its conjugate acid. It has the property that the pH of the solution changes very little when a small amount of acid or base is added to it. Buffer solutions are used as a means of keeping pH at a nearly constant value in a wide variety of chemical applications.

Theory

In a simple buffer solution there is an equilibrium between a weak acid, HA, and its conjugate base, A^-.

$$HA + H_2O \rightarrow H_3O^+ + A^-$$

When hydrogen ions are added to the solution the equilibrium moves to the left, in accordance with Le Chatelier's principle, as there are hydrogen ions on the right-hand side of the equilibrium expression. When hydroxide ions are added the equilibrium moves to the right as hydrogen ions are removed in the reaction $H^+ + OH^- \rightarrow H_2O$. Thus, some of the added reagent is consumed in shifting the equilibrium and the pH changes by less than it would do if the solution were not buffered.

The acid dissociation constant for a weak acid, HA, is defined as

$$K_a = \frac{[H^+][A^-]}{[HA]}$$

Simple manipulation with logarithms gives the Henderson-Hasselbalch equation, which describes pH in terms of pK_a

$$pH = pK_a + \log_{10}\frac{[A^-]}{[HA]}.$$

In this equation $[A^-]$ is the concentration of the conjugate base and [HA] is the concentration of the acid. It follows that when the concentrations of acid and conjugate base are equal, often described as half-neutralization, $pH=pK_a$. In general, the pH of a buffer solution may be easily calculated, knowing the composition of the mixture, by means of an ICE table.

The same considerations apply to a mixture of a weak base, B and its conjugate acid BH^+.

$$B + H_2O \rightleftharpoons BH^+ + OH^-.$$

The pK_a value to be used is that of the acid conjugate to the base.

In general a buffer solution may be made up of more than one weak acid and its conjugate base; if the individual buffer regions overlap a wider buffer region is created by mixing the two buffering agents.

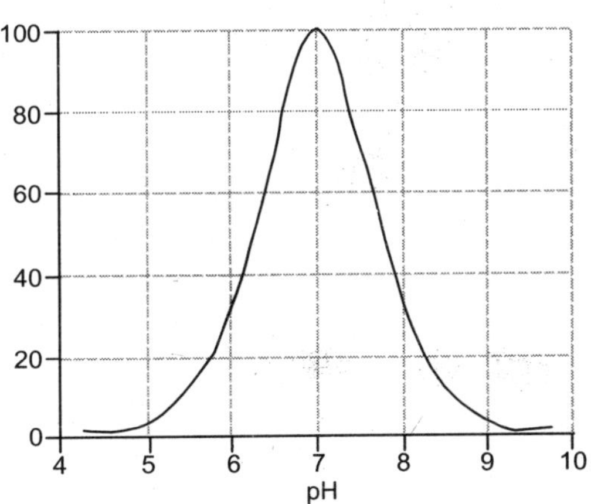

Fig. 4.9 : Buffer capacity for pK_a=7 as percentage of maximum

Buffer Capacity

Buffer capacity is a quantitative measure of the resistance of a buffer solution to pH change on addition of hydroxide ions. It can be defined as follows.

$$\text{buffer capacity} = \frac{dn}{d\,(pH)}$$

where dn is an infinitesimal amount of added base and d(pH) is the resulting infinitesimal change in pH. With this definition the buffer capacity can be expressed as:

$$\frac{dn}{d(pH)} = 2.303\left(\frac{K_w}{[H^+]} + [H^+] + \frac{C_A K_a [H^+]}{(K_a + [H^+])^2}\right)$$

where K_w is the self-ionization constant of water and C_A is the analytical concentration of the acid, equal to $[HA]+[A^-]$. The term $K_w/[H^+]$ becomes significant at pH greater than about 11.5 and the second term becomes significant at pH less than about 2. Both these terms are properities of water and are independent of the weak acid. Considering the third term, it follows that

Buffer capacity of a weak acid reaches its maximum value when pH = pK_a

At pH = $pK_a \pm 1$ the buffer capacity falls to 33% of the maximum value. This is the approximate range within which buffering by a weak acid is effective. Note: at pH = pK_a - 1, The Henderson-Hasselbalch equation shows that the ratio $[HA]:[A^-]$ is 10:1.

Buffer capacity is directly proportional to the analytical concentration of the acid.

Applications

Their resistance to changes in pH makes buffer solutions very useful for chemical manufacturing and essential for many biochemical processes. The ideal buffer for a particular pH has a pK_a equal to that pH, since such a solution has maximum buffer capacity.

Buffer solutions are necessary to keep the correct pH for enzymes in many organisms to work. Many enzymes work only under very precise conditions; if the pH strays too far out of the margin, the enzymes slow or stop working and can denature, thus permanently disabling its catalytic activity. A buffer of carbonic acid (H_2CO_3) and bicarbonate (HCO_3^-) is present in blood plasma, to maintain a pH between 7.35 and 7.45.Industrially, buffer solutions are used in fermentation processes and in setting the correct conditions for dyes used in colouring fabrics. They are also used in chemical analysis and calibration of pH meters.Majority of biological samples that are used in research are made in buffers specially PBS (phosphate buffer saline) at pH 7.4.

Useful Buffer Mixtures

Components	pH range
HCl, sodium citrate	1 - 5
citric acid, sodium citrate	2.5 - 5.6
acetic acid, sodium acetate	3.7 - 5.6
Na_2HPO_4, NaH_2PO_4	6 - 9
Borax, sodium hydroxide	9.2 - 11

"Universal" Buffer Mixtures

By combining substances with pK_a values differing by only two or less and adjusting the pH a wide-range of buffers can be obtained. Citric acid is a useful component of a buffer mixture because it has three pK_a values, separated by less than two. The buffer range can be extended by adding other buffering agents. The following two-component mixtures have a buffer range of pH 3 to 8.

0.2M Na_2HPO_4 /mL	0.1M Citric Acid /mL	pH...
20.55	79.45	3.0
38.55	61.45	4.0
51.50	48.50	5.0
63.15	36.85	6.0
82.35	17.65	7.0
97.25	2.75	8.0

A mixture containing citric acid, potassium dihydrogen phosphate, boric acid, and diethyl barbituric acid can be made to cover the pH range 2.6 to 12.

Table 4.3 : Common Buffer Compounds Used in Biology

Common Name	pK_a at 25°C	Buffer Range	Temp Effect d pH/d T in (1/K) **	Mol. Weight	Full Compound Name
TAPS	8.43	7.7–9.1	–0.018	243.3	3-{[tris(hydroxymethyl) methyl] amino} propanesulfonic acid
Bicine	8.35	7.6–9.0	–0.018	163.2	N,N-bis(2-hydroxyethyl) glycine
Tris	8.06	7.5–9.0	–0.028	121.14	tris (hydroxymethyl) methylamine
Tricine	8.05	7.4–8.8	–0.021	179.2	N-tris (hydroxymethyl) methylglycine
HEPES	7.48	6.8–8.2	–0.014	238.3	4-2-hydroxyethyl-1-piperazineethanesulfonic acid
TES	7.40	6.8–8.2	–0.020	229.20	2-{[tris(hydroxymethyl) methyl] amino} ethanesulfonic acid
MOPS	7.20	6.5–7.9	–0.015	209.3	3-(N-morpholino) propanesulfonic acid
PIPES	6.76	6.1–7.5	–0.008	302.4	piperazine-N,N2 -bis (2-ethanesulfonic acid)
Cacodylate	6.27	5.0–7.4		138.0	dimethylarsinic acid
MES	6.15	5.5–6.7	–0.011	195.2	2-(N-morpholino) ethanesulfonic

The kidneys and the lungs work together to help maintain a blood pH of 7.4 by affecting the components of the buffers in the blood. Therefore, to understand how these organs help control the pH of the blood, we must first discuss how buffers work in solution.

Acid-base buffers confer resistance to a change in the pH of a solution when hydrogen ions (protons) or hydroxide ions are added or removed. An acid-base buffer typically consists of a **weak acid**, and its **conjugate base** (salt) (see Equations 2-4). Buffers work because the concentrations of the weak acid and its salt are large compared to the amount of protons or hydroxide ions added or removed. When protons are added to the solution from an external source, some of the base component of the buffer is converted to the weak-acid component (thus using up most of the protons added); when hydroxide ions are added to the solution (or, equivalently, protons are removed from the solution; see Equations 8-9), protons are dissociated from some of the weak-acid molecules of the buffer, converting them to the base of the buffer (and thus replenishing most of the protons removed). However, the change in acid and base concentrations is small relative to the amounts of these species present in solution. Hence, the **ratio** of acid to base changes only slightly. Thus, the effect on the pH of the solution is small, within certain limitations on the amount of H^+ or OH^- added or removed.

The Carbonic-Acid-Bicarbonate Buffer in the Blood

By far the most important buffer for maintaining acid-base balance in the blood is the carbonic-acid-bicarbonate buffer. The simultaneous equilibrium reactions of interest are

$$H^+(aq) + HCO_3^-(aq) \rightleftharpoons H_2CO_3(aq) \rightleftharpoons H_2O(I) + CO_2(g) \quad (1)$$

We are interested in the change in the pH of the blood; therefore, we want an expression for the concentration of H^+ in terms of an equilibrium constant (see blue box, below) and the concentrations of the other species in the reaction (HCO_3^-, H_2CO_3, and CO_2).

Recap of Fundamental Acid-Base Concepts

An **acid** is a chemical species that can donate a proton (H^+), and a **base** is a species that can accept (gain) a proton, according to the common Bronstead-Lowry definition. (A subset of the Brønstead-Lowry definition for aqueous solutions is the Arrhenius definition, which defines an acid as a proton producer and a base as a hydroxide (OH^-) producer.) Hence, the conjugate base of an acid is the species formed after the acid loses a proton; the base can then gain another proton to return to the acid. In solution, these two species (the acid and its conjugate base) exist in equilibrium.

$$pH = -\log\left[H^+\right],_{-2} \quad (2)$$

where $[H^+]$ is the molar concentration of protons in aqueous solution. When an acid is placed in water, free protons are generated according to the general reaction shown in Equation 3. [**Note**: HA and A^- are generic symbols for an acid and its deprotonated form, the conjugate base.]

$$\underset{\text{acid}}{HA} + \underset{\text{base}}{H_2O} \rightleftharpoons \underset{\text{conjugatedacid}}{H_3O^+} + \underset{\text{conjugatedbase}}{A} \quad (3)$$

Equation 3 is useful because it clearly shows that HA is a Brønstead-Lowry acid (giving up a proton to become A^-) and water acts as a base (accepting the proton released by HA). However, the nomenclature H_3O^+ is somewhat misleading, because the proton is actually solvated by many water molecules. Hence, the equilibrium is often written as Equation 4, **where H_2O is the base**:

$$HA \rightleftharpoons H^+ + A^- \tag{4}$$

The Law of Mass Action and Equilibrium Constants

Using the Law of Mass Action, which says that for a balanced chemical equation of the type

$$aA + bB \overset{K}{\rightleftharpoons} cC + dD \tag{5}$$

in which A, B, C, and D are chemical species and a, b, c, and d are their stoichiometric coefficients, a constant quantity, known as the **equilibrium constant** (K), can be found from the expression:

$$K = \frac{[C]^c [D]^d}{[A]^a [B]^b} \tag{6}$$

where the brackets indicate the concentrations of species A, B, C, and D **at equilibrium**.

Equilibrium Constant for an Acid-Base Reaction

Using the Law of Mass Action, we can also define an equilibrium constant for the acid dissociation equilibrium reaction in Equation (4). This equilibrium constant, known as K_a, is defined by Equation 7:

$$K_a = \frac{\left[H^+\right]\left[A^-\right]}{[HA]} \tag{7}$$

Equilibrium Constant for the Dissociation of Water

One of the simplest applications of the Law of Mass Action is the dissociation of water into H^+ and OH^- [Equation (8)].

$$H_2O(I) \rightleftharpoons H^+(aq) + OH^-(aq) \tag{8}$$

The equilibrium constant for this dissociation reaction, known as K_w, is given by

$$K_w = \left[H^+\right]\left[OH^-\right] \tag{9}$$

(H_2O is not included in the equilibrium-constant expression because it is a pure liquid.) Hence, we can see that increasing the OH^- concentration of an aqueous solution has the effect of decreasing the H^+ concentration, because the product of these two concentrations must remain constant at a given temperature. Thus, in water, the equilibrium in Equation (8) underlies the equivalency of the Brønstead-Lowry definition of a base (an H^+ acceptor) and the Arrhenius definition of a base (an OH^- producer).

To more clearly show the two equilibrium reactions in the carbonic-acid-bicarbonate buffer, Equation 1 is rewritten to show the direct involvement of water:

acid-base reaction

$$H_3O + (aq) + HCO_3^- (aq) \underset{}{\overset{K_1}{\rightleftharpoons}} H_2CO_3(aq) + H_2O\ (l) \overset{K_2}{\rightleftharpoons} 2\ H_2O(l) + CO_2\ (g) \quad (10)$$

not an acid-base reaction

The equilibrium on the left is an acid-base reaction that is written in the reverse format from Equation 3. Carbonic acid (H_2CO_3) is the acid and water is the base. The conjugate base for H_2CO_3 is HCO_3^- (bicarbonate ion). Carbonic acid also dissociates rapidly to produce water and carbon dioxide, as shown in the equilibrium on the right of Equation (10). This second process is not an acid-base reaction, but it is important to the blood's buffering capacity, as we can see from Equation (11), below:

$$pH = pK - \log\left(\frac{[CO_2]}{HCO_3^-}\right) \quad (11)$$

The derivation for this equation is shown in the yellow box, below. Notice that Equation (11) is in a similar form to the **Henderson-Hasselbach equation** presented in the introduction to the Experiment (Equation 16 in the lab manual). Equation 11 does not meet the strict definition of a Henderson-Hasselbach equation, because this equation takes into account a non-acid-base reaction (*i.e.,* the dissociation of carbonic acid to carbon dioxide and water), and the ratio in parentheses is not the concentration ratio of the acid to the conjugate base. However, the relationship shown in Equation (11) is frequently referred to as the Henderson-Hasselbach equation for the buffer in physiological applications.

In Equation (11), pK is equal to the negative log of the equilibrium constant, K, for the buffer [Equation (12)].

where $K=K_a/K_2$ [from Equation (10)].

$$pK = -\log K \quad (12)$$

This quantity provides an indication of the degree to which HCO_3^- reacts with H^+

[or with H_3O^+ as written in Equation (10)] to form H_2CO_3, and subsequently to form CO_2 and H_2O. In the case of the carbonic-acid-bicarbonate buffer, pK=6.1 at normal body temperature.

Derivation of the pH Equation for the Carbonic-Acid-Bicarbonate Buffer

We may begin by defining the equilibrium constant, K_1, for the left-hand reaction in Equation 10, using the Law of Mass Action:

$$K_1 = \frac{[H_2CO_3]}{[H^+][HCO_3^-]} \quad (13)$$

K_a (see Equation (9), above) is the equilibrium constant for the acid-base reaction that is the reverse of the left-hand reaction in Equation (10). It follows that the formula for K_a is

$$K_a = \frac{1}{K_1} = \frac{[H^+][HCO_3^-]}{[H_2CO_3]} \tag{14}$$

The equilibrium constant, K_2, for the right-hand reaction in Equation (10) is also defined by the Law of Mass Action:

$$K_2 = \frac{[CO_2]}{[H_2CO_3]} \tag{15}$$

Because the two equilibrium reactions in Equation (10) occur simultaneously, Equations 14 and 15 can be treated as two simultaneous equations. Solving for the equilibrium concentration of carbonic acid gives

$$[H_2CO_3] = \frac{[H^+][HCO_3^-]}{K_a} = \frac{[CO_2]}{K_2} \tag{16}$$

Rearranging Equation 16 allows us to solve for the equilibrium proton concentration in terms of the two equilibrium constants and the concentrations of the other species:

$$[H^+] = \left(\frac{K_a}{K_2}\right)\left(\frac{[CO_2]}{[HCO_3]}\right) \tag{17}$$

Because we are interested in the pH of the blood, we take the negative log of both sides of Equation (17):

$$-\log[H^+] = -\log(K) - \log\left(\frac{[CO_2]}{[HCO_3^-]}\right) \tag{18}$$

Recalling the definitions of pH and pK (Equations (2) and (12), above), Equation (18) can be rewritten using more conventional notation, to give the relation shown in Equation (11), which is reproduced below:

$$pH = pK - \log\left(\frac{[CO_2]}{[HCO_3^-]}\right)$$

As shown in Equation (11), the pH of the buffered solution (*i.e.,* the blood) is dependent only on the **ratio** of the amount of $\mathbf{CO_2}$ present in the blood to the amount of $\mathbf{HCO_3^-}$ (bicarbonate ion) present in the blood (at a given temperature, so that **pK remains constant**). This ratio remains relatively constant, because the concentrations of both buffer components (HCO_3^- and CO_2) are very large, compared to the amount of H^+ added to the blood during normal activities and moderate exercise. When H^+ is added to the blood as a result of metabolic processes, the amount of HCO_3^- (relative to the amount of CO_2) decreases; however, the amount of the change is tiny compared to the amount of HCO_3^- present in the blood. This optimal buffering occurs when the pH is within approximately 1 pH unit from the pK value for the buffering system, *i.e.,* when the pH is between 5.1 and 7.1.

However, the normal blood pH of 7.4 is outside the optimal buffering range; therefore, the addition of protons to the blood due to strenuous exercise may be too great for the buffer alone to effectively control the pH of the blood. When this happens, other organs must help control the amounts of CO_2 and HCO_3^- in the blood. The lungs remove excess CO_2 from the blood (helping to raise the pH via shifts in the equilibria in Equation 10), and the kidneys remove excess HCO_3^- from the body (helping to lower the pH). The lungs' removal of CO_2 from the blood is somewhat impeded during exercise when the heart rate is very rapid; the blood is pumped through the capillaries very quickly, and so there is little time in the lungs for carbon dioxide to be exchanged for oxygen.

BIOLOGICAL OXIDATION

Introduction

The oxidative degradation of carbohydrates, fats and amino acids at cellular level needs oxygen and any metabolism after complete oxidation forms CO_2 and H_2O. Hence, any biological oxidation taking place at tissue level is associated with the uptake of oxygen and release of carbon dioxide and rapidly liberates energy. This biological oxidation accompanied by specific enzymes and coenzymes in a step wise fashion involves the union between hydrogen atoms with oxygen atom to form water. During the electron transport, the electrons are transferred from organic substrates to oxygen yielding energy in the generation bond energy in the form of Adenosine triphosphates (ATP) from Adenosine diphosphates (ADP).

ATP and ADP are known as high energy phosphates as the cleavage of phosphate bond in them yield energy and inorganic phosphate. This energy is utilized for the anabolic and catabolic processes. The oxidative phosphorylation enables the aerobic living organisms to capture a far greater proportion of available free energy of the oxidizing substrates inthe form of ATP. Oxidation involving phosphorylation is a very vital process and it is a continuous process and any disturbance of its function is incompatible with life.

Redox Couple

Always every ***oxidation is accompanied by a reduction process***. All such reactions are termed as ***oxidation-reduction reactions*** and shortly referred as ***redox.*** These redox reactions are associated with movements of electron. The electron donor is called as reductant or reducing agent and the electron acceptor, the oxidant or oxidizing agent. The system which transfer its electron is changed into oxidant form while the system which accepts electrons gets converted to the reductant form.

Oxidation reduction system is simplified and shown below (Fig. 4.10)

$$\text{Electron donor} \longrightarrow e^- + \text{electro acceptor}$$

A specific example is oxidation of ferrous iron to ferric iron indicates the removal of electron (e^-) from ferrous iron.

$$\underset{\text{(Ferrous ion)}}{Fe^{++}} \longrightarrow \underset{\text{(Ferric ion)}}{Fe^{+++}} + e^-$$

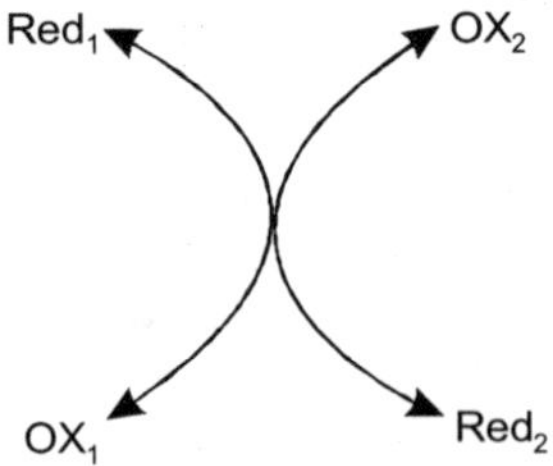

Fig. 4.10 : Oxidation-Reduction System

Since, the electron is not stable in the free form it gets attached to another molecule and thus every oxidation is followed by a reduction. Always these redox reactions are energy yielding. A direct transfer of electrons from substrate to the oxidant would liberate a sudden burst of energy and most of it will be wasted by dissipation. Normally when the electrons of hydrogen combine with oxygen results in explosion. In biological system this oxidation reduction process takes place smoothly without increasing the temperature because the transfer of hydrogen pairs

occurs in a step by step process till it reacts with oxygen. This permits the liberation of energy in small amounts so that it can be captured and saved.

Redox Potential

In oxidation and reduction reactions the free energy exchange is proportionate to the tendency of reactants to donate or to accept electrons. The affinity for the electron by the oxidant is called the electron affinity or redox potential. In biochemistry, the oxygen has the highest redox potential or electron affinity (E0) and therefore the electron pass from the systems of hydrogen donors which have lower potentials. It is usual to compare the redox potential of a system (E0) against the potential of the hydrogen electrode, which is at pH 0 designated as 0.0 volts. However, in a biological system it is normal to express the redox potential (E0) at pH 7.0 at which the electrode potential of hydrogen electrode is - 0.42volts. In the biological system, the enzymes concerned with this oxidation reduction processes are designated as oxidoreductases. Some of the redox potentials are given in the table below:

Some of the redox potential of special interest shown in Table 4.4.

Table 4.4

System	E_0 (volts)
H^+/H_2	– 0.42
Oxygen/water	+ 0.82
Cytochrome a Fe^{3+}/Fe^{2+}	+ 0.29
Cytochrome b Fe^{3+}/Fe^{2+}	+ 0.08
Cytochrome c Fe^{3+}/Fe^{2+}	+ 0.22
NAD^+ / NADH + H +	– 0.32

A positive value for the standard reduction potential means that a compound in question preferentially is reduced when involved in a redox reaction with hydrogen. A negative value means that a compound in question is preferentially oxidized. However, listings of standard reduction potentials are always given in the form of a reduction reaction.

Electron Transport System

The electron transport chain (Fig. 4.11) consists of series of proteins which tightly bound involves the passage of a pair of electrons from one chemical to the next, whereby each chemical in the sequence has less reduced energy than the previous. The electron transport chain oxidizes (i.e. "burns") the $NADH^+H^+$ and FADH2 cofactors, using molecular oxygen as the final electron acceptor. In the electron transport chain electron carriers and hydrogen-electron acceptors are positioned alternatively to carry the function. There are three different regions in the electron transport chain, where energy is released. In each region there is a formation of one ATP. All these reactions and capturing of energy takes place in mitochondria.

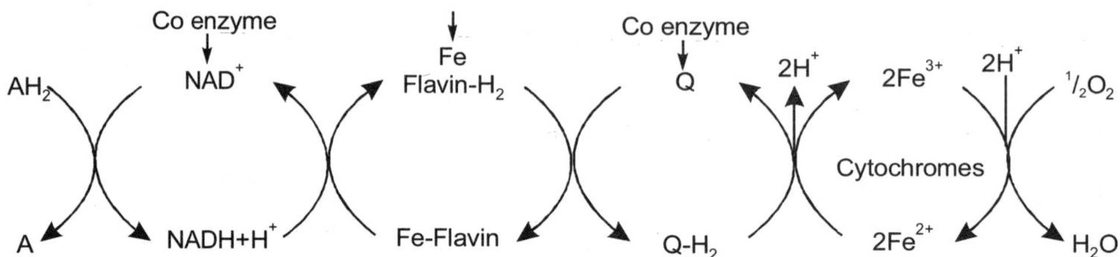

Fig. 4.11 : The Electron Transport Chain in the Enzyme Reaction System

Components of Mitochondria with Marker Enzymes

The histochemical and ultra centrifugation studies clearly established that the major site of cellular oxidation is mitochondria. These are sub cellular organelles and quite vary in size and shape. Ellipsoidal, spherical or rod shaped structures measuring about 0.5-5 μ in length and 0.1-0.6μ in width. Since, the energy released in the oxidation process is converted into chemical energy (ATP). Mitochondria otherwise called as **power house** of the cell.

Hence the number of mitochondria in a cell depends on it's metabolic activity. All the reducing equivalence that can release energy during oxidation of carbohydrates, fatty acids and proteins are available in the mitochondria. In mitochondria, a series of catalysts referred as respiratory chain that collects these reducing equivalents and direct them towards oxygen to form water. The electron microscopic picture of the mitochondria shows a double membrane, an ***outer and inner membrane which*** consists of different specific enzymes. The folding of the inner membrane produces a number of partitions called ***cristae*** that extend into the matrix. The inner membrane encloses the matrix and it is very selective in its permeability. Inner membrane is highly complex in its structure and function. The space between the inner and the outer membrane is called as ***Inter membrane space*** which is surrounded by **matrix.** The mitochondria contains its own circular DNA and ribosomes. Some mitochondrial proteins are thus coded for and produced by the mitochondria itself. Other mitochondrial proteins are coded by nuclear DNA, synthesized by cytosolic ribosomes, and subsequently transported to the mitochondria. The structure of mitochondria (Fig. 4.12) and the location of various essential enzymes are given in the form of diagram.

The electron transport chain is initiated by the reaction of an organic metabolite (intermediate in metabolic reactions) with the coenzyme NAD^+ (nicotinamide adenine dinucleotide is a coenzyme containing the B-vitamin, nicotinamide). This is an oxidation reaction where 2 hydrogen atoms (or 2 hydrogen ions and 2 electrons) are removed from the organic metabolite. (The organic metabolites are usually from the citric acid cycle and the oxidation of fatty acids—details in following pages). The reaction can be represented simply where M = any metabolite.

$$MH2 + NAD^+ \rightarrow NADH + H^+ + M + energy$$

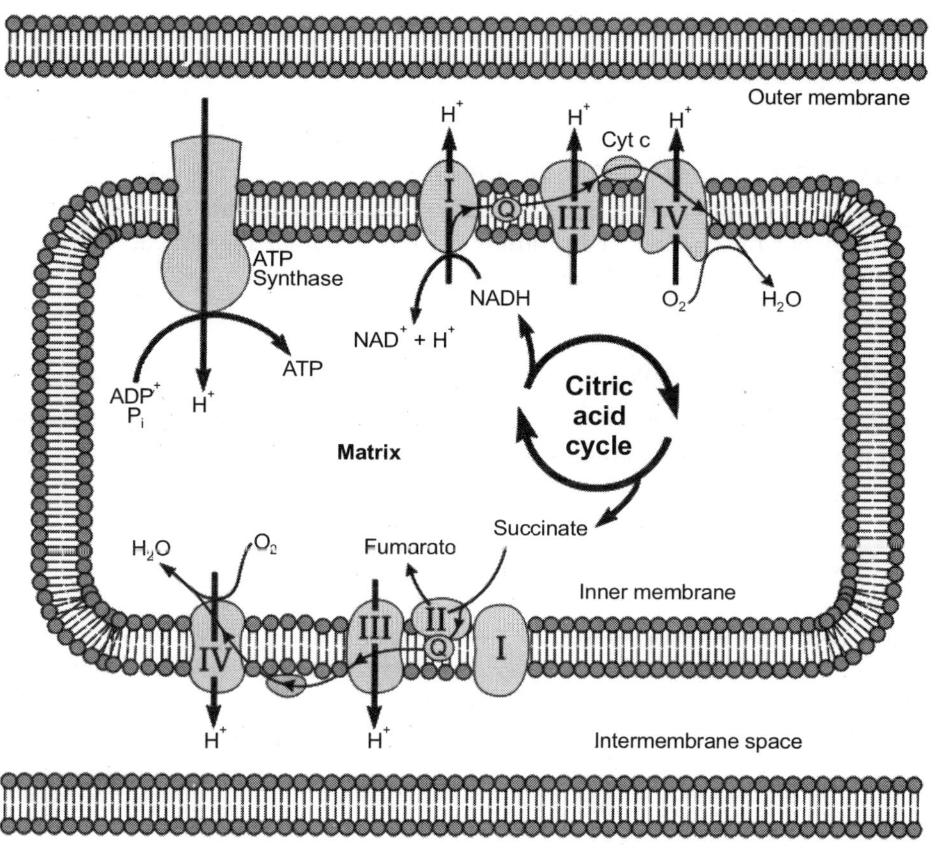

Fig. 4.12 : The Enzymes used Inside the Mitochondria

Complex I—*NADH dehydrogenase*, also called NADH coenzyme Q reductase located in the inner mitochondrial membrane and also contains non heme iron atoms. These dehydrogenase enzyme does not react with oxygen instead an electron carrier is interposed between the metabolite and next member in the chain. These enzymes consist of a protein part and a non protein part which is a coenzyme. The co enzyme NAD^+ or $NADP^+$ are utilized as the prime carriers of hydrogen.

Complex II—*Coenzyme Q* (Q for Quinone) or cytochrome c reductase is a Ubiquinone. It is in the inner membrane in the free form or protein bound form. Coenzyme Q occupies the position between metalloflavoproteins and cytochrome in the chain. At the point of coenzyme the H^+ ion dissociate and go into solution, leaving the electrons to the cytochromes.

Complex III—Cytochrome c oxidase. Cytochromes are very similar to the structure of myoglobin or hemoglobin. The significant feature is the heme structure containing the iron (Fe) ions, initially in the +3 state andchanged to the +2 state by the addition of an electron. Cytochrome molecules accept only the electron from each hydrogen, not the entire atom. The several types of cytochromes hold electrons at slightly different energy levels. Electrons are passed along from one cytochrome to the next in the chain, losing energy as they go. Finally, the last cytochrome in the chain, cytochrome a3, passes two electrons to molecular oxygen. These cytochromes are proteins that carry a prosthetic group that has an embedded metal atom. The protein 'steals' the ability of the metal atom to accept and release electrons.

Complex IV—ATP synthase, also known as the F0 F1 particle has two components F0 and F1 (F - indicates the factor). F1 protruding into matrix from the inner membrane and F0 embedded and extend across the inner membrane. The protruding F1 is essential for the energy coupling to ATP molecule. Careful removal of this component (experimentally) leads to impairment in ATP production though the intact respiratory chain is present.

Reactions of Electron Transport

The electron acceptors in the electron transport chain include FMN, ubiquinone (CoQ), and a group of closely related proteins called cytochromes. Figure 4.14 shows arrangement of the protein

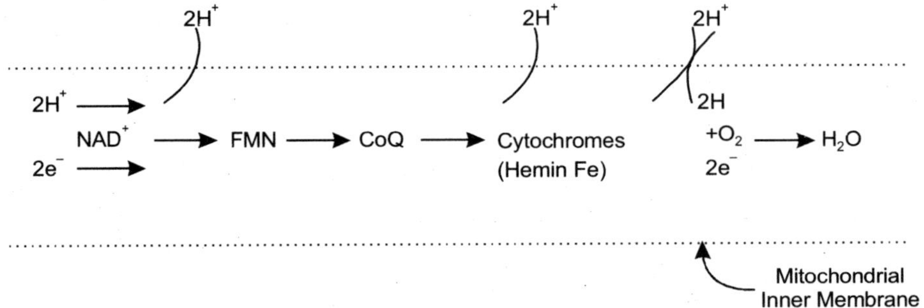

Fig. 4.13 : The Simplified Electron Transport

Oxidative Phosphorylation in Electron Transport Chain consist of the electron donors

NADH+H+

FADH2

The coupled oxidation/reduction reaction is:

$$NADH + H^+ + 1/2O_2 \longrightarrow H_2O + NAD^+$$
$$FADH_2 + 1/2O_2 \longrightarrow H_2O + FAD$$

This coupled reactions yield free energy NADH+H+ yields 52 Kcal/mole as the electrons from NADH+H+ transfer to oxygen consist of three pumps yield 3 ATP molecule at 3 sites FADH2 yields 36 Kcal/mole as the electron from FADH2 transfer to oxygen there are two pumps yield 2 ATP molecule at 2 sites. The position at which the energy capture occurring as ATP are given in Figs. 4.14 and 4.16.

Spontaneous flow of electrons through each of the respiratory chaincomplexes I, III, and IV is coupled to **ejection of H+** from the mitochondrial matrix to inner membrane space. The ejection of proton gradient is done through inner membrane protein, **ATPase that uses released energy to drive the synthesis of ATP from ADP.** The terminal acceptor of electrons is molecular oxygen and it is reduced to water. However not all the energy released are captured as high energy phosphate bond and liberated as heat. In warm blooded animals this heat is used for the maintenance of body temperature. The important ***respiratory control of electron transport chain is the availability of ADP,*** the substrate for the ATP Synthase.

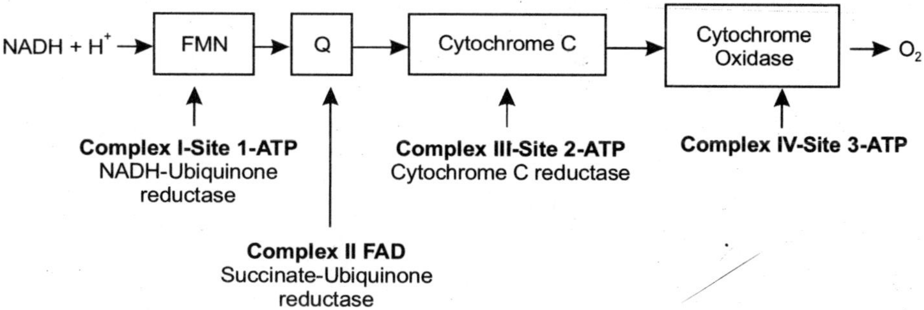

Fig. 4.14 : Arrangement of Proteins in Electron Transport Chain

The use of inhibitors gives much information about the electron transport chain.

They are classified as:

(*a*) **Inhibitors that arrest respiration** are barbiturates like amobarbital, antibiotic like piericidin A, antimycin A and fish poison retinone. **The carbon monoxide and cyanide inhibit cytochrome oxidase** so that it cannot transport electrons to oxygen. This blocks the further passage of electrons through the chain, halting ATP production and life.

(*b*) **Inhibitors of oxidative phosphorylation** are oligomycin and atrctyloside.

(*c*) **Uncouplers:** dissolve in the membrane, and function as **carriers for H+ or it can be an ionophores.** Uncouplers block oxidative phosphorylation by dissipating the H^+ electro chemical gradient by an un coupling the essential linkage between electron transport and ATP synthesis. Un couplers are 2,4 dinitro phenol, dinitrocresol, pentacholorophenol.

Ionophores (ion carriers) are lipid soluble substance capable of carrying specific ions through the membrane. They slightly differ in their action from the uncouplers as they also transport cation other than H+ through the membrane. Valiomycin forms a lipid complex through which the K^+ ion readily pass through. The ionophore gramicidin induces penetration to H^+, K^+ or Na^+ and uncouples the oxidative phosphorylation.

Oxidative phosphorylation Hydrogens or their electrons, pass down the electron transport chain in a series of redox reaction. The electrons entering the electron transport system have relatively high energy content. As they pass along the chain of electron acceptors, they lose much of their energy, some of which is used to pump the protons across the inner mitochondrial membrane. The flow of electrons in the electron transport is usually coupled tightly to the production of ATP with the help of the enzyme ATP synthetase, and it does not occur unless the phosphorylation of ADP can also proceed.

This prevents a waste of energy, because high-energy electrons do not flow unless ATP can be produced. Because the phosphorylation of ADP to form ATP is coupled with the oxidation of electron transport components, this process of making ATP is referred to as oxidative phosphorylation. The electron transport and oxidative phosphorylation depends upon the availability of ADP and Pi and it is referred as acceptor control of respiration

Chemiosmotic theory Peter Mitchell got the Nobel prize in 1978 for his theory of chemiosmosis. **The chemiosmotic theory of Mitchell claims that oxidation of components in respiratory chain generates hydrogen ion and ejected across the inner membrane.** The electrochemical potential difference resulting from the asymmetric distribution of the hydrogen ion is used as the driving force (potential energy). This consist of a **chemical concentration gradient of protons across the membrane (pH gradient) also provides a charge gradient.** The inner mitochondrial membrane is impermeable to the passage of protons, which can flow back into the matrix of the mitochondrion only through special channels in the inner mitochondrial membrane. In these channels, the enzyme ATP synthetase is present. As the protons move down the energy gradient **(proton motive force = chemiosmotic energy)**, the energy releases is used by ATP synthetase to produce ATP. The chemiosmotic model explains that this electrochemical potential difference across the membrane is used to drive a membrane located ATP synthetase which couple the energy to ADP, to form ATP. Protons are pumped across the inner mitochondrial membrane by three electron transfer complexes, each associated with particular steps in the electron transport system. As electrons are transferred along the acceptors in the electron transport chain, sufficient energy is released at three points to convey protons across the inner mitochondrial membrane and ultimately to synthesize ATP.

Role of $F_0 - F_1$ ATPase ATP synthetase or F_0 F_1 ATPase, has two major components, F_0 and F_1 (F for factor). **F_1** consists of 5 polypeptides, with stoichiometry **a3, b3, g, d, e.** The F_1 component resembles a doorknob protruding into the matrix from the inner membrane. It is attached to Fo by a stalk, which is embedded in the inner membrane and extend across it. Fo is a complex of integral membrane proteins. When F_1 is carefully extracted (from inside out vesicles prepared) from the inner mitochondrial membrane, the vesicles still contain intact respiratory chains. However, since it no longer contain the F_1 knobs, as confirmed by electron microscopy, they cannot make ATP. When a preparation of isolated F_1 is added back to such depleted vesicles under appropriate conditions, to reconstitute the inner membrane structure, with F_1 knobs, the capacity of the inner membrane vesicles to carry out energy coupling between electron transport and ATP formation is restored. This shows the precise arrangement of these F_0 and F_1 make the ATP synthetase to form complexes called respiratory assemblies. It is proposed that an irregularly shaped "**shaft**" linked to Fo was able to produce conformational changes as follows:

1. A **loose** conformation in which the active site can loosely bind ADP + Pi
2. A **tight** conformation in which substrates are tightly bound and ATP is formed

3. An **open** conformation that favours ATP release.As the protons move down the energy gradient, the energy releases is used by ATP synthetase to produce ATP.

High energy compounds The high energy compound is the ATP. The other high energy compounds include ADP,1,3-diphospho glycerate, phosphoenol pyruvate and also creatine phosphate.

The phosphate group of the high energy phosphate may transfer directly to another organic compound. For this reason the term **group transfer potential** is preferred by some high energy bond. However, the phosphorylated compound may or may not have high energy phosphate though the total energy content of the molecule is higher than a nonphosphorylated compound.

Storage form of high energy compounds They are called as **phosphogens** and help to store the high energy. The example for this is the creatine phosphate present in the vertebrate muscles, the reaction works in both directions it is a reversible reaction form ATP when ATP is required. When ATP is more, creatine reacts with ATP and forms the phosphocreatine.

$$\begin{matrix} CO-O-PO_3^{2-} \\ | \\ H-C-OH + ADP \\ | \\ CH_2-O-PO_3^{2-} \end{matrix} \underset{\text{Phosphoglycerate kinase}}{\overset{Mg^{2+}}{\rightleftharpoons}} \begin{matrix} CO-O^- \\ | \\ H-C-OH + ATP \\ | \\ CH_2-O-PO_3^{2-} \end{matrix}$$

1, 3-diphosphoglycerate 3-phosphoglycerate

One of the phosphate groups undergoes hydrolysis to form the acid and a phosphate ion, giving off energy. This first energy producing reaction is coupled with the next endothermic reaction making ATP. The phosphate is transferred directly to an ADP to make ATP and this is catalysed by phosphoglycerate Kinase enzyme. Since one molecule of glucose yield 2 molecule of Glyceraldehyde 3-phosphate, 2 high energy ATP are produced for one molecule of glucose.

Role of Phosphoenol pyruvate Phosphoenol pyruvate, which is formed during breakdown of glucose to lactic acid, donates its phosphate group to ADP in a reaction catalyzed by Pyruvate kinase. One of the phosphate groups undergoes hydrolysis to form the acid and a phosphate ion, giving off energy. This first energy producing reaction is coupled with the next endothermic reaction making ATP. The phosphate can only exist as the high energy enol form. Thus, when the phosphate group is removed, the pyruvate can revert back to the stable, low-energy keto form and the surplus energy is released. Production of ATP in this reaction is controlled by pyruvate kinase.

$$\begin{matrix} CO-O^- \\ | \\ C-O-PO_3^{2-} \\ \| \\ CH_2 \end{matrix} \xrightarrow[\text{ADP} \rightarrow \text{ATP}]{\text{Pyruvate kinase}} \begin{matrix} CO-O^- \\ | \\ C=OH \\ \| \\ CH_2 \end{matrix} \xrightarrow{\text{Spontaneously}} \begin{matrix} CO-O^- \\ | \\ C=O \\ | \\ CH_3 \end{matrix}$$

Phospho enol pyruvate Enol pyruvate Pyruvate

Actually, this reaction takes place in two steps. First the enolate form of pyruvate is formed, then the transfer of the phosphate group to ADP occur as second step. The keto pyruvic acid may reduced to lactic acid in the lack of oxygen. ***Mitochondria is not involved*** . Since one molecule of glucose yield 2 molecule of Glyceraldehyde 3 phosphate, 2 high energy ATP are produced for one molecule of glucose.

Phosphocreatine

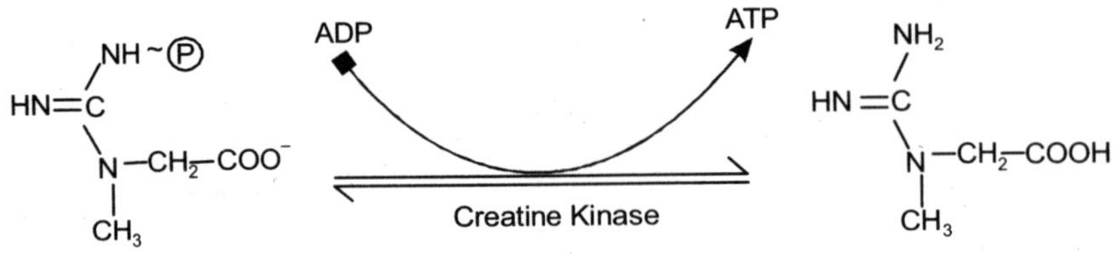

ADP ATP

$HN{=}C(NH{\sim}P)-N(CH_3)-CH_2-COO^-$ ⇌ (Creatine Kinase) $HN{=}C(NH_2)-N(CH_3)-CH_2-COOH$

Phosphocreatine

Phosphocreatine is a phosphogen and it interacts with ADP to form ATP. When ATP is more creatine reacts with ATP and forms phosphocreatine. The enzyme involved is creatine kinase. This energy transfer from creatine phosphate to ADP helps to produce ATP molecule to provide energy during muscle contraction.

ATP as high energy compound ATP is the most widely distributed high-energy compound within the human body. Adenosine triphosphate (ATP) is a useful free-energy currency because the dephosphorylation reaction or hydrolysis, yield an **unusually largeamount ofenergy;** *i.e.,* it releases a large amount of free energy. "High energy" bonds are often represented by the "~" symbol (squiggle), with ~**P** representing a phosphate group with a high free energy on hydrolysis. The terminal phosphate group is then transferred by hydrolysis to another compound, a process called *phosphorylation*, producing ADP, phosphorylated new compound and energy. Thus, the dephosphorylation reaction of ATP to ADP and inorganic phosphate is often coupled with non spontaneous reactions. Generally, ATP is connected to another reaction—a process called *coupling* which means the two reactions occur at the same time and at the same place, usually utilizing the same enzyme complex. Release of phosphate from ATP is exothermic (a reaction that gives off heat) and this reaction is connected to an endothermic reaction (requires energy input in order to occur). The free energy yielded can be coupled to endothermic reaction and useful for the works such as:

Chemical work: ATP energy is consumed to synthesize macromolecules that make up the cell.

Transport work: ATP energy is utilized to pump substances across the plasma membrane.

Mechanical work: ATP provides energy to contract the muscles of the body.

Some time the phosphate group can be transferred to an acceptor molecule and such group transfer potential are associated with some high energy compound. Thus, ATP act as a common intermediate that serves as a vehicle for transfer of chemical energy.

Structure of ATP: ATP is an abbreviation for *adenosine triphosphate*, a complex molecule that contains the nucleoside *adenosine*, ribose and a tail consisting of three phosphates.

Adenosine Triphosphate

The bond is known as a "high-energy" bond and is depicted in the Equation above by a wavy line. The bond between the first and the second phosphate is also "high-energy" bond.

$$\text{ATP} + H_2O \longrightarrow \text{ADP} + \text{Pi} + \text{energy released}$$

ATP

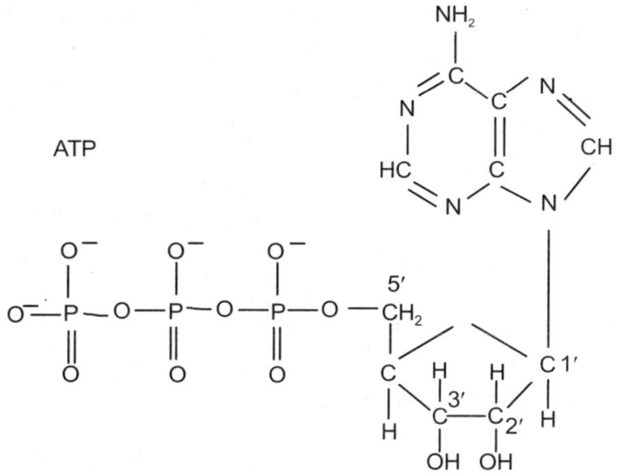

ADP + Pi ⇌ ATP + H_2O
requires energy: 7.3 kcal/mole

$$\Delta G^0 = 7{,}300 \text{ calories/mol} = -7.3 \text{ kcal/mol} = \text{-30.5 kJ/mol } (\Delta C^0 \text{measured at } 37°C)$$

ATP is sometimes referred to as a "High Energy" compound. High energy in this case does **not** refer to total energy in compound, rather just **toenergyof hydrolysis. Thus ATPhasa larger negative** DG for hydrolysis. For biochemistry *High Energy* is defined in terms of ATP: if a compound's free energy for hydrolysis is equal to or greater than ATP's then it is "High Energy," if its free energy of hydrolysis is less than ATP's then it is not a "high energy" compound. Note that ATP has twohighenergy anhydride bonds (AMP **~P ~P).** DG of ATP hydrolysis also depends on the local environments it varies with pH, divalent metal ion concentration, ionic strength and Consumption of ATP. An EATP of -7.3 kcal /mol requires ATP, ADP, and phosphate to be present at equal concentrations. In cells, however the concentration of ATP is often 5 to 10 times that of ADP. As a result, the free energy of ATP hydrolysis is about -12 kcal /mol. One must be clear that the bond energy generally meant by physical chemist is the energy required to break a covalent bond between two atoms. Since relatively a large amount of energy is required to break a covalent bond, the phosphate bond energy is totally a different one.

Phosphate bond energy specifically denotes the difference in the free energyof the reactants when phosphorylated compound undergoes hydrolysis.

Mono(ortho) phosphate cleavage and Pyrophosphate cleavage ATP may under go either an orthophosphate or pyrophosphatecleavage during it's utilization in biosynthetic pathways. In an ATP molecule, when the terminal phosphate is cleaved it is called as mono phosphate or ortho phosphate cleavage.

Adenosine — O — P(=O)(OH) ~ O — P(=O)(OH) ~ O — P(=O)(OH) — OH + H_2O ⟶ ADP + Pi + energy

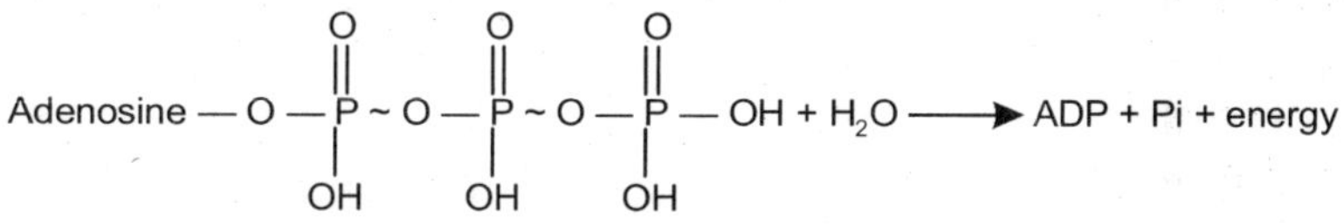

However, in many ATP utilizing reactions instead of one terminal phosphate two terminal phosphate groups are enzymatically hydrolyzed to give a pyro phosphate molecule and a large amount of energy which is greater than the mono phosphate or ortho phosphate cleavage.

$$\text{Adenosine}-O-\overset{\overset{O}{\|}}{\underset{\underset{OH}{|}}{P}}\sim O-\overset{\overset{O}{\|}}{\underset{\underset{OH}{|}}{P}}\sim O-\overset{\overset{O}{\|}}{\underset{\underset{OH}{|}}{P}}-OH + H_2O \longrightarrow AMP + Pi \sim PI + \text{energy}$$

Pyrophosphate (PPi) is often the product of a reaction that needs a driving force. Its spontaneous hydrolysis, catalyzed by Pyrophosphatase enzyme, drives the reaction for which PPi is a substrate. The DG (free energy) for this pyrophosphate cleavage is 10.0 Kcal./mol and thus an extra thermodynamic push is given to certain enzymatic reaction which require more energy than that of a mono phosphate cleavage and assure the completeness of certain biosynthetic reactions.

GENERATION OF SUPER OXIDE FREE RADICALS

Superoxide is the anion O_2^-. It is important as the product of the one-electron reduction of dioxygen, which occurs widely in nature. With one unpaired electron, the superoxide ion is a free radical, and, like dioxygen, it is paramagnetic.

Synthesis, Basic reactions and Structure

Superoxides are compounds in which the oxidation number of oxygen is –½. The O-O bond distance in O_2^- is 1.33 Å, vs. 1.21 Å in O_2 and 1.49 Å in O_2^{2-}. The salts CsO_2, RbO_2, KO_2, and NaO_2 are prepared by the direct reaction of O_2 with the respective alkali metal. The overall trend corresponds to a reduction in the bond order from 2 (O_2), to 1.5 (O_2^-), to 1 (O_2^{2-}).

The alkali salts of O_2^- are orange-yellow in color and quite stable, provided they are kept dry. Upon dissolution of these salts in water, however, the dissolved O_2^- undergoes disproportionation (dismutation) extremely rapidly:

$$2\,O_2^- + 2\,H_2O \rightarrow O_2 + H_2O_2 + 2\,OH^-$$

In this process O_2^- acts as a strong Brønsted base, initially forming HO_2. The pKa of its conjugate acid, hydrogen superoxide (HO_2, also known as "hydroperoxyl" or "perhydroxy radical"), is 4.88 so that at neutral pH 7 the vast majority of superoxide is in the anionic form, O_2^-.

Salts also decompose in the solid state, but this process requires heating:

$$2NaO_2 \rightarrow Na_2O_2 + O_2$$

This reaction is the basis of the use of potassium superoxide as an oxygen source in chemical oxygen generators, such as those used on the space shuttle and on submarines.

Superoxide in Biology

Superoxide is biologically quite toxic and is deployed by the immune system to kill invading microorganisms. In phagocytes, superoxide is produced in large quantities by the enzyme NADPH oxidase for use in oxygen-dependent killing mechanisms of invading pathogens. Mutations in the gene coding for the NADPH oxidase cause an immunodeficiency syndrome called chronic granulomatous disease, characterized by extreme susceptibility to infection. In turn, micro-organisms genetically engineered to lack superoxide dismutase (SOD), lose virulence. Superoxide is also deleteriously produced as a byproduct of mitochondrial respiration (most notably by Complex I and Complex III), as well as several other enzymes, for example xanthine oxidase.

Despite being chemically rather benign, superoxide is so toxic that intracellular levels above 1nM are lethal. The biological toxicity of superoxide is not entirely understood, but derives in part from its capacity to inactivate iron-sulfur cluster containing enzymes (which are critical in a wide variety of metabolic pathways), thereby liberating free iron in the cell, which can undergo Fenton chemistry and generate the highly reactive hydroxyl radical. In its HO_2 form (hydroperoxyl radical), superoxide can also initiate lipid peroxidation of polyunsaturated fatty acids. It also reacts with carbonyl compounds and halogenated carbons to create toxic peroxy radicals. Superoxide can also react with nitric oxide (NO) to form $ONOO^-$. Superoxide can also form tyrosine peroxides as a result of reaction with enzymes containing tyrosyl radicals (such as ribonucleotide reductase). Superoxide can also oxidize hemoglobin (forming the non-oxygen carrying met-hemoglobin), and possibly other low-potential heme proteins. Finally, superoxide can oxidize low potential thiols. As such, superoxide is one of the main causes of oxidative stress.

Because superoxide is toxic, nearly all organisms living in the presence of oxygen contain isoforms of the superoxide scavenging enzyme, superoxide dismutase, or SOD. SOD is an extremely efficient enzyme; it catalyzes the neutralization of superoxide nearly as fast as the two can diffuse together spontaneously in solution. Other proteins, which can be both oxidized and reduced by superoxide, have weak SOD-like activity (e.g. hemoglobin). Genetic inactivation ("knockout") of SOD produces deleterious phenotypes in organisms ranging from bacteria to mice and have provided important clues as to the mechanisms of toxicity of superoxide in vivo.

Yeast lacking both mitochondrial and cytosolic SOD grow very poorly in air, but quite well under anaerobic conditions. Absence of cytosolic SOD causes a dramatic increase in mutagenesis and genomic instability. Mice lacking mitochondrial SOD (MnSOD) die around 21 days after birth due to neurodegeneration, cardiomyopathy and lactic acidosis. Mice lacking cytosolic SOD (CuZnSOD) are viable but suffer from multiple pathologies, including reduced lifespan, liver cancer, muscle atrophy, cataracts, thymic involution, haemolytic anemia and a very rapid age-dependent decline in female fertility.

Superoxide may contribute to the pathogenesis of many diseases (the evidence is particularly strong for radiation poisoning and hyperoxic injury), and perhaps also to aging via the oxidative damage that it inflicts on cells. While the action of superoxide in the pathogenesis of some conditions is strong, for instance, mice and rats overexpressing CuZnSOD or MnSOD are more resistant to strokes and heart attacks, the role of superoxide in aging, must be regarded as unproven for now. In model organisms (yeast, the fruit fly Drosophila and mice), genetically knocking out CuZnSOD shortens lifespan and accelerates certain features of aging (cataracts, muscle atrophy, macular degeneration, thymic involution), but the converse, increasing the levels of CuZnSOD, does not seem (except perhaps in *Drosophila*), to consistently increase lifespan. The most widely accepted view is that oxidative damage (derived amongst other factors, from superoxide) is but one of several factors limiting lifespan.

CYTOCHROME 450

Cytochrome P450 proteins in humans are drug metabolizing enzymes and enzymes that are used to make cholesterol, steroids and other important lipids such as prostacyclins and thromboxane A2. These last two are metabolites of arachidonic acid. Mutations in cytochrome P450 genes or deficiencies of the enzymes are responsible for several human diseases. Induction of some P450s is a risk factor in several cancers since these enzymes can convert procarcinogens to carcinogens. P450 enzymes play a major role in drug interactions. The name cytochrome P450 derives from the fact that these proteins have a heme group, and an unusual spectrum. Mammalian cytochrome P450s are membrane bound. They were originally discovered in rat liver microsomes. Microsomes are turbid suspensions made by grinding up cells and isolating the membrane fraction that is still in suspension after the cell debris and mitochondria have been pelleted. These mixtures are very opaque to standard spectroscopy, because they scatter light so badly. The only way to measure a spectrum on turbid samples like these was to make a special instrument with the light detector very close to the cuvette, and to use dual beams and do difference spectroscopy. In this way all the interfering substances and the light scattering could be subtracted out. With this setup, microsomes treated with dithionite (reduced microsomes) and with carbon monoxide gas added to one cuvette only give a very strong absorption band at 450 nm, thus P450 (P is for pigment). This is called a reduced CO difference spectrum. The CO binds tightly to the ferrous heme, giving a difference between the absorbance of the two cuvettes. This spectrum was first observed in 1958. Other heme containing proteins don't absorb at 450 nm. The reason why cytochrome P450 absorbs in this range is the unusual ligand to the heme iron.

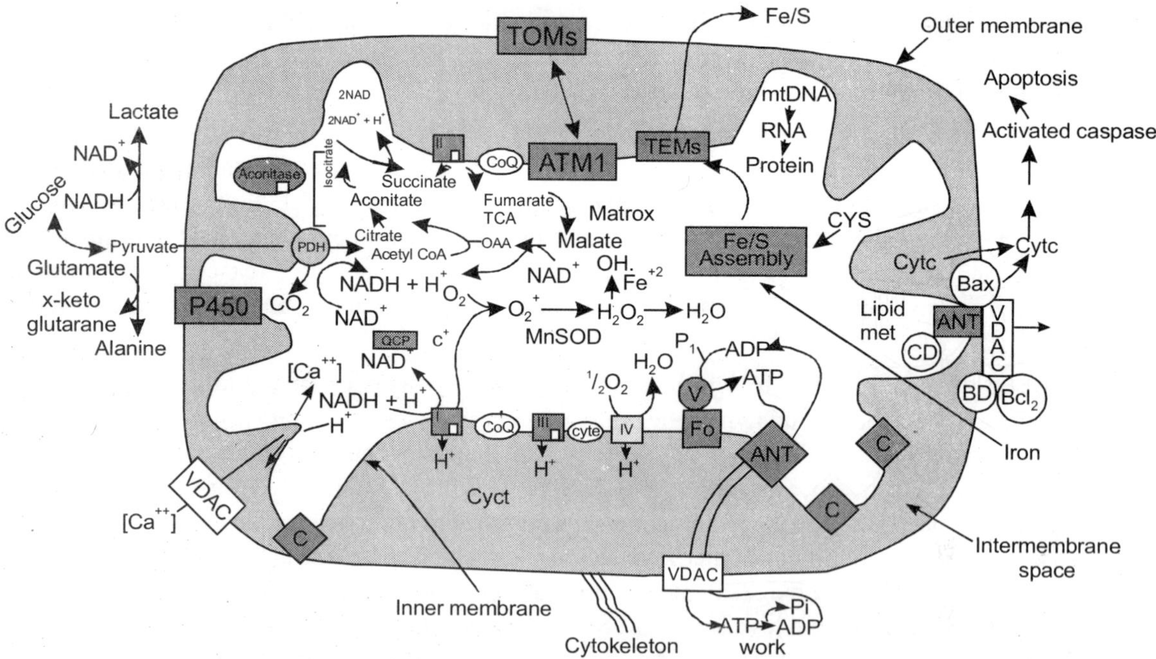

Fig. 4.15 : The Mitochondrial Respiratiory chain action with p450 Cytochrome

Four ligands are provided by nitrogens on the heme ring. Above and below the plane of the heme, there is room for two more ligands, the 5th and 6th ligands. In cytochrome P450s, the 5th ligand is a thiolate anion, a sulphur with a negative charge, S(-). The sulfur comes from a conserved cysteine at the heme binding region of the active site.

X-ray crystal structures are now available for more than ten different bacterial P450s (CYPs 51 [Mycobacterium], 55A1, 101A1, 102A1, 107A1 [eryF], 111A1, 119A1, 121A1 [P450Mt2], 152, 175A1). These proteins are soluble as compared to the eukaryotic P450s that are membrane bound. The structures are similar and they probably form a good model for membrane bound versions, except for the membrane anchoring parts. Here is a picture of the first P450 crystallized. This is P450 cam from Pseudomonas putida, a bacterium that can use camphor as its sole carbon source. This bug is found growing in soil under camphor trees. The protein is shaped like a triangle with the heme buried deep inside. In this structure, there is no access channel for substrate or products, even water, to enter or leave the active site. Therefore, we must assume that the structure breathes when it functions, so a channel will be open at some point in the catalytic cycle. Half of the enzyme is rich in alpha-helix and the other half is beta-sheet or non-repetetive structure. The mammalian P450s are similar to this fold, but with an N-terminal membrane anchor. A cartoon of one possible view is given here. This model shows a single transmembrane segment, but membrane attachment is more complex than that. When the N-terminal sequence is removed the protein still sticks to membranes. The mammalian CYP2C5 protein has been crystallized after removal of the N-terminal anchor peptide and replacement of an internal hydrophobic sequence with a more water soluble sequence from a related enzyme. The X-ray structure has been solved. A brief description of the main features had appeared earlier. The structure is similar to the soluble bacterial enzymes, but there are significant differences.

The phamaceutical industry is very interested in P450 crystal structures with drugs bound, so they can do improvement of drug design. Here we see a press release from May 3, 2001 about a research agreement between Astex and AstraZeneca to determine crystal structures of human P450s with AstraZeneca drugs bound in the active site. Look at the outlined sections. "Cytochrome P450 enzymes are the most prominent group of drug-metabolising enzymes in humans, and consequently are of great importance to the pharmaceutical industry Astex has now solved the crystal structures of CYP2C9 and CYP3A4, two of the most significant drug metabolising enzymes in humans. This structure can now be used to modify existing drugs to make them poorer substrates for 3A4 (or better substrates). Poorer P450 substrates would last longer in the body before elimination, which is desirable for the pharmaceutical industry.

P450s catalyze many types of reactions, but the one that is most important for us is hydroxylation. These enzymes are called mixed function oxidases or monooxygenases, because they incorporate one atom of molecular oxygen into the substrate and one atom into water. They differ from dioxygenases that incorporate both atoms of molecular oxygen into the substrate. Foreign chemicals or drugs are also called xenobiotics. Cytochrome P450s play an important role in xenobiotic metabolism, especially for lipophilic drugs. The metabolism of these compounds takes place in two phases. Phase I is chemical modification to add a functional group that can be used to attach a conjugate. The conjugate makes the modified compound more water soluble so it can be excreted in the urine. Many P450s add a hydroxyl group in a Phase I step of drug metabolism. The hydroxyl then serves as the site for further modifications in Phase 2 drug metabolism. For cytochrome P450s to function, they also need a source of electrons. The addition of two electrons (reduction) to the heme iron makes the difficult chemistry of breaking the oxygen-oxygen bond possible. The electrons are donated by another protein that binds briefly to the P450 and passes an electron from a prosthetic group. This handoff of electrons between proteins is called an electron transfer chain, and it is similar to the electron transfers that go on in complexes I to IV of the electron transfer chain in mitochondria. (However, this is not the same electron transfer chain.)There are two different kinds of electron transfer chains for cytochrome P450s. These depend on the location of the enzyme in the cell. Some P450s are found in the mitochondrial inner membrane and some are found in the endoplasmic reticulum (ER). Both types of P450s are membrane bound proteins. The protein that donates electrons to P450s in the ER is called NADPH cytochrome P450 reductase. It is also membrane bound by an N-terminal tail that crosses the ER membrane once. The bulk of this protein is on the cytosolic side of the ER membrane. This protein has two domains that each contain one flavin. Two electrons are acquired from NADPH and migrate from FAD to FMN, then to the P450 heme iron.In the mitochondria, the electron transfer chain is a little longer.

Ferredoxin (called adrenodoxin in the adrenals, but exactly the same gene codes for both proteins) is the immediate donor of electrons to the P450s in mitochondria (CYP11A1, CYP11B1, CYP11B2, CYP24, CYP27A1, CYP27B1, CYP27C1). Ferredoxin has an iron sulfur cluster instead of a flavin, however, ferredoxin is reduced by ferredoxin reductase (or adrenodoxin reductase in the adrenals) that does contain a flavin. NADPH is the source of electrons that flow from ferredoxin reductase to ferredoxin and then to P450. A few P450s also can accept electrons from cytochrome b5. This is a small membrane bound heme containing protein that gets its reducing equivalents (electrons) from NADH.

The families of human P450s The P450 proteins are categorized into families and subfamilies by their sequence similarities. Sequences that are greater than 40% identical at the amino acid level belong to the same family. Sequences that are greater than 55% identical are in the same subfamily. There are now more than 2500 cytochrome P450 sequences known.

Humans have 18 families of cytochrome P450 genes and 43 subfamilies:

CYP1 drug metabolism (3 subfamilies, 3 genes, 1 pseudogene)

CYP2 drug and steroid metabolism (13 subfamilies, 16 genes, 16 pseudogenes)

CYP3 drug metabolism (1 subfamily, 4 genes, 2 pseudogenes)

CYP4 arachidonic acid or fatty acid metabolism (5 subfamilies, 11 genes, 10 pseudogenes)

CYP5 Thromboxane A2 synthase (1 subfamily, 1 gene)

CYP7A bile acid biosynthesis 7-α hydroxylase of steroid nucleus (1subfamily member)

CYP7B brain specific form of 7-α hydroxylase (1 subfamily member)

CYP8A prostacyclin synthase (1 subfamily member)

CYP8B bile acid biosynthesis (1 subfamily member)

CYP11 steroid biosynthesis (2 subfamilies, 3 genes)

CYP17 steroid biosynthesis (1 subfamily, 1 gene) 17-α hydroxylase

CYP19 steroid biosynthesis (1 subfamily, 1 gene) aromatase forms estrogen

CYP20 Unknown function (1 subfamily, 1 gene)

CYP21 steroid biosynthesis (1 subfamily, 1 gene, 1 pseudogene)

CYP24 vitamin D degradation (1 subfamily, 1 gene)

CYP26A retinoic acid hydroxylase important in development (1 subfamily member)

CYP26B probable retinoic acid hydroxylase (1 subfamily member)

CYP26C probable retinoic acid hydroxylase (1 subfamily member)

CYP27A bile acid biosynthesis (1 subfamily member)

CYP27B Vitamin D3 1-α hydroxylase activates vitamin D3 (1 subfamily member)

CYP27C Unknown function (1 subfamily member)

CYP39 7 α hydroxylation of 24 hydroxy cholesterol (1 subfamily member)

CYP46 cholesterol 24-hydroxylase (1 subfamily member)

CYP51 cholesterol biosynthesis (1 subfamily, 1 gene, 3 pseudogenes) lanosterol 14-α demethylase

Humans have 57 sequenced CYP genes and 58 pseudogenes.

only full length functional genes are named below:

1A1, 1A2, 1B1, 2A6, 2A7, 2A13, 2B6, 2C8, 2C9, 2C18, 2C19, 2D6, 2E1, 2F1, 2J2, 2R1, 2S1, 2U1, 2W1, 3A4, 3A5, 3A7, 3A43, 4A11, 4A22, 4B1, 4F2, 4F3, 4F8, 4F11, 4F12, 4F22, 4V2, 4X1, 4Z1 5A1, 7A1, 7B1, 8A1, 8B1, 11A1, 11B1, 11B2, 17, 19, 20, 21A2, 24, 26A1, 26B1, 26C1, 27A1, 27B1, 27C1, 39, 46, 51,

Induction of P450 enzymes. P450 enzymes have a variety of gene regulatory mechanisms. Many of these genes can be turned on or induced by a chemical signal. The steroid hormones are under strict endocrine control. Their levels are tightly regulated. One example is the induction of steroid biosynthetic P450s by ACTH adrenocorticotropic hormone. ACTH stimulates production of cAMP that presumably activates a protein kinase that phosphorylates some unidentified protein, leading to an increase in gene transcription. Another type of P450 gene regulation is that shown by peroxisome proliferators like clofibrate. These drugs act through a binding protein called the PPAR or peroxisome proliferator activated receptor. When drug is bound to this protein it migrates to the nucleus, heterodimerizes with retinoid X receptor (RXR) and binds to specific DNA sequences in the regulatory region of genes that are needed for peroxisome generation. The CYP4A1 gene is turned on by this mechanism. Peroxisomes oxidize fatty acids and the 4A1 P450 is a known fatty acid hydroxylase.

The members of the CYP1 family are induced by aromatic hydrocarbons. The activation involves a specialized receptor called the Ah receptor. Ah stands for aryl hydrocarbon. This receptor protein binds the aromatic hydrocarbon, but it cannot reach the nucleus to activate gene transcription without another protein called arnt for Ah receptor nuclear translocator. These two proteins bind and together they then bind DNA and activate transcription. Other chemicals also induce P450s. Ethanol induces the CYP2E enzymes. Phenobarbital induces the rat CYP2B enzymes 40-50 fold, through a phenobarbital receptor called CAR. This receptor also dimerizes with RXR as seen above with the PPAR receptor. The heterodimer binds to a phenobarbital response element in the DNA to activate the gene. For details on these receptor mediated induction mechanisms. The general feature that many P450 enzymes are inducible is probably related to P450's role in detoxification of foreign chemicals found in plants.

Non-invasive markers for measuring levels of P450 enzymes in humans P450 enzymes catalyze specific reactions that can be monitored by sampling the urine, blood or breath of patients given a noninvasive marker. Caffeine is a marker for CYP1A2. It is demethylated, and the rate at which it is demethylated is related to the amount of CYP1A2 in a person's liver. By administering caffeine and measuring the rate of demethylation, it is possible to estimate the level of CYP1A2 in a human. This can show if a person has been induced by exposure to polycyclic aromatic hydrocarbons (PAHs). There are a variety of non-invasive markers for different P450s. Assays of CYP1A enzymes from fish livers can also be used to monitor water pollution levels, since certain types of pollutants will induce the enzyme. This is also being done in soil using nematodes like C. elegans.

Functions of Human P450s and Diseases caused by Defects in P450s

The **CYP1 family** of P450s can hydroxylate estrogen (CYP1A2 and 1B1)and oxidize uroporphyrinogen to uroporphyrin (CYP1A2) in heme metabolism,but they may have additional undiscovered endogenous substrates. These enzymes are inducible by some polycyclic hydrocarbons, some of which are found in cigarette smoke and charred food. These enzymes are of interest, because in assays, they can activate compounds to carcinogens. High levels of CYP1A2 have been linked to an increased risk of colon cancer. Since the 1A2 enzyme can be induced by cigarette smoking, this links smoking with colon cancer.

The **CYP1B1** gene has been linked to primary congenital glaucoma. The normal substrate in mammals is not known, but it is speculated that this P450 may be required to eliminate a signaling molecule. Defects in the gene could lead to chronic high concentrations of the signaling molecule that lead to glaucoma. The molecule affected may be a steroid. As you can see from the table of human P450s, the 2 family is the largest family in humans. About one third of human P450s are in this family. Many of these proteins can hydroxylate steroids, and some of them are expressed in a sex specific manner. This would be expected for enzymes that only act on sex specific steroids. Some of these may also be drug metabolism enzymes that are defensive, to protect us from toxins in our food. Plants especially make many toxic components that are probably defensive for the plants.

CYP2B is inducible by barbiturates in rodents. It was one of the first P450s to be purified from mammals, but its role in humans is not understood.

CYP2C8 is known to catalyze the 6-α hydroxylation of taxol. This is a drug used in treating breast cancer.

CYP2C9 is one of two human P450s that has a known crystal structure. The other is CYP3A4 (still confidential). CYP2C9 structure was published in Nature this summer.

CYP2C19 metabolizes omeprazole, a common ulcer medication.Polymorphisms in this gene cause a higher incidence of poor metabolizer phenotypes in Asians (23%).

Drug metabolism differences caused by polymorphisms in P450s. A polymorphism is a difference in DNA sequence found at 1% or higher in a population. These differences in DNA sequence can lead to differences in drug metabolism, so they are important features of P450 genes in humans. CYP2C19 has a polymorphism that changes the enzyme's ability to metabolize mephenytoin (a marker drug). In Caucasians, the polymorphism for the poor metabolizer phenotype is only seen in 3% of the population. However, it is seen in 20% of the asian population. Because of this difference, it is important to be aware of a person's race when drugs are given that are metabolized differently by different populations. Some drugs that have a narrow range of effective dose before they become toxic might be overdosed in a poor metabolizer. Very recently, Roche has marketed a CYP450 DNA chip to detect major known polymorphisms in human CYP2D6 and CYP2C19. For about $400 you can test a person to see if they are a poor metabolizer, normal metabolizer or ultra metabolizer, for a large number of drugs. Since 1A2, 2C9, 2C19, 2D6 and 3A4 are responsible for oxidizing more than 90% of currently used drugs (2C9 paper above), this is a significant beginning to characterizing risk of adverse drug reactions in people.

CYP2D6 is perhaps the best studied P450 with a drug metabolism polymorphism. This enzyme is responsible for more than 70 different drug oxidations. Since there may be no other way to clear these drugs from the system, poor metabolizers may be at severe risk for adverse drug reactions. There are at least 72 named alleles identified in CYP2D6.

CYP2D6 Substrates

- **Antiarrhythmics:** Flecainide, Mexiletine, Propafenone
- **Antidepressants:** Amitriptyline, Paroxetine, Venlafaxine, Fluoxetine (Prozac), Trazadone
- **Antipsychotics:** Clorpromazine, Haloperidol, Thoridazine
- **Beta-Blockers:** Labetalol, Timolol, Propanolol, Pindolol, Metoprolol
- **Analgesics:** Codeine, Fentanyl, Meperidine, Oxycodone, Propoxyphene

oxycodone is oxycontin, a favorite drug of abuse.

CYP2E1 is induced in alcoholics. There is a polymorphism associated with this gene that is more common in Chinese people. The mutation correlates with a 2-fold increased risk of nasopharyngeal cancer linked to smoking. This is the second P450 enzyme that may be related to smoking induced cancer (see 1A2).

The **CYP3A subfamily** is one of the most important drug metabolizing families in humans. The crystal structure of 3A4 is known, but confidential and heavily patented. CYP3A4 is "the most abundantly expressed P450 in human liver". (Arch. Biochem. Biophys. 369, 11-23 1999) The color of perfused liver is due to this protein. CYP3A4 is known to metabolize more than 120 different drugs. Some of these are well known and I give a list here of some of the recognizable ones.

CYP3A4 Substrates

- **Acetominophen** (Tylenol)
- **Codeine** (narcotic)
- **Cyclosporin A** (an immunosuppresant),
- **Diazepam** (Valium)
- **Erythromycin** (antibiotic)
- **Lidocaine** (anaesthetic),
- **Lovastatin** (HMGCoA reductase inhibitor, a cholesterol lowering drug),
- **Taxol** (cancer drug),
- **Warfarin** (anticoagulant).

Poisoning by acetominophen overdose is caused by P450 enzymes in the liver and kidney that convert acetominophen into a very toxic intermediate that can react with cellular macromolecules to damage cells and eventually kill them. This intermediate normally reacts with glutathione, a naturalantioxidant in cells. It is only when the glutathione is depleted that cell death can occur. That's why acetominophen overdoses don't have any serious symptoms until 3-4 days later. This problem is worse in alcoholics, since they have induced P450 enzymes that make more of the toxic intermediate.

There are common drugs given for special purposes that inhibit P450 enzymes. These include erythromycin (an antibiotic), ketoconazole, and itraconazole (both antifungals that inhibit the fungal CYP51 and unintentionally they also inhibit CYP3A4). If these drugs are given with other drugs that arenormally metabolized by P450 enzymes, the lifetime of these other drugs will be prolonged, and plasma levels will be increased, since they won't be cleared as fast. If these drugs affect heart rhythms or other critical systems, the result can be fatal.

For example, inhibition of CYP3A4 in a patient taking warfarin can cause bleeding. This is called a drug interaction. Drug interactions are one of the major causes of death in hospitalized patients. Another factor in drug dosage is interfering substances from food. Grapefruit juice contains a CYP3A4 inhibitor that causes about a 12-fold increase in some drug concentrations. And the effect lasts for several days. It is advisable to discourage your patients from drinking grapefruit juice while on medication metabolized by CYP3A4.Now we will leave the drug metabolizing enzymes behindand talk about P450s that are very specific in their reactions, just the opposite of CYP3A4.

These enzymes tend to be in families with one or two members and they have only one substrate. Most of these enzymes use steroids or steroid precursors as their substrates.

CYP5 is the thromboxane A2 synthase. Thromboxane A2 is a fatty acid in the arachidonic acid cascade. Arachidonic acid can be metabolized in two pathways, the linear pathway that leads to leukotrienes, and the cyclic pathway that leads to prostaglandins and thromboxanes. The first enzymes leading to cyclic products of arachidonic acid are cyclooxygenases 1 and 2.

These enzymes are inhibited by aspirin and non-steroidal antiinflammatory drugs (NSAIDS). Aspirin acetylates a serine in the enzyme that blocks the binding of arachidonic acid. Current research shows that COX2 is inducible and is found to be induced in inflammation. COX1 is constitutive. This difference suggests that COX2 specific inhibitors would block inflammation while not interfering with the beneficial effects of COX1, such as maintaining the stomach lining. These drugs are now on the market. Mouse knockouts have been made, but the full analysis of these COX1 and COX2 knockouts is not finished yet. After this step the pathway branches. Two of the branches include cytochrome P450 reactions. One leads to thromboxane A2 (CYP5) and the other to prostacyclin (CYP8A1). Thromboxane A2 causes platelet aggregation and that is why aspirin prevents platelet aggregation. Prostacyclin acts in opposition to thromboxane A2. It is a vasodilator and an inhibitor of platelet aggregation. The acetylation of COX1 and COX2 in platelets is critical since the platelets have no nucleus and cannot resynthesize the inhibited enzymes.

CYP7A is the first and rate limiting step of bile acid synthesis. This pathway is the only means the body has of eliminating cholesterol in liver. CYP39 can substitute for it in brain. As we will see later, CYP51 is a key enzyme in cholesterol biosynthesis, so P450s are active at both ends of cholesterol metabolism. In 2002 patients were found with defects in this gene. They had elevated levels of cholesterol, decreased levels of bile acids and increased triglycerides, as a compensation for the reduced bile acids. John Kane et al. Journal of Clin. Invest. July 2002.

CYP7B a novel brain cytochrome P450, catalyzes the synthesis of neurosteroids 7-alpha hydroxy dehydroepiandrosterone and 7-alpha hydroxy pregnenolone Proc. Natl. Acad. Sci. USA 94, 4925-4930 (1997).

CYP8A is prostacyclin synthase (prostaglandin I2). It is part of a regulatory component of hemostasis that opposes CYP5 that makes thromboxane A2.

CYP8B is the 12-alpha hydroxylase needed in bile acid biosynthesis.

CYP11A1 is the side chain cleavage enzyme that converts cholesterol to pregnenolone. This is the first step in steroid biosynthesis.

CYP11B1 is the 11-β hydroxylase enzyme that can act on 11-deoxycortisol to make cortisol or it can hydroxylate 11-deoxycorticosterone to make corticosterone.

CYP11B2 is aldosterone synthase that hydroxylates coricosterone at the 18 position Mitochondrial.

CYP17 is the 17 α hydroxylase and 17-20 lyase (two enzymes in one).

CYP19 is aromatase that makes estrogen by aromatizing the A ring of the steroid nucleus. Lack of this enzyme causes a lack of estrogen and failure of women to develop at puberty. An interesting defect found in a male was an overactive CYP19 enzyme with about 50 times normal activity. This boy developed breasts at a young age.

CYP20 is a new P450 found only on chordates so far, it may be chordate specific and be involved in development. Nothing is known yet.

CYP21 is the C21 steroid hydroxylase. Defects in this gene cause congential adrenal hyperplasia due to lack of cortisol synthesis. Since cortisol is not made, the precursor 17 hydroxy progesterone builds up and this causes excessive androgen (testosterone) biosynthesis resulting in virilization.

CYP24 is a 25-hydroxyvitamin D(3) 24-hydroxylase used in the degradation or inactivation of vitamin D metabolites. [mitochondrial]

CYP26A1 is an all trans retinoic acid hydroxylase. It does not recognize 9-cis or 13-cis retinoic acid. CYP26A1 has been mutated in zebrafish and it causes a developmental defect. The human and mouse cDNAs have been cloned, but the effects of a mutation in mammals is not yet determined. Retinoic acid is known to be an important molecule in vertebrate development. It operates through several retinoic acid receptors. The hydroxylase may be a means of degrading the retinoic acid signal and thus turning off a developmental switch.

CYP26B1 is a recently discovered human P450. It metabolizes retinoic acid and its expression is induced by retinoic acid during development in chickens (and probably all vertebrates).

CYP26C1 is only known from genomic DNA sequencing. The function is not known.

Use of a P450 for Gene Therapy in Cancer

As mentioned earlier that CYP1A2 can activate procarcinogens to carcinogens. The induction of this enzyme may be a cancer risk. The activation of a prodrug to an active form by a P450 mediated reaction has been exploited to fight cancer. A vector with a P450 gene on it (and a P450 reductase gene) can be injected into cancer tumors. Some of these cells take up the vector and express The P450 and its reductase. Then a non-toxic prodrug is administered that is converted by the P450 into a toxic compound that kills the cells. Since the cancer cells have cellular connections, the toxin gets shared around and the tumor dies.

ATP/ADP CYCLE

An **ATP synthase** (EC 3.6.3.14) is a general term for an enzyme that can synthesize adenosine triphosphate (ATP) from adenosine diphosphate (ADP) and inorganic phosphate by using some form of energy. This energy is often in the form of protons moving down an electrochemical gradient, such as from the lumen into the stroma of chloroplasts or from the inter-membrane space into the matrix in mitochondria. The overall reaction sequence is:

$$ADP + P_i \rightarrow ATP$$

These enzymes are of crucial importance in almost all organisms, because ATP is the common "energy currency" of cells.

The antibiotic oligomycin inhibits the F_0 unit of ATP synthase.

Structure and Nomenclature

In mitochondria, the F_1, F_0 ATP synthase has a long history of scientific study.

- the F_0 portion is within the membrane.
- The F_1 portion of the ATP synthase is above the membrane.

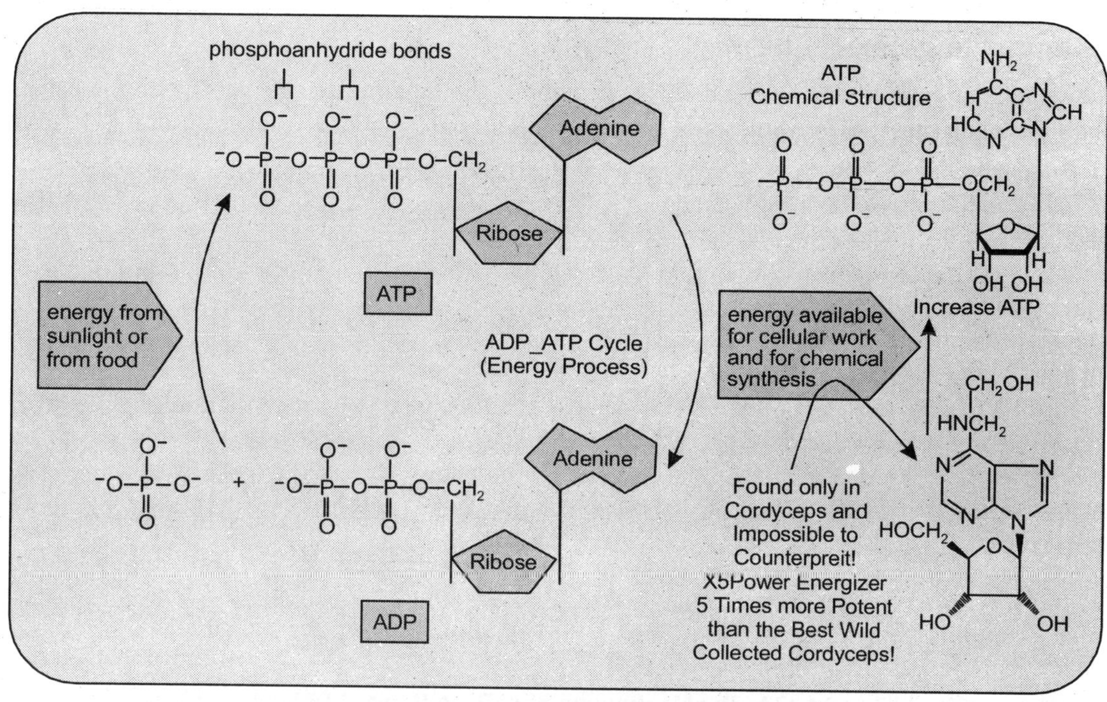

Fig. 4.16 : The Complete ATP – ADP cycle

The nomenclature of the enzyme suffers from a long history. The F_1 fraction derives it name from the term "Fraction F_1 and F_0 (written as a subscript "O", not "zero") derives its name from being the oligomycin binding fraction.

Taking as an example the nomenclature of subunits in the bovine enzyme, many subunits have alphabet names:

- Greek letters: α, β, γ, δ, ε
- Roman letters: a, b, c, d, e, f, g, h

Others have more complex names:

- F_6 (from "Fraction 6")
- OSCP (the oligomycin sensitivity conferral protein), *ATP5O*
- A6L (named for the gene that codes for it in the mitochondrial genome)
- IF_1 (inhibitory factor 1), *ATPIF1*

The F_1 particle is large and can be seen in the transmission electron microscope by negative staining. These are particles of 9 nm diameter that pepper the inner mitochondrial membrane. They were originally called elementary particles and were thought to contain the entire respiratory apparatus of the mitochondrion, but through a long series of experiments were able to show that this particle is correlated with ATPase activity in uncoupled mitochondria and with the ATPase activity in submitochondrial particles created by exposing mitochondria to ultrasound. This ATPase activity was further associated with the creation of ATP by a long series of experiments in many laboratories.

Binding change Mechanism

In the 1960s through the 1970s, Paul Boyer developed the binding change, or flip-flop, mechanism, which postulated that ATP synthesis is coupled with a conformational change in the ATP synthase

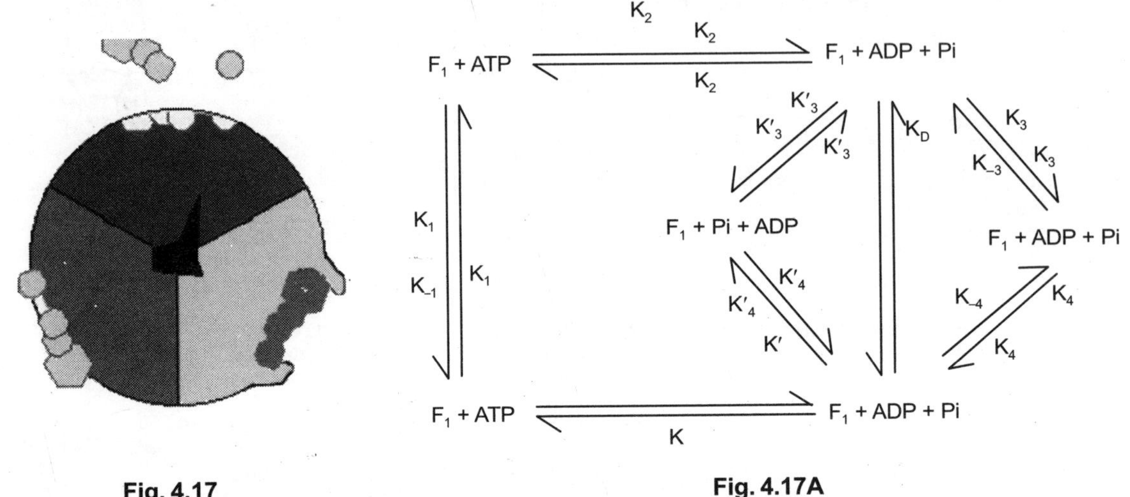

Fig. 4.17 **Fig. 4.17A**

Fig. 4.17 and 4.17A: The Mechanism of ATP synthase. ATP is shown in red, ADP and phosphate in pink and the rotating g subunit in black

generated by rotation of the gamma subunit. The research group of John E. Walker, then at the MRC Laboratory of Molecular Biology in Cambridge but now at the MRC Dunn Human Nutrition Unit (also in Cambridge) crystallized the F_1 catalytic-domain of ATP synthase. The structure, at the time the largest asymmetric protein structure known, indicated that Boyer's rotary-catalysis model was essentially correct. For elucidating this Boyer and Walker shared half of the 1997 Nobel Prize in Chemistry. Jens Christian Skou received the other half of the Chemistry prize that year "for the first discovery of an ion-transporting enzyme, Na+, K+ -ATPase"

The crystal structure of the F_1 showed alternating alpha and beta subunits (3 of each), arranged like segments of an orange around an asymmetrical γ subunit. According to the current model of ATP synthesis (known as the alternating catalytic model), the proton-motive force across the inner mitochondrial membrane, generated by the electron transport chain, drives the passage of protons through the membrane via the F_O region of ATP synthase. A portion of the F_O (the ring of c-subunits) rotates as the protons pass through the membrane. The c-ring is tightly attached to the asymmetric central stalk (consisting primarily of the γ subunit) which rotates within the $\alpha_3\beta_3$ of F_1 causing the 3 catalytic nucleotide binding sites to go through a series of conformational changes that leads to ATP synthesis. The major F_1 subunits are prevented from rotating in sympathy with the central stalk rotor by a peripheral stalk that joins the $alpha_3beta_3$ to the non-rotating portion of F_O. The structure of the intact ATP synthase is currently known at low-resolution from electron cryo-microscopy (cryo-EM) studies of the complex. The cryo-EM model of ATP synthase suggests that the peripheral stalk is a flexible structure that wraps around the complex as it joins F_1 to F_O. Under the right conditions, the enzyme reaction can also be carried out in reverse, with ATP hydrolysis driving proton pumping across the membrane.

The binding change mechanism involves the active site of a β subunit cycling between three states. In the "open" state, ADP and phosphate enter the active site, in the diagram to the right this is shown in brown. The protein then closes up around the molecules and binds them loosely - the "loose" state (shown in red). The enzyme then undergoes another change in shape and forces these molecules together, with the active site in the resulting "tight" state binding the newly-produced ATP molecule with very high affinity. Finally, the active site cycles back to the open state, releasing ATP and binding more ADP and phosphate, ready for the next cycle of ATP production.

Physiological Role

Like other enzymes, the activity of F_1F_O ATP synthase is reversible. Large enough quantities of ATP cause it to create a transmembrane proton gradient, this is used by fermenting bacteria which do not have an electron transport chain, and hydrolyze ATP to make a proton gradient, which they use for flagella and transport of nutrients into the cell.

In respiring bacteria under physiological conditions, ATP synthase generally runs in the opposite direction, creating ATP while using the protonmotive force created by the electron transport chain as a source of energy. The overall process of creating energy in this fashion is termed oxidative phosphorylation. The same process takes place in the mitochondria, where ATP synthase is located in the inner mitochondrial membrane (so that F_1-part sticks into mitochondrial matrix, where ATP synthesis takes place).

ATP Synthase in Different Organisms

Plant ATP Synthase

In plants ATP synthase is also present in chloroplasts (CF_1F_O-ATP synthase). The enzyme is integrated into thylakoid membrane; the CF_1-part sticks into stroma, where dark reactions of photosynthesis (Also called the light-independent reactions or the Calvin cycle) and ATP synthesis take place. The overall structure and the catalytic mechanism of the chloroplast ATP synthase are almost the same as those of the mitochondrial enzyme. However, in chloroplasts the proton motive force is generated not by respiratory electron transport chain, but by primary photosynthetic proteins.

E. coli ATP Synthase

E. coli ATP synthase is the simplest known form of ATP synthase, with 8 different subunit types.

Yeast ATP Synthase

Yeast ATP synthase is one of the best-studied eukaryotic ATP synthases and five F_1, eight F_O subunits and seven associated proteins have been identified. Most of these proteins have homologues in other eukaryotes.

Human ATP Synthase

The following is a list of humans genes that encode components of ATP synthases:

- ATP5A1, ATP5AL1
- ATP5B, ATP5BL1
- ATP5C2, ATP5D, ATP5E, ATP5F1, ATP5G1, ATP5G2, ATP5G3, ATP5H, ATP5HP1, ATP5I, ATP5J, ATP5J2, ATP5L, ATP5L2, ATP5O, ATP5S
- ATP6, ATP6AP1, ATP6AP2
- ATPSBL1, ATPSBL2
- MT-ATP6, MT-ATP8

SHUTTLES AND MECHANISM

Shuttles are systems of enzymes and transporters. The enzymes convert molecules into metabolites that are capable of crossing membranes via the transporters, a process that is frequently followed by reformation of the original molecule. The electrons of NADH produced in the cytoplasm must be transported into the mitochondria for conversion to ATP by the electron transport pathway. Because the NADH itself cannot cross the mitochondrial membrane, one important function of shuttle mechanisms is the transport of reducing equivalents across the mitochondrial membrane. Two separate methods are used for this purpose:

The Glycerophosphate Shuttle Malate-Aspartate Shuttle

The glycolytic intermediate dihydroxyacetone phosphate can be converted to glycerol-3-phosphate by glycerol-3- phosphate dehydrogenase; this process also results inconversion of NADH to NAD. Glycerol-3-phosphate can then be converted back to dihydroxyacetone phosphateby flavoprotein dehydrogenase (a different glycerol-3-phosphate dehydrogenase); this second enzyme is an FAD-dependent enzyme located in the mitochondrial inner membrane. Like Complex II of the electron transport chain, flavoprotein

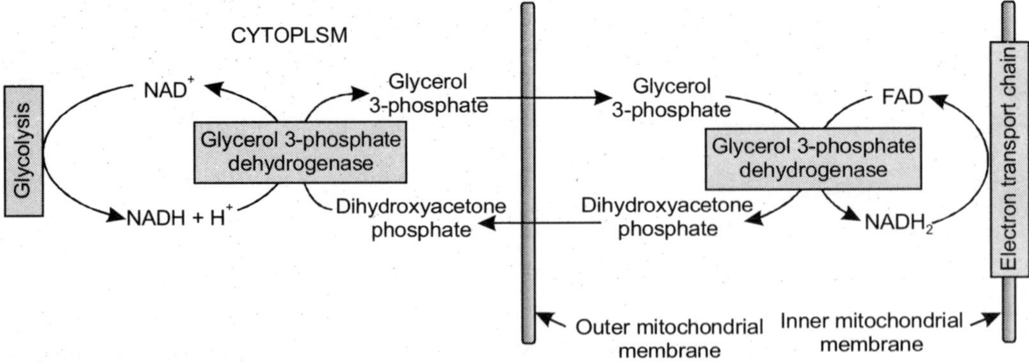

Fig. 4.18 : Glycerophosphate shuttle

dehydrogenase donates electrons directly to Coenzyme Q without pumping protons. The glycerophosphate shuttle is essentially irreversible, and therefore can be used under essentially all conditions. Because the electrons using the glycerophosphate shuttle enter the electron transport pathway at the level of Coenzyme Q, theelectrons can only be used to synthesize a maximum of two ATP, instead of the maximum of three ATP derived from NADH formed inside the mitochondria. The glycerophosphate shuttle is heavily used in insect flight muscle (a tissue heavily specialized for conversion of chemical energy to mechanical energy). Some mammalian tissues use the glycerophosphate shuttle also, but tend to prefer a more energy efficient shuttle system that uses malate and aspartate.

Malate-Aspartate Shuttle

Mammalian tissues can use a shuttle system involving malate and aspartate to transport electrons across the mitochondrial inner membrane. Oxaloacetate in the cytoplasm is converted to malate by malate dehydrogenase, oxidizing NADH to NAD. The malate enters the mitochondria using an exchanger protein that must also transport α-ketoglutarate in the opposite direction. The malate is then oxidized to oxaloacetate by the mitochondrial malate dehydrogenase, resulting in formation of NADH, which can then enter the electron transport pathway. Return of the oxaloacetate to the cytoplasm requires a separate transporter, which exchanges aspartate for glutamate. (Note that this conserves the nitrogens present in these amino acids; oxaloacetate and α-ketoglutarate are the α-keto acid counterparts of aspartate and glutamate, respectively.) This separate exchanger is necessary to allow net movement of electrons from one side of the membrane to the other.Endogenous creatine, which is already in skeletal muscle, assists in recharging adenosine triphosphate (ATP), an energy-carrying molecule, by transferring a high-energy phosphate group to adenosine diphosphate (ADP) (Fig. 4.19).

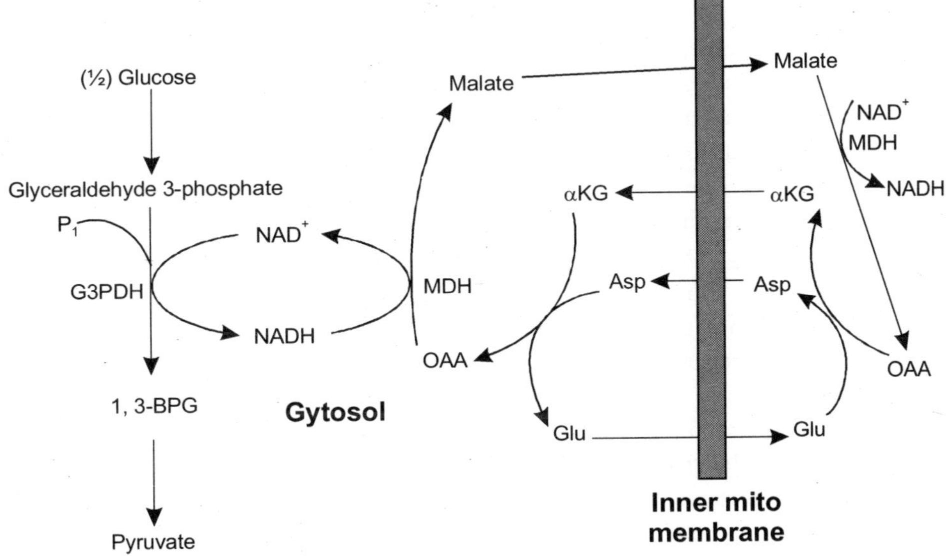

Fig. 4.19 : The Malate-Aspartate Shuttle Regeneration of Phosphocreatine at Rest

Although there has been a considerable amount of research on the properties of creatine, there is no conclusive answer as to exactly how creatine supplements work. However, there are several proposed mechanisms:

1. Increased intramuscular creatine may help maintain a high ATP/ADP ratio.
2. Creatine phosphate may buffer protons accumulating in the acidic environment of anaerobic exercise.
3. Creatine may help energy transfer from the mitochondria to sites of ATP use in a process described as the creatine phosphate energy shuttle.
4. Excess creatine in the muscle cell may osmotically draw water into the cell and stimulate protein synthesis.
5. Exogenous creatine may be converted back into amino acids and serve as precursors for muscle protein synthesis.

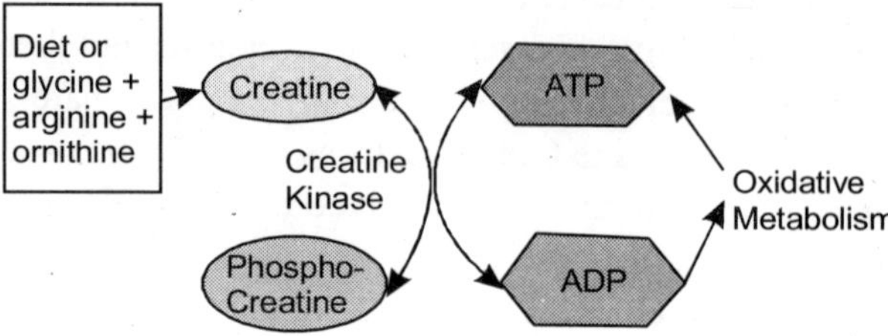

Fig. 4.20 : The Creatine Phosphate Shuttle

As described in subsequent sections, the mechanism of action for creatine may actually be a combination of these proposed mechanisms.

Citrate-pyruvate Shuttle

A related shuttle is used to move reducing equivalents out of the mitochondria. Citrate can exchange for malate. The citrate in the cytoplasm can act as a regulator of phosphofructokinase, or as a substrate for ATP-citrate lyase, an enzyme that reverses the citrate synthase reaction, and produces acetyl-CoA and oxaloacetate. (Note that ATP is required to generate the high-energy bond present in acetyl-CoA.) The oxaloacetate produced in the ATP-citrate lyase reaction can be converted tomalate. Both the oxaloacetate and malate can be used for a variety of reactions; for malate one additional reaction, catalyzed by malic enzyme, results in the formation of NADPH and pyruvate and carbon dioxide. The pyruvate produced by malic enzyme can return to the mitochondria to complete the cycle.

> Note that pyruvate has a proton-dependent pump, and generally cannot leave the mitochondria directly.

The net reaction for this shuttle reveals that two ATP are being used to transport acetyl-CoA out of the mitochondria and to transfer electrons from NADH toNADPH. The citrate-pyruvate shuttle as drawn here is irreversible, because four of the enzymes (pyruvate carboxylase, citrate synthase, ATP-citrate lyase, and malic enzyme) and the pyruvate pump mediate irreversible processes. Note, however, that the citrate and malate are transported across the membrane by exchangers that allow movement in either direction, and that malate has several mechanisms for both entering and leaving the mitochondria. The shuttle mechanisms presented here are examples of commonly used processes. The discussion above does not include a complete list of the exchangers and pumps used in the mitochondria membrane. In addition, note that malate is used in both the citrate-pyruvate and malate-aspartate shuttles; cells can use mixtures of theseprocesses to move a variety of carbon units or electrons from one side of the mitochondrial membrane to the other.

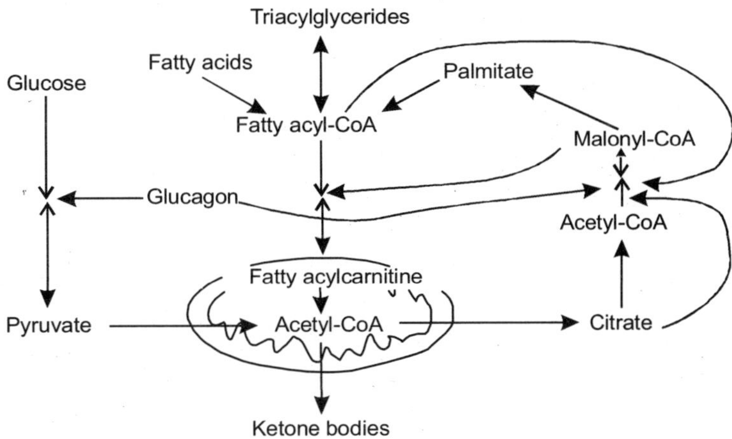

Fig. 4.21 : The Citrate-pyruvate shuttle

TRANSPORT MECHANISMS ACROSS BIOMEMBRANE

Active Transport and Passive Transport

Active transport is the pumping of molecules or ions through a membrane **against** their concentration gradient. It requires:

- a transmembrane protein (usually a complex of them) called a transporter and
- energy. The source of this energy is **ATP**.

The energy of ATP may be used directly or indirectly.

- **Direct Active Transport.** Some transporters bind ATP directly and use the energy of its hydrolysis to drive active transport.
- **Indirect Active Transport.** Other transporters use the energy already stored in the gradient of a directly-pumped **ion**. Direct active transport of the ion establishes a concentration gradient. When this is relieved by facilitated diffusion, the energy released can be harnessed to the pumping of some other ion or molecule.
- **Passive Transport.** Requires no energy from the cell. Examples include the diffusion of oxygen and carbon dioxide, osmosis of water, and facilitated diffusion.

The Na^+/K^+ ATPase

The cytosol of animal cells contains a concentration of potassium ions (K^+) as much as 20 times higher than that in the extracellular fluid. Conversely, the extracellular fluid contains a concentration of sodium ions (Na^+) as much as 10 times greater than that within the cell.

These concentration gradients are established by the active transport of both ions. And, in fact, the same transporter, called the Na^+/K^+ ATPase, does both jobs. It uses the energy from the hydrolysis of ATP to:

- actively transport 3 Na^+ ions out of the cell; and
- for each 2 K^+ ions pumped into the cell.

This accomplishes several vital functions:

- It helps establish a net charge across the plasma membrane with the interior of the cell being negatively charged with respect to the exterior. This **resting potential** prepares nerve and muscle cells for the propagation of action potentials leading to nerve impulses and muscle contraction.
- The accumulation of sodium ions outside of the cell draws water out of the cell and thus enables it to maintain osmotic balance (otherwise it would swell and burst from the inward diffusion of water).
- The gradient of sodium ions is harnessed to provide the energy to run several types of indirect pumps.

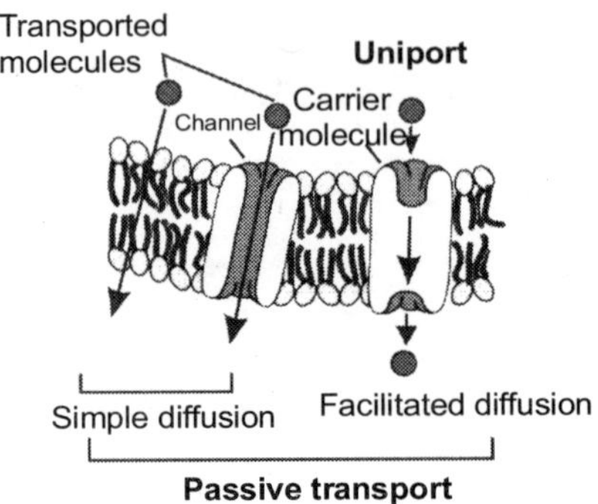

Fig. 4.22 : The Passive Transport

The crucial roles of the Na^+/K^+ ATPase are reflected in the fact that almost one-third of all the energy generated by the mitochondria in animal cells is used just to run this pump.

Passive Transport

Requires no energy from the cell. Examples include the diffusion of oxygen and carbon dioxide, osmosis of water, and facilitated diffusion.

Symport Pumps

In this type of indirect active transport, the driving ion (Na^+) and the pumped molecule pass through the membrane pump in the **same** direction.

Examples:

- **The Na^+/glucose transporter**. This transmembrane protein allows sodium ions and glucose to enter the cell together. The sodium ions flow **down** their concentration gradient while the glucose molecules are pumped **up** theirs. Later the sodium is pumped back out of the cell by the Na^+/K^+ ATPase.
- The Na^+/glucose transporter is used to actively transport glucose out of the intestine and also out of the kidney tubules and back into the blood.
- All the **amino acids** can be actively transported, for example
  - out of the kidney tubules and into the blood
  - the reuptake of Glu from the synapse back into the presynaptic neuron

by sodium-driven symport pumps.

- **The Na^+/iodide transporter**. This symporter pumps iodide ions into the cells of the thyroid gland (for the manufacture of thyroxine) and also into the cells of the mammary gland (to supply the baby's need for iodide).
- The **permease** encoded by the lac operon of *E. coli* that transports lactose into the cell.

Antiport Pumps

In antiport pumps, the driving ion (again, usually sodium) diffuses through the pump in one direction providing the energy for the active transport of some other molecule or ion in the opposite direction.

Example: Ca^{2+} ions are pumped out of cells by sodium-driven antiport pumps.

Antiport pumps in the vacuole of some plants harness the outward facilitated diffusion of protons (themselves pumped into the vacuole by a H^+ ATPase).

- to the active inward transport of sodium ions. This sodium/proton antiport pump enables the plant to sequester sodium ions in its vacuole. Transgenic tomato plants that overexpress this sodium/proton antiport pump are able to thrive in saline soils too salty for conventional tomatoes.
- to the active inward transport of nitrate ions (NO_3^-).

Some Inherited Ion-channel Diseases

A growing number of human diseases have been discovered to be caused by inherited mutations in genes encoding channels.

Some examples:

- **Chloride-channel** diseases
  - Cystic fibrosis
  - inherited tendency to kidney stones (caused by a different kind of chloride channel than the one involved in cystic fibrosis)
- **Potassium-channel** diseases
  - some inherited life-threatening defects in the heartbeat
  - a rare, inherited tendency to epileptic seizures in the newborn.
  - several types of inherited **deafness**
- **Sodium-channel** diseases
  - inherited tendency to certain types of muscle spasms
  - Liddle's syndrome. Inadequate sodium transport out of the kidneys, because of a mutant sodium channel, leads to elevated **osmotic pressure** of the blood and resulting hypertension (high blood pressure).

ENDOCYTOSIS

In endocytosis, the cell engulfs some of its extracellular fluid (**ECF**) including material dissolved or suspended in it. A portion of the plasma membrane is invaginated and pinched off forming a membrane-bounded **vesicle** called an **endosome**.

Phagocytosis

Phagocytosis ("cell eating"):

- results in the ingestion of particulate matter (e.g., bacteria) from the ECF.
- The endosome is so large that it is called a **phagosome** or **vacuole**.
- Phagocytosis occurs only in certain specialized cells (e.g., neutrophils, macrophages, the amoeba), and
- occurs sporadically.

This micrograph shows a guinea phagocyte ingesting polystyrene beads. Several beads are already enclosed in phagosomes while the others are in the process of being engulfed.

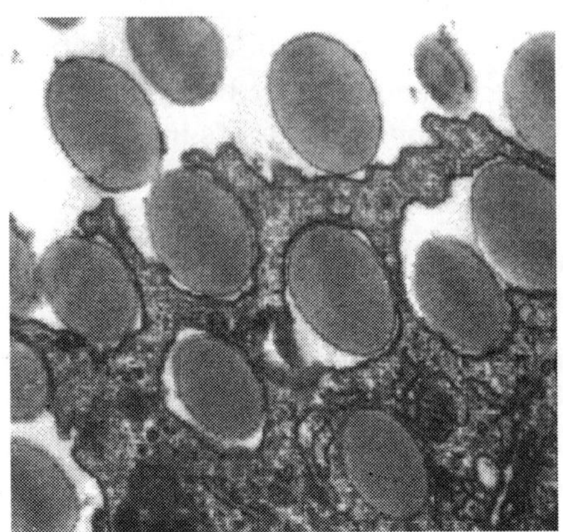

Fig. 4.22A : The Phagocytosis

In due course, phagosomes deliver their contents to lysosomes. The membranes of the two organelles fuse. Once inside the lysosome, the contents of the phagosome, e.g. ingested bacteria, are destroyed by the degradative enzymes of the lysosome.

Games Parasites Play!

Phagocytic cells, like macrophages and neutrophils, are an early line of defense against invading bacteria. However, some bacteria have evolved mechanisms to avoid destruction even after they have been engulfed by phagocytes.

Two examples:

- **Salmonella enterica** is a bacterium that causes food poisoning in humans. Once engulfed by phagocytosis, it secretes a protein that prevents the fusion of its phagosome with a lysosome.
- **Mycobacteria** (e.g., the tubercle bacillus that causes tuberculosis) use a different trick.
  - When the phagosome is first pinched off from the plasma membrane, it is coated with a protein called "TACO" (for tryptophan-aspartate-containing coat protein).
  - This must be removed before the phagosome can fuse with a lysosome.
  - Mycobacteria taken into a phagosome are able, in some way, to keep the TACO coat from being removed.
  - Thus there is no fusion with lysosomes and the mycobacteria can continue to live in this protected intracellular location.

Pinocytosis

In pinocytosis ("cell drinking"), the drop engulfed is relatively small.

Pinocytosis

- occurs in almost all cells
- occurs continuously

A cell sipping away at the ECF by pinocytosis acquires a representative sample of the molecules and ions dissolved in the ECF. But pinocytosis also provides a much more elegant method for cells to pick up critical components of the ECF that may be in scant supply.

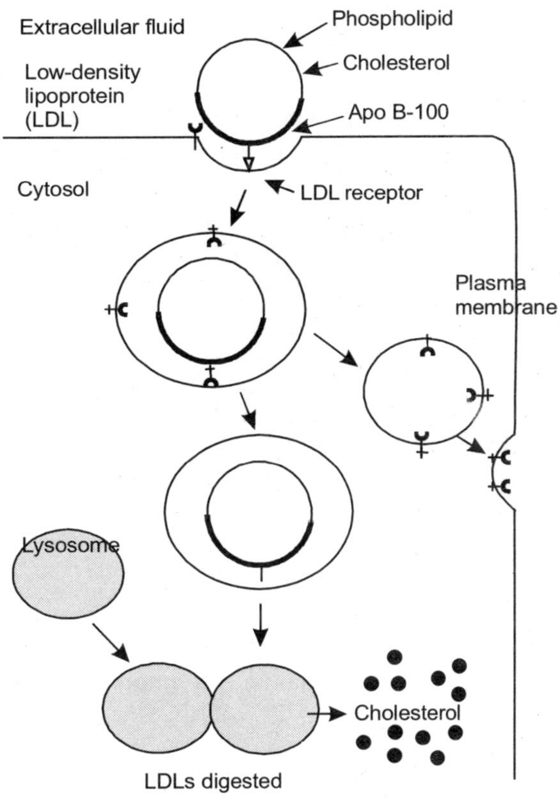

Fig. 4.23 : The diagram showing the Endocytosis in LDL

Receptor-Mediated Endocytosis

Some of the integral membrane proteins that a cell displays at its surface are receptors for particular components of the ECF. For example, iron is transported in the blood complexed to a protein called **transferrin**. Cells have receptors for transferrin on their surface. When these receptors encounter a molecule of transferrin, they bind tightly to it. The complex of transferrin and its receptor is then engulfed by endocytosis. Ultimately, the iron is released into the cytosol. The strong affinity of the transferrin receptor for transferrin (its ligand) ensures that the cell will get all the iron it needs even if transferrin represents only a small fraction of the protein molecules present in the ECF. Receptor-mediated endocytosis is many thousand times more efficient than simple pinocytosis in enabling the cell to acquire the macromolecules it needs.

The Low-Density Lipoprotein (LDL) Receptor

Cells take up cholesterol by receptor-mediated endocytosis. Cholesterol is an essential component of all cell membranes. Most cells can, as needed, either synthesize cholesterol or acquire it from the ECF. Human cells get much of their cholesterol from the liver and, if your diet is not strictly "100% cholesterol-free", by absorption from the intestine.

Cholesterol is a hydrophobic molecule and quite insoluble in water. Thus it cannot pass from the liver and/or the intestine to the cells simply dissolved in blood and ECF. Instead it is carried in tiny droplets of lipoprotein. The most abundant cholesterol carriers in humans are the **low-density lipoproteins** or **LDL**s.

LDL particles are spheres covered with a single layer of phospholipid molecules with their hydrophilic heads exposed to the watery fluid (e.g., blood) and their hydrophobic tails directed into the interior. Over a thousand molecules of cholesterol are bound to the hydrophobic interior of LDL particles. One molecule of a protein, called **apolipoprotein B-100** (Apo B-100) is exposed at the surface of each LDL particle.

The first step in acquiring LDL particles is for them to bind to **LDL receptors** exposed at the cell surface. These transmembrane proteins have a site that recognizes and binds to the apolipoprotein B-100 on the surface of the LDL. The portion of the plasma membrane with bound LDL is internalized by endocytosis. A drop in the pH (from ~7 to ~5) causes the LDL to separate from its receptor. The vesicle then pinches apart into two smaller vesicles: one containing free LDLs; the other containing now-empty receptors. The vesicle with the LDLs fuses with a lysosome to form a **secondary lysosome**. The enzymes of the lysosome then release free cholesterol into the cytosol. The vesicle with unoccupied receptors returns to and fuses with the plasma membrane, turning inside out as it does so (exocytosis). In this way the LDL receptors are returned to the cell surface for reuse.

People who inherit two defective (mutant) genes for the LDL receptor have receptors that function poorly or not at all. This creates excessively high levels of LDL in their blood and predisposes them to atherosclerosis and heart attacks. The ailment is called **familial** (because it is inherited) **hypercholesterolemia**.

Mutations in the **Apo B-100 gene** cause another form of inherited hypercholesterolemia.

Other small hydrophobic molecules are also transported in the blood while bound to soluble proteins:

- the retinoid vitamin A (retinol) bound to the retinol-binding protein
- the steroids
  - 25[OH] vitamin D_3 bound to the vitamin D binding protein
  - cortisol bound to the corticosteroid binding globulin
  - testosterone and estrogens bound to the sex hormone binding globulin

and there is growing evidence that, like cholesterol, they are taken into the cell by receptor-mediated endocytosis.

More Games Parasites Play

Some intracellular parasites exploit receptor-mediated endocytosis to sneak their way into their host cell.

They have evolved surface molecules that serve as decoy ligands for receptors on the target cell surface. Binding to these receptors tricks the cell into engulfing the parasite.

Some examples:

- This is the organism that led to the discovery that genes are DNA
- **Epstein-Barr Virus (EBV)**. This virus causes mononucleosis and is a contributing factor in the development of Burkitt's lymphoma, a cancer of B lymphocytes. It binds to a receptor present on the surface of B cells.
- **Influenza virus**. The hemagglutinin on the surface of the virus binds to carbohydrate on the surface of the target cell tricking the cell into engulfing it.

- **Listeria monocytogenes**. This food-borne bacterium can be dangerous to people with defective immune systems as well as to pregnant women and their newborn babies. It has two kinds of surface molecules each a ligand for a different receptor on the target cell surface.
- **Streptococcus pneumoniae**. Epithelial cells like those in the nasopharynx have receptors that are responsible for transporting IgA and IgM antibodies from the blood to the cell surface. The pneumococcus exploits this receptor for a return trip into the cell.

Coming Full Circle

Endocytosis removes portions of the plasma membrane and takes them inside the cell. To keep in balance, membrane must be returned to the plasma membrane. This occurs by exocytosis.

TRANSPORT THROUGH CELL MEMBRANE

Importance

All cells acquire the molecules and ions they need from their surrounding extracellular fluid (ECF). There is an unceasing traffic of molecules and ions

- in and out of the cell through its plasma membrane
  - Examples: glucose, Na^+, Ca^{2+}
- In eukaryotic cells, there is also transport in and out of membrane-bounded intracellular compartments such as the nucleus, endoplasmic reticulum, and mitochondria.
  - Examples: proteins, mRNA, Ca^{2+}, ATP

Two problems to be considered

Relative concentrations

Molecules and ions move spontaneously down their concentration gradient (i.e., from a region of higher to a region of lower concentration) by diffusion.

Molecules and ions can be moved **against** their concentration gradient, but this process, called **active transport**, requires the expenditure of energy (usually from ATP).

Lipid bilayers are impermeable to most essential molecules and ions

The lipid bilayer is permeable to **water** molecules and a few other small, uncharged, molecules like oxygen (O_2) and carbon dioxide (CO_2). These diffuse freely in and out of the cell. The diffusion of water through the plasma membrane is of such importance to the cell that it is given a special name: **osmosis**.

Lipid bilayers are **not** permeable to:

- **ions** such as
  - K^+, Na^+, Ca^{2+} (called **cations** because when subjected to an electric field they migrate toward the cathode [the negatively-charged electrode])
  - Cl^-, HCO_3^- (called **anions** because they migrate toward the anode [the positively-charged electrode])

- small hydrophilic **molecules** like glucose
- **macromolecules** like proteins and RNA.

This page will examine how ions and small molecules are transported across cell membranes. The transport of macromolecules through membranes is described in Endocytosis.

Solving these Problems

Mechanisms by which cells solve the problem of transporting ions and small molecules across their membranes:

Facilitateddiffusion: Transmembrane proteins create a water-filled pore through which ions and some small hydrophilic molecules can pass by diffusion. The channels can be opened (or closed) according to the needs of the cell.

Activetransport: Transmembrane proteins, called transporters, use the energy of ATP to force ions or small molecules through the membrane **against** their concentration gradient.

Facilitated Diffusion of Ions

Facilitated diffusion of ions takes place through proteins, or assemblies of proteins, embedded in the plasma membrane. These transmembrane proteins form a water-filled channel through which the ion can pass down its concentration gradient.

The transmembrane channels that permit facilitated diffusion can be opened or closed. They are said to be "gated".

Some types of gated ion channels:

- ligand-gated.
- mechanically-gated.
- voltage-gated, and
- light-gated.

Ligand-gated Ion Channels

Many ion channels open or close in response to binding a small signaling molecule or **"ligand"**. Some ion channels are gated by extracellular ligands; some by intracellular ligands. In both cases, the ligand is **not** the substance that is transported when the channel opens.

External Ligands

External ligands bind to a site on the extracellular side of the channel.

***Examples*:**

- **Acetylcholine (ACh)**. The binding of the neurotransmitter acetylcholine at certain synapses opens channels that admit Na^+ and initiate a nerve impulse or muscle contraction.
- **Gamma amino butyric acid (GABA)**. Binding of GABA at certain synapses—designated $GABA_A$—in the central nervous system admits Cl^- ions into the cell and inhibits the creation of a nerve impulse.

Internal ligands

Internal ligands bind to a site on the channel protein exposed to the cytosol.

Examples:

- "Second messengers", like **cyclic AMP (cAMP)** and **cyclic GMP (cGMP)**, regulate channels involved in the initiation of impulses in neurons responding to odors and light respectively.

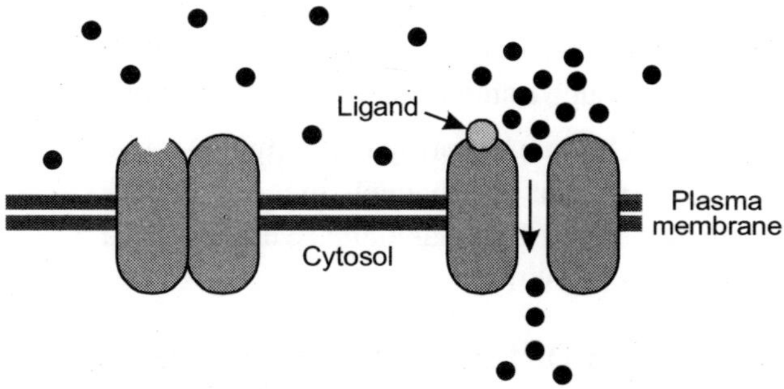

Fig. 4.24 : Diffusion Through a Ligand-gated Channel

- **ATP** is needed to open the channel that allows chloride (Cl^-) and bicarbonate (HCO_3^-) ions out of the cell. This channel is defective in patients with **cystic fibrosis**. Although the energy liberated by the hydrolysis of ATP is needed to open the channel, this is **not** an example of active transport; the ions diffuse through the open channel following their concentration gradient.

Mechanically-gated Ion Channels

Examples:

- Sound waves bending the cilia-like projections on the hair cells of the inner ear open up ion channels leading to the creation of nerve impulses that the brain interprets as sound.
- Mechanical deformation of the cells of stretch receptors opens ion channels leading to the creation of nerve impulses.

Voltage-gated Ion Channels

In so-called "excitable" cells like **neurons** and **muscle cells**, some channels open or close in response to changes in the charge (measured in volts) across the plasma membrane.

Example: As an impulse passes down a neuron, the reduction in the voltage opens sodium channels in the adjacent portion of the membrane. This allows the influx of Na^+ into the neuron and thus the continuation of the nerve impulse.

Some 7000 sodium ions pass through each channel during the brief period (about 1 millisecond) that it remains open. This was learned by use of the patch clamp technique.

The Patch Clamp Technique

The properties of ion channels can be studied by means of the patch clamp technique.

- A very fine pipette (with an opening of about 0.5 μm) is pressed against the plasma membrane of
- either an intact cell or
- the plasma membrane can be pulled away from the cell and the preparation placed in a test solution of desired composition.
- Current flow through a single ion channel can then be measured.

Such measurements reveal that each channel is either fully open or fully closed; that is, facilitated diffusion through a single channel is "all-or-none".

This technique has provided so much valuable information about ion channels that its inventors, Erwin Neher and Bert Sakmann, were awarded a Nobel Prize in 1991).

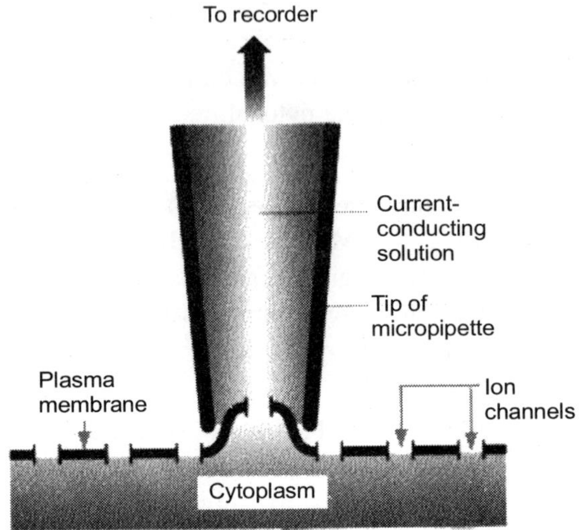

Fig. 4.25 : The Patch Clamp Technique

Facilitated Diffusion of Molecules

Some small, hydrophilic organic molecules, like sugars, can pass through cell membranes by facilitated diffusion.

Once again, the process requires trans-membrane proteins. In some cases, these—like ion channels—form water-filled pores that enable the molecule to pass in or out of the membrane following its concentration gradient.

Example: Maltoporin. This homotrimer in the outer membrane of E. coli forms pores that allow the disaccharide maltose and a few related molecules to diffuse into the cell.

Another example: The plasma membrane of human red blood cells contain transmembrane proteins that permit the diffusion of glucose from the blood into the cell.

Note that in all cases of facilitated diffusion through channels, the channels are selective; that is, the structure of the protein admits only certain types of molecules through.

Whether all cases of facilitated diffusion of small molecules use channels is yet to be proven. Perhaps some molecules are passed through the membrane by a conformational change in the shape of the transmembrane protein when it binds the molecule to be transported.

Hypotonic Solutions

If the concentration of water in the medium surrounding a cell is greater than that of the cytosol, the medium is said to be **hypotonic**. Water enters the cell by osmosis.

A red blood cell placed in a hypotonic solution (e.g., pure water) bursts immediately ("hemolysis") from the influx of water.

Plant cells and bacterial cells avoid bursting in hypotonic surroundings by their strong cell walls. These allow the buildup of **turgor** within the cell. When the turgor pressure equals the osmotic pressure, osmosis ceases.

Isotonic Solutions

When red blood cells are placed in a 0.9% salt solution, they neither gain nor lose water by osmosis. Such a solution is said to be **isotonic**.

The extracellular fluid (ECF) of mammalian cells is isotonic to their cytoplasm. This balance must be actively maintained because of the large number of organic molecules dissolved in the cytosol but not present in the ECF. These organic molecules exert an osmotic effect that, if not compensated for, would cause the cell to take in so much water that it would swell and might even burst. This fate is avoided by pumping sodium ions out of the cell with the Na^+/K^+ ATPase.

Hypertonic Solutions

If red cells are placed in sea water (about 3% salt), they lose water by osmosis and the cells shrivel up. Sea water is **hypertonic** to their cytosol.

Similarly, if a plant tissue is placed in sea water, the cell contents shrink away from the rigid cell wall. This is called **plasmolysis**.

Sea water is also hypertonic to the ECF of most marine vertebrates. To avoid fatal dehydration, these animals (e.g., bony fishes like the cod) must

- continuously drink sea water and then
- desalt it by pumping ions out of their gills by active transport. (Marine reptiles—turtles and snakes—use special salt glands for the same purpose.)

Osmosis is important A report in the 23 April 1998 issue of **The New England Journal of Medicine** tells of the life-threatening complications that can be caused by an ignorance of osmosis.

- Large volumes of a solution of 5% human albumin are injected into people undergoing a procedure called plasmapheresis.
- The albumin is dissolved in physiological saline (0.9% NaCl) and is therefore isotonic to human plasma (the large protein molecules of albumin have only a small osmotic effect).
- If 5% solutions are unavailable, pharmacists may substitute a proper dilution of a 25% albumin solution. Mixing 1 part of the 25% solution with 4 parts of diluent results in the correct 5% solution of albumin.
- **BUT**, in several cases, the diluent used was sterile water, not physiological saline.
- **SO**, the resulting solution was strongly hypotonic to human plasma.
- **The Result:** massive, life-threatening hemolysis in the patients.

ROLE OF G PROTEINS

G proteins, short for **guanine nucleotide-binding proteins**, are a family of proteins involved in second messenger cascades.

G proteins are so called because they function as "molecular switches", alternating between an inactive guanosine diphosphate (GDP) and active guanosine triphosphate (GTP) bound state, ultimately going on to regulate downstream cell processes.

G proteins were discovered when Alfred G. Gilman and Martin Rodbell investigated stimulation of cells by adrenaline. They found that when a hormone like adrenaline bound to a receptor, the receptor did not stimulate enzymes like adenylate cyclase directly. Instead, the receptor stimulated a G protein, which then stimulated the adenylate cyclase to produce a second messenger, cyclic AMP. For this discovery they won the 1994 Nobel Prize in Physiology or Medicine.

G proteins belong to the larger group of enzymes called GTPases.

Functions

G proteins are important signal transducing molecules in cells. In fact, diseases such as diabetes, blindness, allergies, depression, cardiovascular defects and certain forms of cancer, among other pathologies, are thought to arise due to derangement of G protein signaling.

The human genomes encodes roughly 350 G protein-coupled receptors, which detect hormones, growth factors and other endogenous ligands. Approximately 150 of the GPCRs found in the human genome have unknown functions.

Types of G protein Signaling

G protein can refer to two distinct families of proteins. Heterotrimeric G proteins, sometimes referred to as the "large" G proteins that are activated by G protein-coupled receptors and made up of alpha (α), beta (β), and gamma (γ) subunits. There are also *"small" G proteins* (20-25kDa) that belong to the Ras superfamily of small GTPases. These proteins are homologous to the alpha (α) subunit found in heterotrimers, and are in fact monomeric. However, they also bind GTP and GDP and are involved in signal transduction.

Heterotrimeric G Proteins

Heterotrimeric G proteins share a common mode of action, i.e., activation in response to a conformation change in the G-protein-coupled receptor, exchange of GTP for GDP and dissociation in order to activate further proteins in the signal transduction pathway. However, the specific mechanism differs between different types of G proteins.

Common Mechanism

Receptor-activated G proteins are bound to the inside surface of the cell membrane. They consist of the G_α and the tightly associated $G_{\beta\gamma}$ subunits. At the present time, four main families exist for G_α subunits: $G_{\alpha s}$, $G_{\alpha i}$, $G_{\alpha q/11}$, and $G_{\alpha 12/13}$. These groups differ primarily in effector recognition, but share a similar mechanism of activation.

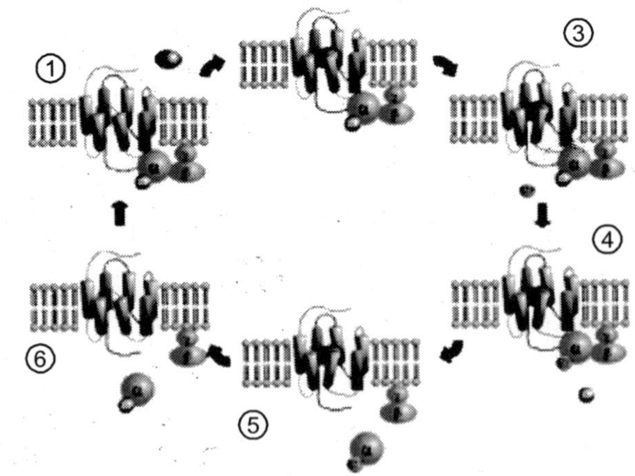

Fig. 4.26 : Activation cycle of G-proteins by G-protein-coupled receptors

Activation

When a ligand activates the G protein-coupled receptor, it induces a conformation change in the receptor (a change in shape) that allows the receptor to function as a guanine nucleotide exchange factor (GEF) that exchanges GTP in place of GDP on the G_α subunit. In the traditional view of heterotrimeric protein activation, this exchange triggers the dissociation of the G_α subunit, bound to GTP, from the $G_{\beta\gamma}$ dimer and the receptor. However, models that suggest molecular rearrangement, reorganization, and pre-complexing of effector molecules are beginning to be accepted. Both G_α-GTP and $G_{\beta\gamma}$ can then activate different *signaling cascades* (or *second messenger pathways*) and effector proteins, while the receptor is able to activate the next G protein.

Termination

The G_α subunit will eventually hydrolyze the attached GTP to GDP by its inherent enzymatic activity, allowing it to re-associate with $G_{\beta\gamma}$ and starting a new cycle. There do exist groups of proteins called RBMs that act as GTPase-activating proteins (GAPs), which are specific for G_α subunits, which act to accelerate hydrolysis and terminate the transduced signal. In some cases the effector itself may possess intrinsic GAP activity, which helps deactivate the pathway. This is true in the case of phospholipase C beta, which possesses GAP activity within its C-terminal region. This is an alternate form of regulation for the G_α subunit.

Specific Mechanisms

- **G_{as}** stimulates the production of cAMP from ATP. This is accomplished by direct stimulation of the membrane-associated enzyme adenylate cyclase. cAMP acts as a second messenger that goes on to interact with and activate protein kinase A (PKA). PKA can then phosphorylate a myriad of downstream targets.
- **G_{ai}** inhibits the production of cAMP from ATP.
- **$G_{aq/11}$** stimulates membrane-bound phospholipase C beta, which then cleaves PIP_2 (a minor membrane phosphoinositol) into two second messengers, IP3 and diacylglycerol (DAG).
- **$G_{a12/13}$** are involved in Rho family GTPase signaling (through RhoGEF superfamily) and control cell cytoskeleton remodeling, thus regulating cell migration.
- **G_{bg}** sometimes also have active functions, e.g., coupling to L-type calcium channels.

Small GTPases

Small GTPases also bind GTP and GDP and are involved in signal transduction. These proteins are homologous to the alpha (α) subunit found in heterotrimers, but exist as monomers. They are small (20-kDa to 25-kDa) proteins that bind to guanosine triphosphate (GTP). This family of proteins is homologous to Ras GTPases and is also called the Ras superfamily GTPases.

Lipidation

In order to associate with the inner leaflet of the plasma membrane, many G proteins are covalently modified with lipid extensions, i.e., they are lipidated.

- Heterotrimeric G protein subunits may be myristolated, palmitoylated, or prenylated.
- Small G proteins may be prenylated.

CYCLIC ADENOSINE MONOPHOSPHATE

Cyclic adenosine monophosphate (**cAMP**, **cyclic AMP** or 3'-5'-cyclic adenosine monophosphate) is a second messenger that is important in many biological processes. cAMP is derived from adenosine triphosphate (ATP) and used for intracellular signal transduction in many different organisms.

Synthesis and Decomposition

cAMP is synthesised from ATP by adenylyl cyclase which is located at the cell membranes. Adenylyl cyclase is activated by a range of signaling molecules through the activation of adenylyl cyclase stimulatory G (G_s)-coupled receptors and inhibited by agonists of adenylyl cyclase inhibitory G (G_i)-protein coupled receptors. Liver adenylyl cyclase responds more strongly to glucagon, and muscle adenylyl cyclase responds more strongly to adrenaline.

cAMP decomposition into AMP is catalyzed by the enzyme phosphodiesterase.

Functions

cAMP is a second messenger, used for intracellular signal transduction, such as transferring the effects of hormones like glucagon and adrenaline, which cannot get through the cell membrane. Its purposes include the activation of protein kinases and regulating the effects of adrenaline and glucagon. It is also used to regulate the passage of Ca^{2+} through ion channels.

In humans

Epinephrine (adrenaline) binds its receptor, that associates with an heterotrimeric G protein. The G protein associates with adenylyl cyclase that converts ATP to cAMP, spreading the signal.

cAMP and its associated kinases function in several biochemical processes, including the regulation of glycogen, sugar, and lipid metabolism.

In humans, cyclic AMP works by activating protein kinase A (PKA, also known as cAMP-dependent protein kinase). This is normally inactive as a tetrameric holoenzyme, consisting of 2 catalytic and 2 regulatory units (C_2R_2), with the regulatory units blocking the catalytic centers of the catalytic units. Cyclic AMP binds to specific locations on the regulatory units of the protein kinase, and causes dissociation between the regulatory and catalytic subunits, thus activating the catalytic units and enabling them to phosphorylate substrate proteins.

The active subunits catalyze the transfer of phosphate from ATP to specific serine or threonine residues of protein substrates. The phosphorylated proteins may act directly on the cell's ion channels, or may become activated or inhibited enzymes. Protein kinase A can also phosphorylate specific proteins that bind to promoter regions of DNA, causing increased expression of specific genes. Not all protein kinases respond to cAMP: several types of protein kinases are not cAMP dependent, for example protein kinase C. Further effects depend on cAMP-dependent protein kinase, which vary based on the type of cell.

AN OVER VIEW

Metabolism is the set of chemical reactions that occur in living organisms in order to maintain life. These processes allow organisms to grow and reproduce, maintain their structures, and respond to their environments. Metabolism is usually divided into two categories. Catabolism breaks down organic matter, for example to harvest energy in cellular respiration. Anabolism, on the other hand, uses energy to construct components of cells such as proteins and nucleic acids. The chemical reactions of metabolism are organized into metabolic pathways, in which one chemical is transformed into another by a sequence of enzymes.

Enzymes are crucial to metabolism because they allow organisms to drive desirable but thermodynamically unfavorable reactions by coupling them to favorable ones, and because they act as catalysts to allow these reactions to proceed quickly and efficiently. Enzymes also allow the regulation of metabolic pathways in response to changes in the cell's environment or signals from other cells. The metabolism of an organism determines which substances it will find nutritious and which it will find poisonous.

For example, some prokaryotes use hydrogen sulfide as a nutrient, yet this gas is poisonous to animals. The speed of metabolism, the metabolic rate, also influences how much food an organism will require. A striking feature of metabolism is the similarity of the basic metabolic pathways between even vastly different species. For example, the set of carboxylic acids that are best known as the intermediates in the citric acid cycle are present in all organisms, being found in species as diverse as the unicellular bacteria *Escherichia coli* and huge multicellular organisms like elephants. These striking similarities in metabolism are most likely the result of the high efficiency of these pathways, and of their early appearance in evolutionary history.

CONTROL OF METABOLIC PATHWAYS

In biochemistry, a metabolic pathway is a series of chemical reactions occurring within a cell. In each pathway, a principal chemical is modified by chemical reactions. Enzymes catalyze these reactions, and often require dietary minerals, vitamins and other cofactors in order to function properly. Because of the many chemicals that may be involved, pathways can be quite elaborate. In addition, many pathways can exist within a cell. This collection of pathways is called the metabolic network. Pathways are important to the maintenance of homeostasis within an organism.

Metabolism is a step by step modification of the initial molecule to shape it into another product. The result can be used in one of three ways:

- Stored by the cell.
- Be used immediately, as a metabolic product.
- Initiate another metabolic pathway, called a flux generating step.

A molecule called a substrate enters a metabolic pathway depending on the needs of the cell and the availability of the substrate. An increase in concentration of anabolic and catabolic end products would slow the metabolic rate for that particular pathway. Each metabolic pathway is composed of a series of biochemical reactions that are connected by their intermediates: the reactants (or substrates) of one reaction are the products of the previous one, and so on. Metabolic pathways are usually considered in one direction (although all reactions are chemically reversible, conditions in the cell are such that it is thermodynamically more favorable for flux to be in one of the directions).

- Glycolysis was the first metabolic pathway discovered:
    1. As glucose enters a cell it is immediately phosphorylated by ATP to glucose 6-phosphate in the irreversible first step. This is to prevent the glucose leaving the cell.
    2. In times of excess lipid or protein energy sources glycolysis may run in reverse (gluconeogenesis) in order to produce glucose 6-phosphate for storage as glycogen or starch.
- Metabolic pathways are often regulated by feedback inhibition, or by a cycle where one of the products in the cycle starts the reaction again, such as the Krebs Cycle.
- Anabolic and catabolic pathways in eukaryotes are separated by either compartmentation or by the use of different enzymes and cofactors.

Cellular Respiration

Several distinct but linked metabolic pathways are used by cells to transfer the energy released by breakdown of fuel molecules to ATP. These occur within all living organisms in some forms:

1. Glycolysis
2. Anaerobic respiration
3. Krebs cycle/Citric acid cycle
4. Oxidative phosphorylation

Other pathways occurring in (most or) all living organisms include:

- Fatty acid oxidation (β-oxidation)
- Gluconeogenesis
- HMG-CoA reductase pathway (isoprene prenylation chains, see cholesterol)
- Pentose phosphate pathway (hexose monophosphate shunt)
- Porphyrin synthesis (or heme synthesis) pathway
- Urea cycle

Creation of energetic compounds from non-living matter:

- Photosynthesis (plants, algae, cyanobacteria).
- Chemosynthesis (some bacteria).

METABOLISM OF CARBOHYDRATE

The chief primary materials for carbohydrate metabolism supplied to the blood from the intestine are the monosaccharides from the digestion of food for the synthesis of tissue proteins in growth and maintenance. The materials absorbed into the portal blood pass to the liver. Where fructose and galactose are converted to glucose and glycogen and some of the amino acids are deaminized to for keto acids which are also converted to these carbohydrate. Lactic and pyruvic acids which are also converted to these carbohydrates. Lactic and pyruvic acids which are also converted to glucose and glycogen. All the process by which the liver converts non-glucose substances into glucose constitutes the gluconeogenetic mechanism of the liver. Although the kidneys have some capacity for gluconeogenesis, it is very limited as compared with that of the liver. Gluconeogenesis is constantly taking place but at widely varying rates. It is increased on high protein diets when large amounts of amino acids are absorbed into the blood, and decreased on high carbohydrate diets. When there is an abundance of preformed glucose. It is increased during exercise, when large amount of lactic acid and pyruvic acids escape from the working muscles and there is need to keep up the blood glucose and replenish the muscle glycogen supply. Under these conditions the liver acts to recover and return to them sources of energy lost by the muscles. During starvation gluconeogenesis from the amino acids of tissue protein is the chief source of blood sugar and tissue glycogen. In diabetes the rate if gluconeogenesis from both food and tissue protein may be greatly increased, contributing to body emaciation.

The liver stores glucose as glycogen when the supplies of blood glucose and glucose precursors are liberal and reconverts this glycogen to glucose for addition to the blood when the blood sugar falls. The liver thus acts as a glucostatic mechanism to maintain the blood glucose within normal physiologic limits. The conversion of glucose to glycogen by the liver (as well as other tissues) requires preliminary phosphorylation of the glucose, for which ATP is essential chemical energy in the form of ATP is required also for many other liver processes. Part of this ATP is provided through oxidation of fatty acids, and a partly by the oxidation of the glycogen. In this latter process the glycogen first undergoes breakdown or glycolysis through a long series of phosphyrelated intermediates of pyruvic acid. In this process some useful energy as ATP is stored up.

The pyruvic acid is then oxidized through the tricarboxylic or citric acid cycle to CO_2 and H_2O, with the formation of most of the ATP which is produced from glycogen, while both lactic and pyruvic acids are formed in the liver cells, essentially none escapes into the blood (in contrast with muscle.

This is due to the efficiency of reconversion to glycogen; in case oxidation of pyruvic acid cannot keep up with the rate of its formation in the glycolytic stage. Lactic acid is first converted to pyruvic acid is then converted to glycogen. These processes thus from lactic acid and pyruvic acid. It will be noted from the fig 136 that glycerol is formed in the glycolytic phase of glycogen breakdown in the liver. Fatty acids are synthesized from pyruvic acid, which upon combination with the glycerol from the glycerides of fat form the carbohydrates. The liver also synthesizes cholesterol from the pyruvic acid of glycogen breakdown. This synthesis proceeds through acetic acid (acetyl CoA) as an intermediate stage as also does the synthesis of fatty acids from the pyruvic acid. Acetic acid as acetyl coenzyme CoA formed from pyruvic acid is used by the liver in acetylating amines in the process of detoxication.

The chemical mechanism involved in liver gluconeogenesis from non glucose monosaccharides and amino acids will be discussed latter.

Carbohydrate metabolism in muscle is highly specialized and designed primarily for the production of ATP as a source of energy for the concentrated process. Muscle is limited essentially to blood glucose for its carbohydrate supply. In muscle, glucose is converted to glycogen through the same phosphorylating reactions involved in the formation of liver glycogen yields energy as ATP through glycolysis and oxidation of pyruvic acid in the tricarboxylic acid cycle just as Liver glycogen does.

Muscle also forms glycogen from lactic acid through pyruvic acid and reversal of the glycolytic reactions. A part of lactic acid is reconverted to glycogen, in the case when rigorous exercise the rate of production of pyruvic and lactic acid in muscle greatly exceeds the rate at which they can be oxidized or reconverted to glycogen, and much escapes into the blood stream. Any condition such as asphyxia which causes tissue anoxia or which limits the oxidative processes in tissue likewise causes the rapid breakdown of muscle glycogen and escape of these acids into the blood most of the acids are not picked up from the blood and utilized by the muscles and other peripheral tissues are transported to the liver and converted to glycogen, which is passed back into the blood as glucose when the blood sugar level falls. However, the conservation by the liver of lactic and pyruvic acids rise appreciable amounts may be lost in the urine.

While pyruvic acid is the primary product of glycolysis it is reversely convertible to lactic acid under the influence of the enzyme lactic acid and the blood of normal resting adults generally contains 10 mg of lactic acid and 1 mg of pyruvic acid per 100ml. the lactate–pyruvate ratio then is about 10:1 These values vary with exercise and disturbances in normal tissue oxidative process.

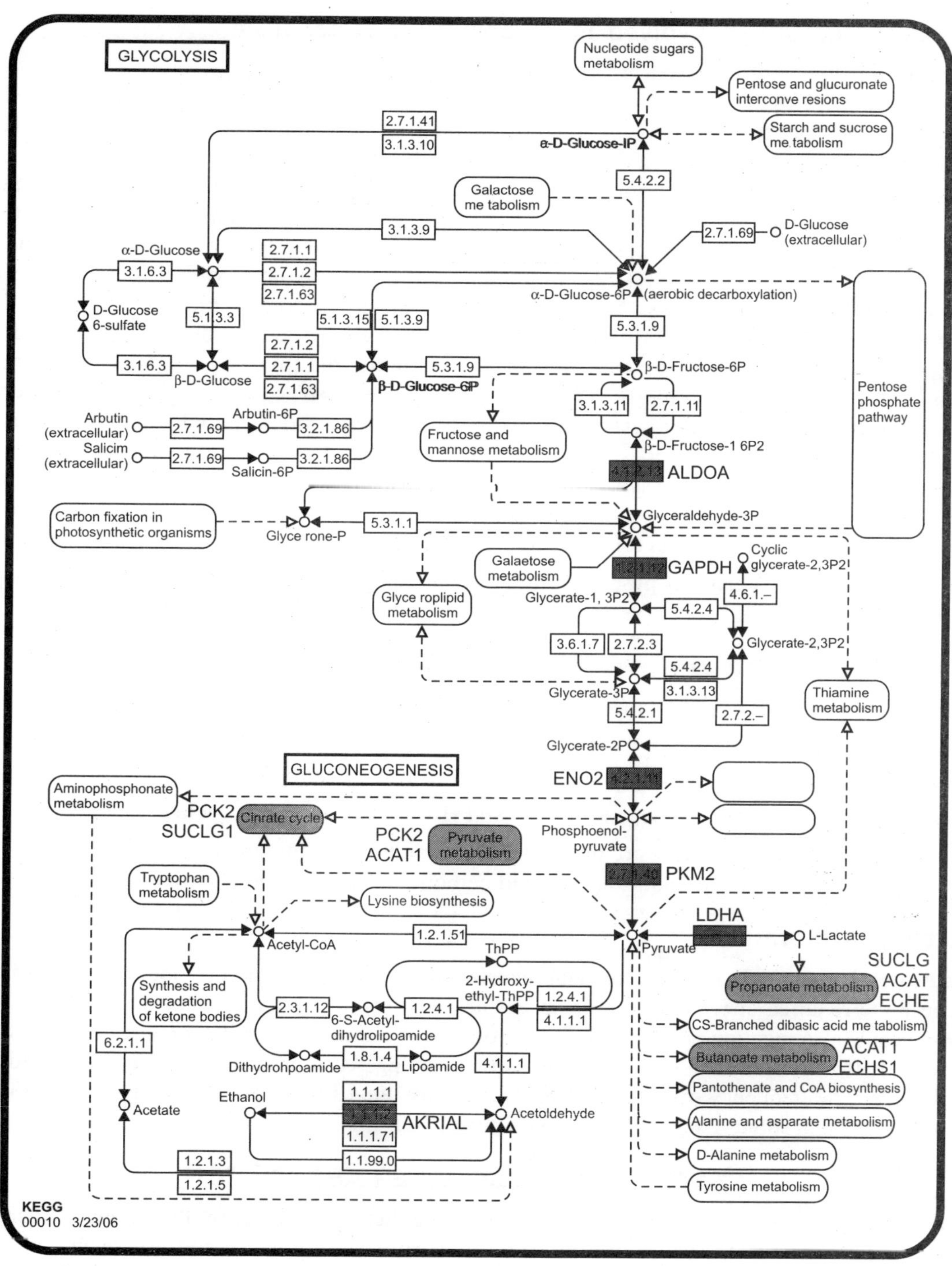

Fig. 5.1 : The Carbohydrate Metabolism

Chemcial Mechanism involved in Carbohydrate Metabolism

The chemical reactions involved in carbohydrate metabolism in the body represent complex groups sequences and cycles of reactions which integrate at various points with the reactions concerned in the metabolism. So that the reactions involved in one stage of metabolism. For example the glycolytic breakdown of glucose or glycogen through many phosphorylated intermediates to pyruvic acid and the subsequent oxidation of this pyruvic acid through the sequence of reactions in the tricarboxylic acid. Now we can see that the reactions of glycolysis must take place preliminary to the reactions of carbohydrate oxidation. The synthesis of fat from carbohydrate involves synthesis of fatty acids from pyruvic acids formed by glycolysis and combinations of these fatty acids with glycerol, which is also a product of glycolytic reactions. Another interesting aspect of carbohydrate metabolism is that the reactions involved in one phase of metabolism may represent essentially the reverse of the reaction concerned in a second phase of metabolism. For example, the process of glucogenesis from the pyruvic acid in liver consists chiefly in reversal of the glycolysis reactions by which the liver converts glucose to pyruvic acid. Thus it is seen that while we may differentiate carbohydrate metabolism into various phases such as glucogenesis glycolysis (breakdown of carbohydrate generally glycogen or glucose), glycogenesis (glycogen synthesis), glycogenolysis (breakdown of glycogen synthesis specially) carbohydrate oxidation, and fat synthesis from carbohydrate it is impossible to segregate into different distinct groups the chemical reactions concerned with these phases of carbohydrate metabolism.

The major types chemical processes involved in carbohydrate metabolism may be grouped as follows:

(*i*) The reversible process glucose $\rightleftharpoons$ glycogen.

(*ii*) The process by which sugar such as fructose mannose, and galactose, are converted to glucose.

(*iii*) The reversible process by which glucose is converted to pyruvic acid.

(*iv*) The oxidation of pyruvic acid to CO_2 and H_2O in the citric acid cycle.

(*v*) Reduction of —CHOH groups to the conversion of carbohydrate carbon (–CHOH) to fatty acid carbon ($-CH_2$) and of amino acid carbon ($-CH_2$) to carbohydrate carbon (–CHOH).

Lactose Intolerance

Lactose intolerance is the inability to metabolize lactose, a sugar found in milk and other dairy products, because the required enzyme lactase is absent in the intestinal system or its availability is lowered. Some people also mention pasteurized dairy products as a cause (raw milk contains small amounts of lactase). It is estimated that 75% of adults show some decrease in lactase activity during adulthood worldwide. The frequency of decreased lactase activity ranges from nearly 5% in northern Europe to more than 90% in some Asian and African countries.

FATE OF GLUCOSE AFTER ABSORPTION

Glucose is primarily the carbohydrate supplied by the diet and utilized by the tissues, Other sugars, such as fructose, galactose, and mannose are convertible into glucose and glycogen and undergo metabolism largely through such conversion.

The interconversions of sugars and their conversion to glycogen are effected through enzymatic reactions of the sugar phosphates.

For orientation purposes, general relations of the sugars in metabolism are shown in Fig. 137.

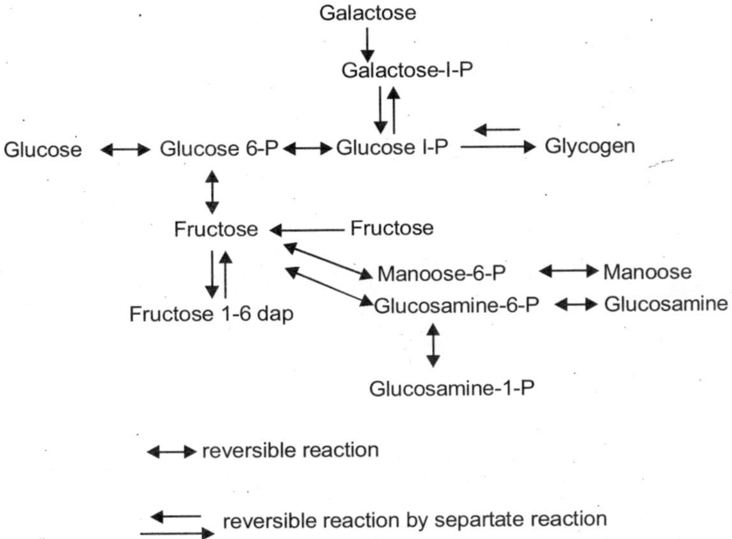

Fig. 5.2 : The Conversion Reaction of Different Glucoses after Absorption

The metabolism of sugars takes place through their phosphate to pyruvic (and lactic) acid in process called glycolysis which may proceed anaerobically. Pyruvic acid is then converted to acetyl CoA which is oxidized to carbon dioxide and water in the citric acid cycle. Discussion in this subsection will deal particularly with the metabolism of sugars relating to their conversion to sugar phosphates, and with the relation of these phosphates to each other after absorption.

The Hexokinase Reaction The first reaction involved in the metabolism of sugar in phosphorylation by ATP, which is catalysed by a Hexokinase enzyme.

$$\text{Sugar} + \text{ATP} \xrightarrow{\text{Hexokinase}} \text{Sugarphosphate} + \text{ATP}$$

The Hexokinase reaction was first discovered in yeast. The enzyme catalyzes the phosphorylation of glucose mannose, fructose, glucosamine and 2 dexoxyglucose to the 6-Phosphate. A glucokinase specific for glucose is present in animal tissues and bacteria. It forms glucose 6-P Liver and muscle contain a fructokinase which forms fructose-1-P instead of the 6-phosphate. Galactokinase present present in yeast and animal tissue forms galactose-1-P. Liver galactokinase also acts upon galactosamine.

The free energy liberated in the Hexokinase reaction is relatively large (about 4 K/cal for glucose)

$$\text{Glucose} + \text{ATP} \xrightleftharpoons{\text{Glucokinase}} \text{Glucose 6 P} + \text{ATP}$$

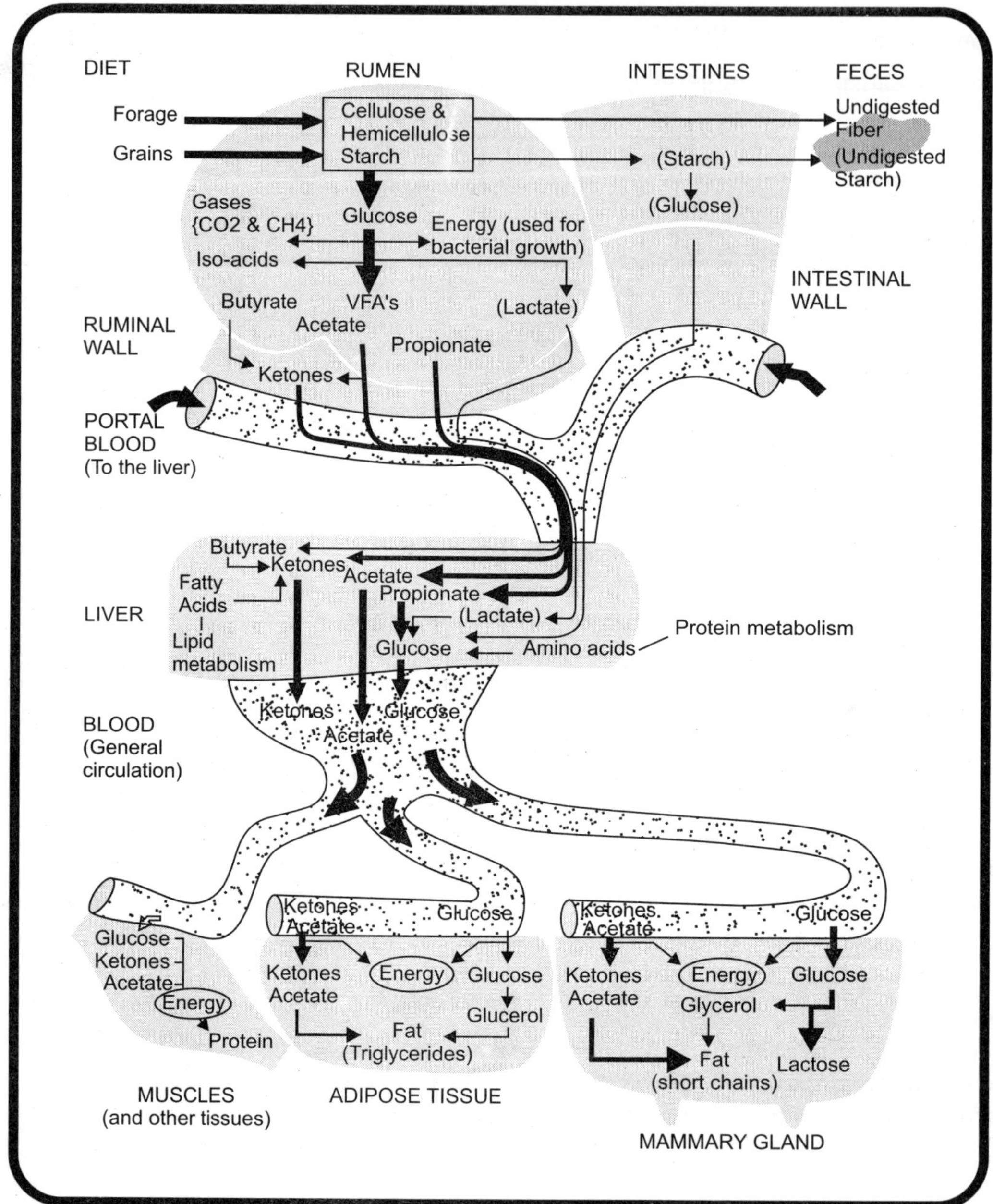

Fig. 5.3 : The Total Absorption Diagram of the Carbohydrate after Absorption

Though the reaction is revesible it is slight and of no appreciable metabolic importance. Glucose-6-P in the liver is hydrolyzed to glucose and Pi by a specific enzyme glucose-6-phosphatase. While glucose 6-P is formed in the muscle it is not converted to glucose because of the absence of glucose-

6-phosphatase. Hydrolysis of glucose-6-P in liver represents that final reaction is the formation of glucose from glycogen and other sugar phosphates. The enzyme is deficient in Von Gieke's disease which leads to massive accumulation of glycogen in liver.

Interconversion of Glucose-1-P Glucose-6-P Phosphoglucomutase

As pointed out glycogen is synthesized from glucose-1-P and broken down into glucose-6 P in the Hexokinase reaction it is reversely converted to glucose-1-P by the phosphoglucomutase is a phosphoprotein and the phosphate group of the enzyme participate in the reaction.

$$\text{Glucose-1-P + Enzyme 1P} \rightleftharpoons \text{Glucose 1-6dip + Enzyme}$$

$$\text{Glucose 1-6 dip + Enzyme} \rightleftharpoons \text{Glucose-6-P + Enzyme P}$$

Muscle and yeast contain and enzyme glucose-1-Phosphate kinase, which catalyses the phosphorylation of glucose 1-P to glucose-1-6-diP by ATP

Also muscle and bacteria contain glucose-1-phosphate transphophorylase which forms glucose – 1-6 diP by the relation.

$$\text{2 Glucose-IP} \rightleftharpoons \text{Glucose-1-6diP+Glucose}$$

These processes supply the amounts of glucose-1-6 diP necessary for the phosphoglucomutase reaction. Glucose-6-P is reversely converted to fructose 6-P by a phosphohexose isomerizes enzyme.

$$\text{Glucose-6-P} \rightleftharpoons \text{Fructose-6-P}$$

At equilibrium (30°C pH 8) there is about 30% fructose-6-P and 70% glucose-6-P. The reaction appears to proceed through an enol intermediate

$$R\text{-}CH(OH)\text{-}CH{=}O \rightleftharpoons R\text{-}C(OH){=}CH\text{-}OH \rightleftharpoons R\text{-}C({=}O)\text{-}CH_2\text{-}OH$$

The phosphorylation of fructose-6-P to fructose-1-6 disphosphate by ATP is catalysed by the enzyme phosphofructokinase, and resembles the hexokynase reaction.

$$\text{Frucose-6-P + ATP} \rightarrow \text{Fructose-1-6diP}$$

The reaction like the hexokynase reaction is inappreciably reversible, and fructose-1-6 disphosphate is converted back to fructose-6-P may be through hydrolysis by a specific enzyme.

$$\text{Frucose-1-6-diP} + H_2O \rightarrow \text{Fructose-6-P + Pi}$$

It was proved that fructose-6-P may be phosphorylated to fructose-1-6 diphosphate also by isosine triphosphate (ITP) uridine triphosphate (UTP)

Metabolism of fructose: Fructose is phosphorylated by ATP through the action of two different fructokinase. As previously indicated, glucose forms fructose-6-P.

$$\text{Fructose + ATP} \rightarrow \text{Fructose-6-P + ATP}$$

The fructose-6-P may then be converted to glucose-1-P through glucose, or to glucose-1-P through glucose-6-P, and then to glycogen. It may also be phosphorylated to fructose-1-6- dip, which undergoes conversion of pyruvic acid (by oxidation etc) to acetylCoA, which is oxidized to CO_2 and H_2O is the citric acid cycle. Other sugars are metabolized similarly by being converted meeting point in the metabolic stream of sugar.

Another fructokinase in animal tissues as previously indicated, with ATP phosphrylates fructose to fructose-1-P

Fructose + ATP $\rightarrow$ Fructose-1-P

Fructose-1-P can be converted to fructose-1-6-Disphosphate as follows:

$$\text{Fructose-1-P} \xrightleftharpoons{\text{Aldolase}} \text{Dihydroxoyacetone-P+Glyceraldehyde}$$

$$\text{Glyceraldehyde-3-P+Dihydroxyacetone-P} \xrightleftharpoons{\text{Aldolase}} \text{Fructose-1, 6 Diphosphate}$$

Metabolism of Mannose

Mannose enters the carbohydrate stream through phosphorylation in the hexokinase reaction and then conversion to fructose-6-P

$$\text{Manmose+ATP} \xrightarrow{\text{Glucokinase}} \text{Manmose-6P+ADP}$$

$$\text{Manmose-6P} \xrightleftharpoons[\text{Isomerase, Muscle}]{\text{Phosphomannose}} \text{Fructose-6P}$$

Metabolism of Galactose

Galactose enters the blood stream from the intestine after digestion of carbohydrate rather lactose. It is converted into glucose-P by the following reaction:

$$\text{Galactose+ATP} \xrightarrow{\text{Galactokinase}} \text{Galactose-1-P+ADP}$$

$$\text{Galactose-1-P+UDP-Glucose} \xleftrightarrow[\text{Uridyltransferase}]{\text{Phosphogalactase}} \text{UDP-Galactase+Glucose-1P}$$

$$\text{UDP-Galactase} \xleftrightarrow[\text{Epimerase}]{\text{UDP-Galactose}} \text{UDP-Glucose}$$

Galactose + ATP $\rightarrow$ Glucose-1-P+ADP

The Glucose 1-P then may undergo the transformations. It has also indicated that congenital absence of the enzyme phosphogalactase uridyl transferase is the cause of the disease galactosemia in the children. This means that in galactosemia the conversion of the galactose-1-P to glucose-1-P definitely is executed.

Under these conversion of galactose-1-P accumulates in erythrocytes and other tissues, causing damage particularly to the liver brain, and optic lens. Treatment consists in feeding galactose (lactose free diets. UDP-galactose is required in forming tissue glycolipids and lactose in the adult female, the galactosemia this is formed form glucose through the epimerize reaction.

$$\text{UDPGlucose} \xleftrightarrow{\text{Epimerase}} \text{UDP-galactose}$$

An alternate route for the metabolism of galactose is due to the presence of the enzyme uridine diphosphate galactose pyrophosphorylase, which catalyses the formation of UDP-galactose from galactose-1-P

$$\text{Galactose-1-P+UTP} \rightleftharpoons \text{UDP + galactosePPi}$$

UDP-galactose pyrophosphorylase is present in low activity in infant liver but increases with age, so the galactosemic adult to metabolise appreciable quantities of galactose.

Summary

The Monosaccharaides produced by the action of oligosaccharides and disaccharides of the brush boders of the intestine cells by several types of transport mechanism. Fructose moves into the cell on a membrane but is not actively transported against a concentration gradiant i.e. facilitated diffusion. Glucose and galactose are actively transported and this is accomplished by a sodium symport system (Fig. 5.4). This transport process ensures that the intracellular glucose concentration is maintained at a vascular side of the cell, thus ensuring that glucose will be transport passively into the bloodstream. The carrier-bound sodium ions and glucose are internalized along an electrochemical gradient that results from the low intracellular sodium concentration. On the luminal side, one molecule of glucose and two molecules of sodium bind to the membrane carrier. Presumably the binding of the sodium ions produces a conformational change such that glucose now binds with greater affinity to the carrier.

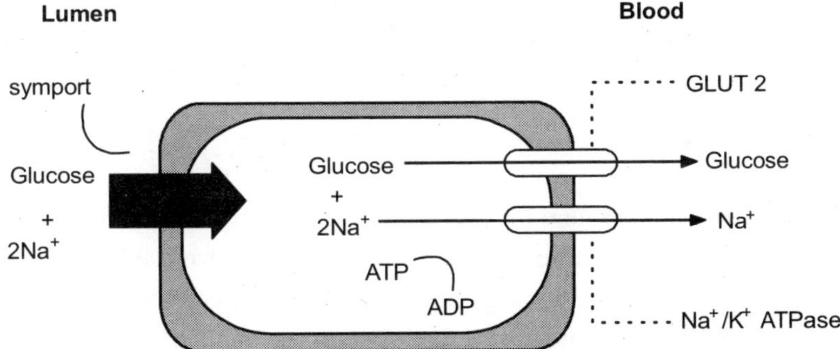

Fig. 5.4 : The Glucose Symport

Inside the cell, the sodium ions are released from the carrier, and diminished affinity allows the glucose to dissociate. A Na^+/K^+ ATPase allows the sodium ions to be transported into the lateral intercellular spaces against a concentration gradiant using the free energy of ATP hydrolysis. Since glucose transport does not involve ATP directly, it can be considered as secondary active transport. Glucose is transported from the mucosal cell into the intercellular space by a high-capacity glucose transporter (GLUT2)

GLYCOLYTIC PATHWAYS

We can now start our consideration of the glycolytic pathway. This pathway is common to virtually all cells, both prokaryotic and eukaryotic cells. In eukaryotic cells the glycolysis takes place in the cystol.

This pathway can be thought of as consisting of three major stages; Stage 1 includes the conversion of glucose into fructose 1p-diphosphate through three intermediate processes first of all the phosphorylation then the isomerization and followed by a second phosphorylation reaction.

The strategy of these individual steps in glycolysis is to trap the glucose in the cell and then from a compound that can be readily changed into phosphorylated three carbon units

Stage 2 Includes the transformation of fructose-1-6 diphosphate into two three carbon fragments. Thus resulting three carbon units which are readily interconvertable. In stage three ATP is produced when three carbon fragments are converted to pyruvic acid by oxidation. (Fig. 5.5 and 5.5A).

Glucose enters cell through specific transport proteins and has one principal fate, it is phosphorylated by ATP to form glucose 6-phosphate. This step is notable for two reasons (1) glucose 6-phosphate cannot diffuse through the membrane because of its negative charges and (2) the addition of the phosphoryl group begins to destabilize glucose, thus facilitating its furdoxyl group on carbon 6 of glucose is catalyzed by hexokynase.

Glucose + ATP $\xrightarrow{\text{Hexokinase}}$ Glucose 6 - Phosphate + ADP + H^+

Phosphoryl transfer is a fundamental reaction in biochemistry and is one that has been discussed. Kinases are enzymes that catalyze the transfer of a phosphoryl group from ATP to an acceptor. Hexokinase then catalyzes the transfer of a phosphoryl group from ATP to a variety of six carbon sugar (hexose) such as glucose and mannose, Hexokinase like adenylatekinase. And all other kinase requiring Mg^{2+} (or another divalent metal ion such as Mn^{2+}) for activity.

The divalent metal ions form complex with ATP. The result of X-Ray crystallographic study of the yeast hexokinase revealed that the binding of glucose induces a large conformational change on the enzyme, analogous to the conformational change under gone by NMP kinases on substrate binding. Hexokinase consists of two parts which move toward each other when glucose is bound (Fig. 5.5). On glucose binding one part rotates 12 degree angle with the respect of the other part, resulting in the movements of the polypeptide backbone of as much as 8Å. The cleft between the parts closes and the bond glucose becomes surrounded by protein, except of the hydroxyl group of carbon 6, which will accept the phosphoryl group ATP. The closing of the cleft in hexokinase is striking example of the role of induced fit in enzyme action.

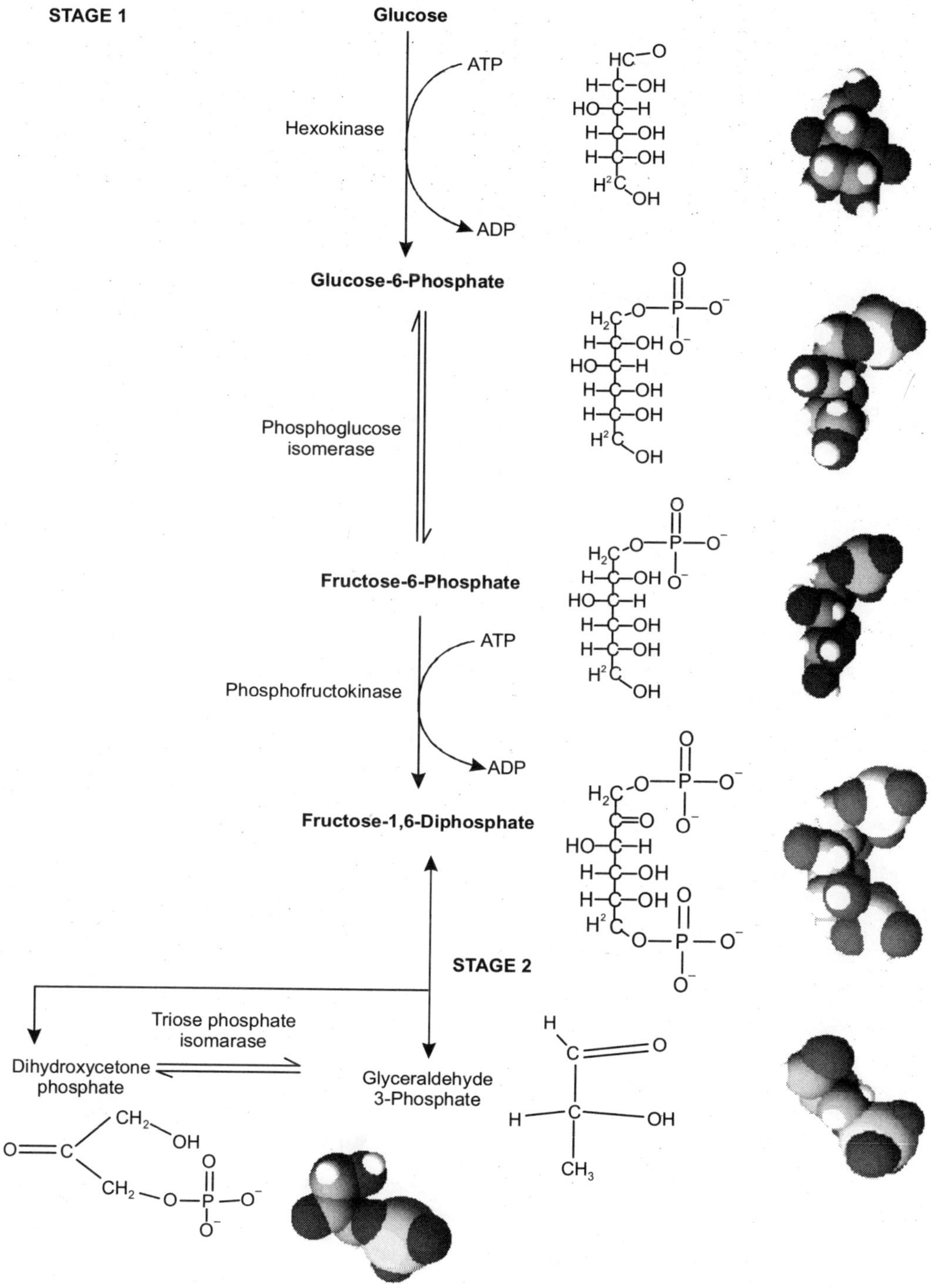

Fig. 5.5 : The Glycolytic Pathway

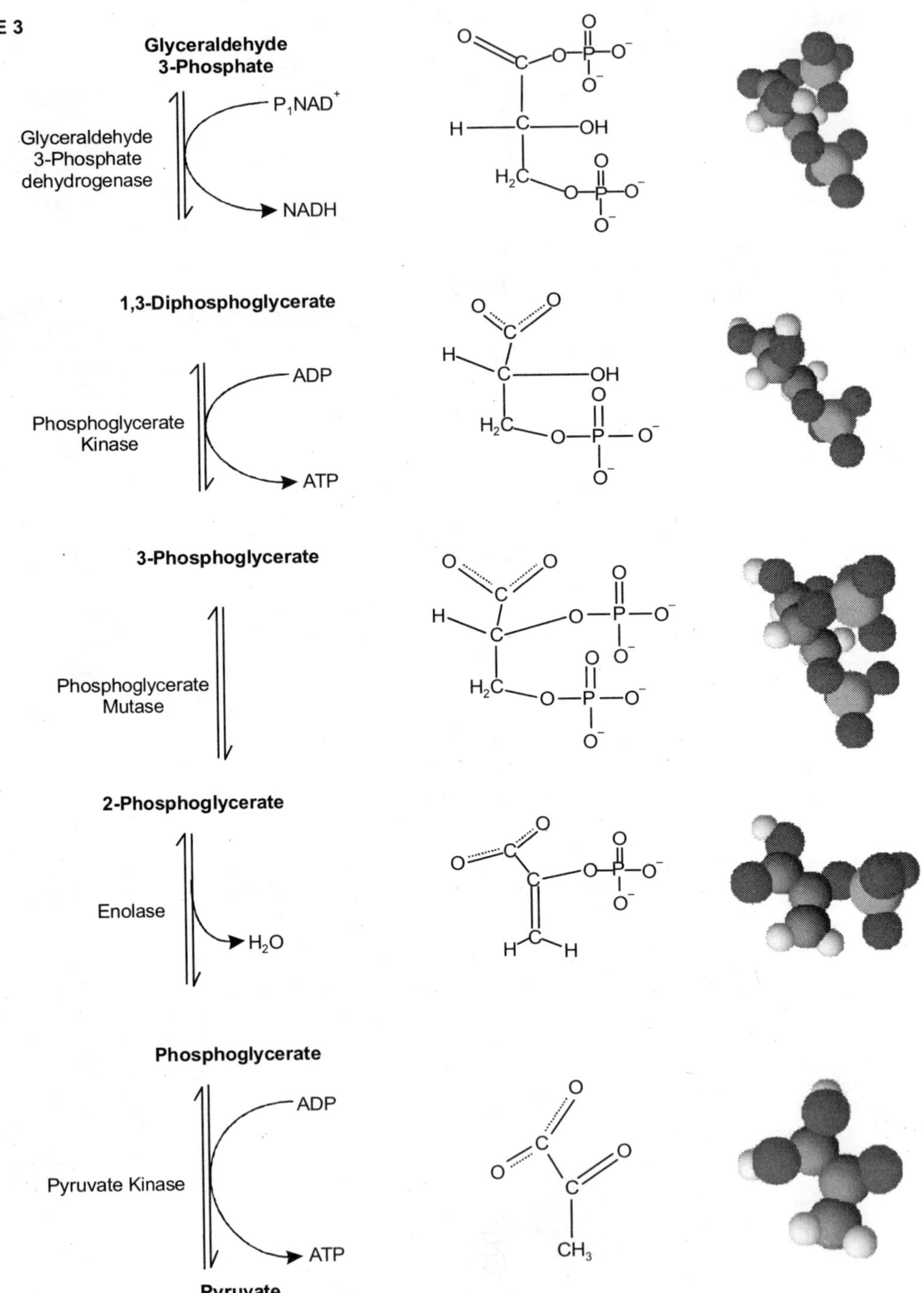

Fig. 5.5A : The Glycolytic Pathway

The glucose included structural changes are significant in two respects. First the environment around the glucose become much more non-polar, which favours the donation of the terminal phosphoryl group of ATP. Secondly as noted the substrate induced conformational changes within the kinase enables it to discriminate against H_2O as a substrate. If hexokinase was rigid a molecule of water occupying the binding site for the $-CH_2OH$ of glucose would attract the phosphoryl group of ATP, forming ADP and P_1. In other words, a rigid kinase would necessarily also is an ATPase. It is interesting to note that other kinase taking part in glycolysis-pyruvate kinase, phosphoglycerate kinase and phosphofructokinase also contain clefts between parts that close when substrate is bound although the structures of these enzymes are different in other regards Substrate-introduced cleft closing is a general feature of kinase.

The formation of Fructose 1, 6 Diphosphate from Glucose 6-Phosphate The next step in glycolysis is the isomerization of glucose-6-phosphate to fructose-6-phosphate. Remember that the open chain from of glucose has an aldehyde group at carbon 1, where as the open chain form the fructose has a keto group at carbon 2. Thus the isomerization of glucose 6-phosphate to fructose-6-phosphate is a conversion of an aldose into a ketose. The reaction catalyzed by phosphoglucose isomerizes include addition steps because both glucose-6-phosphate pair is present primarily in the cyclic froms. The enzyme must first not open the six membered ring of glucose-6-phosphate catalyzes the isomerization, and then promote the formation of the five membered ring of fructose 6-phosphate.

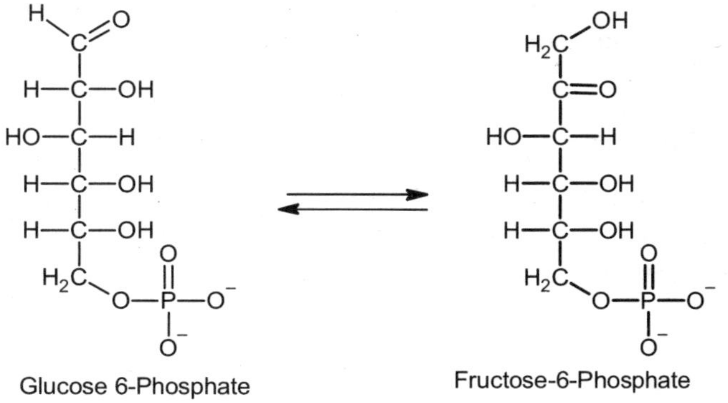

Glucose 6-Phosphate ⇌ Fructose-6-Phosphate

A second phosphorylation reaction follows the isomerization step. Fructose-6-phosphate is phosphorylated by ATP to fructose 1-6 diphosphate.

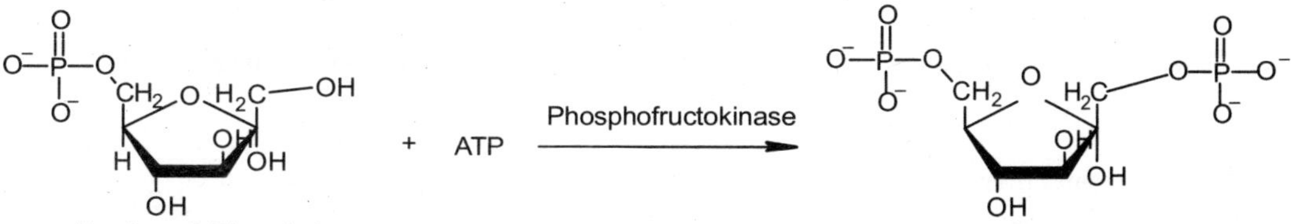

Fructose-6-Phosphate + ATP —Phosphofructokinase→ Fructose-1,6 Diphosphate

The reaction is catalyzed by the enzyme phosphofructokinase; it is allosteric[‡] in nature, that states the pace of glycolysis. As we all learn, this enzymes plays a central role in the integration very much of metabolism.

The Six-Carbon Sugar is cleaved into Two Three Carbon Fragments by Aldose. The second stage of glycolysis begins with the splitting of fructose 1, 6 disphosphate into glyceraldehyde 3-phosphate and dihydroxyacetone phosphate. The products of the remaining steps in glycolysis consists of three carbon units rather than six-carbon units.

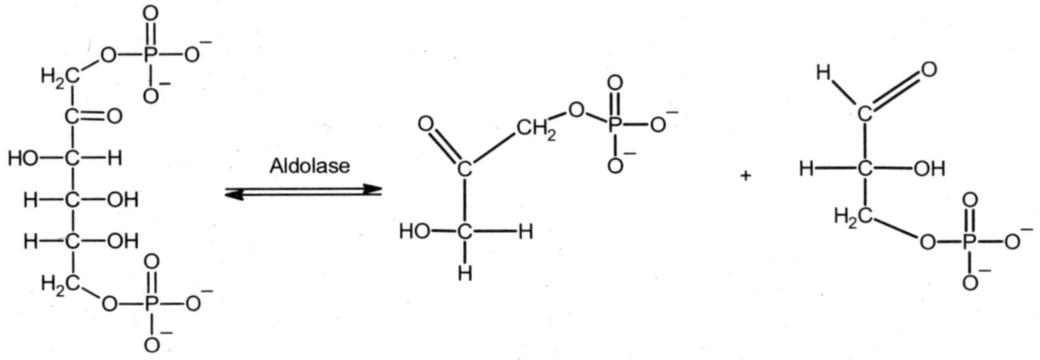

Aldolase

Fructose - 6 - diphosphate   Dihydooxyacetone phosphate   Glyceraldehyde

This reaction is catalyzed by aldolase. This enzyme derives its name from the nature of the reverse reaction, an aldol condensation. The reaction catalyzed by aldolase is readily reversible under intermolecular conditions.

The formation of ATP from 1, 3 Diphosphoglycerate

The final stage in glycolysis generation of ATP from the phosphorylated three carbon metabolites of glucose. Phosphoglycerate kinase catalyzes the transfer of the phosphoryl group from the acyl phosphate of 1, 3 diphosphoglycerate are the products.

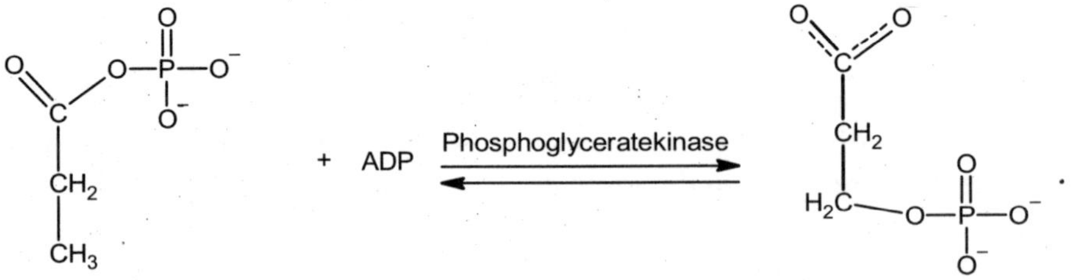

+ ADP   Phosphoglyceratekinase

1, 3 - Phosphoglycerate   3 - Phosphoglycerate

‡ The term *allostery* comes from the Greek *allos*, "other", and *stereos*, "solid (object) Allosteric enzymes are enzymes that change their conformation upon binding of an effector. An allosteric enzyme is an oligomer whose biological activity is affected by *altering* the conformation(s) of its tertiary structure. Allosteric enzymes tend to have several subunits.

Thus the outcomes of the reactions catalyzed by glyceraldehyde–3-phosphate dehydrogenese and phosphoglyceratekinase are:

1. Glyceraldehyde 3-Phosphate and aldehyde is oxidized to 3 – phosphoglycerate a carboxylic acid.
2. NAD^+ is concomitantly[§] reduced to NADH
3. ATP^+ is formed from P_1 and ADP at the enzyme expense of carbon oxidation energy.

Keep it in mind, that because of the actions of aldolase and triose phosphate isomerase, two molecules of ATP were generated. These ATP molecules of ATP consumed in the first stage of glycolysis.

The Generation of Additional ATP and the Formation of Pyruvate: In the remaining steps of glycolysis 3-phosphoglycerate is converted into pyruvate with the concomitant conversion of ADP to ATP.

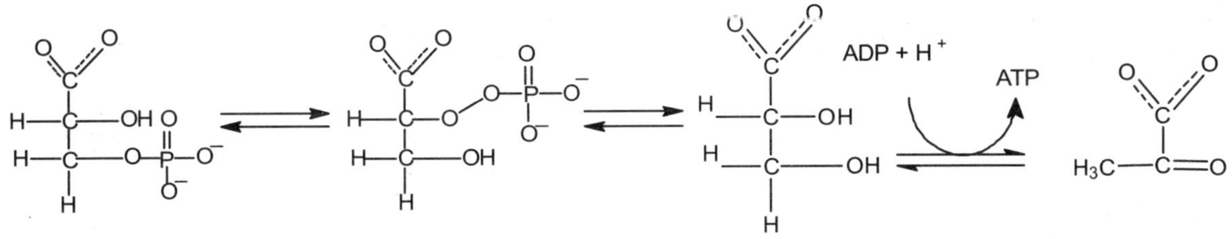

ADP + H^+ ATP

3 - Phosphoglycerate 2 - Phosphoglycerate Phosphoglycerate Pyruvate

The first reaction is a rearrangement; the position of the phosphoryl group shifts in the conversion of 3 phosphoglycerate into 2-phosphoglycerate, a reaction catalyzed Phsophoglycerate Mutase. In general, a mutase is an enzyme that catalyzes the intramolecular or shift of a chemical group, such as a phosphoryl group. The phosphoglycerate mutase reaction has an interesting mechanism; the phosphoryl group is not simply moved from one carbon to another.

This enzyme requires catalytic amounts of 2, 3-diphosphoglycerate to maintain an active site hystidine residue in a phosphorylated form.

Enz-His-phosphate+2,3 Diphosphoglycerate ⇌ Enz-His-phosphate+2 phosphoglycerate

Enz-His-phosphate+3Phosphoglycerate ⇌ Enz-His+2,3 Diphosphoglycerate

The sum of these reaction equation yields mutase reaction

3-Phosphoglycerate ⇌ 2-Phosphoglycerate

Examination of the first partial reaction reveals that the mutase functions as a phosphate, i.e. converts 2, 3 disphosphoglycerate into 2, phosphoryl group remains linked to the enzyme. This phosphoryl group is then transferred to 3-phosphoglycerate to reform 2, 3 diphosphoglycerate.

In the next reaction an enol is formed by the dehydration of 2 phosphoglycerate. Enolase catalyzes to formation of phosphoenolpyruvate (PEP). This dehydration distingusihly displays the transfer potential of the phosphoryl group. An enol phosphate has a high phosphoryl transfer potential where as the phosphate ester such as 2-phosphoglycerate of an ordinary alcohol is low one. It is a fact that phosphoenol pyruvate has a very high phosphoryl transfer potential. The phosphoryl group traps the molecule in its unstable enol form, when the phosphoryl group has been donated to ATP the enol undergoes a conversion into the more stable ketone namely pyruvate.

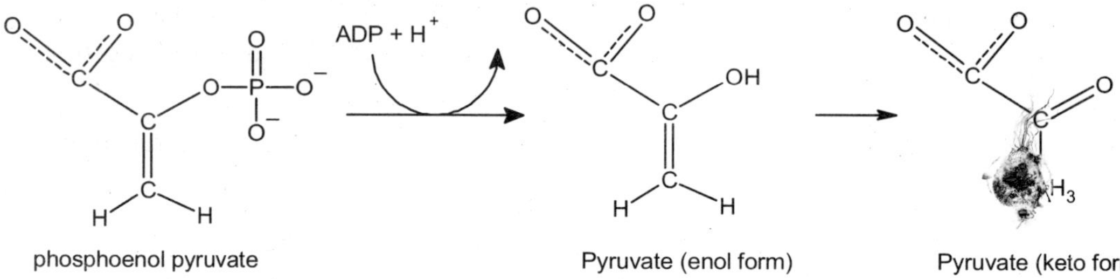

ADP + H^+

phosphoenol pyruvate

Pyruvate (enol form)

Pyruvate (keto form)

Thus, the high phosphoryl transfer potential of phosphoenolpyruvate arises primarily from the large driving force of the subsequent enol ketone conversion. Hence pyruvate is formed and ATP is generated concomitantly. The virtually irreversibly transfer of a phosphoryl group from the phosphoenolpyruvate to ADP is catalyzed by pyruvatekinase. Because the molecules of ATP used in forming fructose 1,6 diphosphate have already been regenerated the two molecules of ATP generated from phosphoenolpyruvate are " profit".

Enzyme	Reaction Type	ΔG° in kcal §mol^{-1}	ΔG° in kcal mol^{-1}
Hexokinase	Phosphoryl transfer	–4.0	–8.0
Phosphoglucose isomerase	Isomerization	+0.4	–0.6
Phosphofructokinase	Phosphoryl Transfer	–3.4	–5.3
Aldolase	Aldol Cleavage	+5.7	–0.3
Triose phosphate isomerase	Isomerization	+1.8	+0.6
Glyceraldehyde 3-phosphate Dehydrogenese	Phosphorylation coupled to oxidation	+1.5	–0.4
Phosphoglycerate kinase	Phosphoryl-transfer	–4.5	+0.3
Phosphoglycerate mutase	Phosphoryl shift	+1.1	+0.2
Enolase	Dehydration	+0.4	–0.8
Pyruvate Kinase	Phosphoryl transfer	–7.5	–4.0

§**Note:** ΔH, the normal free energy change has been calculated from ΔG, and known concentrations of reactants under typical physiologic conditions. Glycolysis can proceed only if the ΔG values of all reactions are negative. The small ΔG values of the above reactions indicate that the concentrations of metobolities in vivo in cells undergoing glycolysis are not precisely known.

GLYCOGENESIS AND GLYCOGENOLYSIS

Glycogenesis Introduction

Glycogen is found in many cell types in the body but only in high concentration in liver and muscle. A fed man weighing 70 kg will have about 1.6 kg liver containing about 100 g glycogen and 35 kg of muscle containing approximately 400 g glycogen. At a caloric value of 17 kJ/g, the stores of glycogen represent about 8500 kJ of fuel. Relative to the fat stores, this is a small reserve yet it has great functional significance. Glycogen is stored in the fed state and utilized during fasting and exercise. Synthesis *(glycogenesis)* and breakdown *(glycogenolysis)* occur by separate but related pathways (Fig. 5.6) and are controlled in an integrated fashion via allosteric and covalent mechanisms; hormonal control is very critical. The enzymes of glycogen metabolism are associated with the glycogen granules in cells.

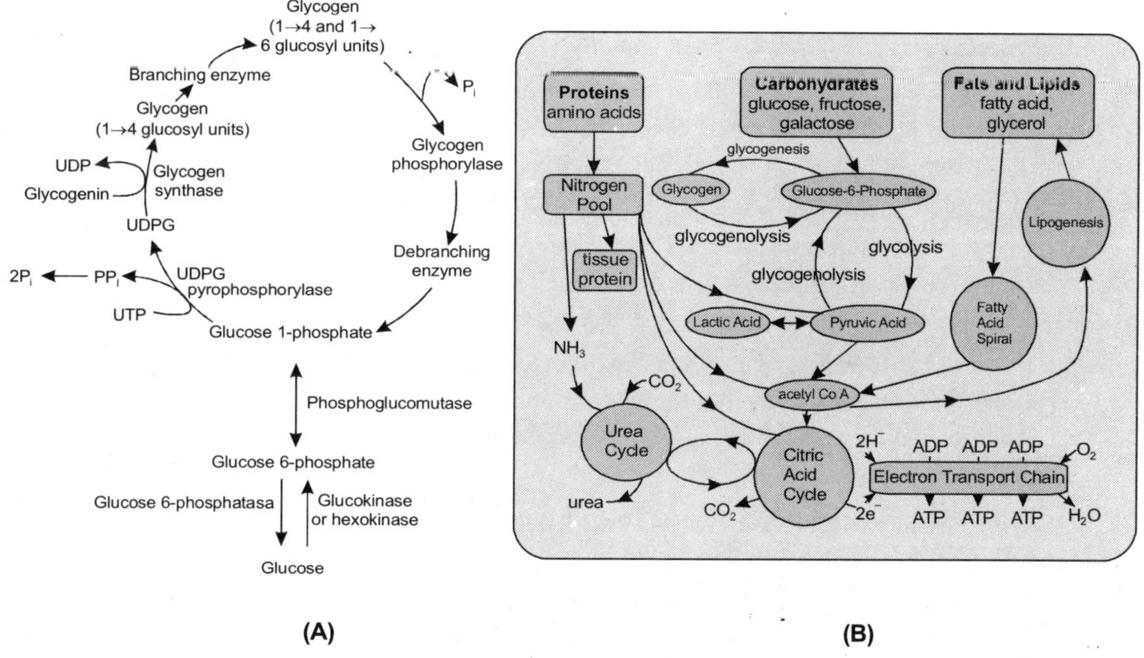

Fig. 5.6 : The Glycogenesis and Glycogenolysis Process

GLYCOGENESIS

The key points about the storage of glycogen are:

1. A step unique to glycogenesis is the formation of UDP-glucose from G-I-P and the pyrimidine nucleotide, UTP
2. The glucose moieties added to form glycogen come directly from UDP-glucose

3. Each glycogen molecule contains a protein, *glycogen in.* The identification of this protein explained how the synthesis of the large glycogen molecule is initiated. Glycogenin has enzymatic activity catalyzing the addition of the first 4-8 glucose moieties to a tyrosine in glycogenin, using UDPG as the source of the glucoses.
4. Glycogen synthase transfers the glucose moiety of UDPG to the non-reducing end of the primer, giving a polymer with a-l,4-linkages
5. Glycogen is branched and the *branching enzyme* involved is a 4:6-transferase.

GLYCOGENOLYSIS

Key points about the breakdown of stored glycogen are:

1. *Glycogen phosphorylase* catalyses the interaction of inorganic phosphate (Pj) with terminal a-l:4-glycosidic bonds at the multiple non-reducing ends of glycogen to yield G-l-P
2. Phosphorylase contains the coenzyme pyridoxal phosphate.
3. The branching of glycogen means that there are many sites (ends) for phosphorolysis and this allows for rapid production of G-l-P, which is beneficial in both liver and muscle.
4. *Debranching enzyme* has two distinct catalytic sites and is important for complete utilization of glycogen. Glucoses are removed from a branch by phosphorylase action until there are only four glucoses attached. Then, a 4:4 *glucan transferase* transfers the trisaccharide attached to the glucose in a l,6-linkage to a different chain. This leaves a single glucose attached at the branch point and the a-l,6-linkage is hydrolyzed by *a-l:6-glucosidase* to yield free glucose.
5. There are different end-products in liver compared with muscle. Muscle lacks *glucose 6-phosphatase (G-6-Pase)* so the end-products of increased muscle glycogenolysis will be pyruvate and lactate (following glycolysis); liver contains G-6-Pase which means that the end product is glucose.
6. Clearly this is consistent with the role of glycogen in muscle, which is to supply that tissue with ATP and the role of glycogen in liver to maintain blood glucose levels.

It is essential to consider the control of the synthetic breakdown pathways together (see Fig. 5.7)

(*b*) The fed state: high insulin levels, low glucagon levels.

The control steps in glycogenesis and glycogenolysis are catalyzed by glycon synthase and glycogen phosphorylase, respectively. They are both subject to control by allosteric and covalent modification.

ATP and G-6-P levels high	Glycogen synthase active, phosphorylase inactive
Amp levels High	Glycogen synthase inactive, phosphorylase active

High ATP in muscle is an indication of high energy charge and is a signal that there is less need for glycogen breakdown. In contrast, increased AMP is a signal that ATP utilization is high.

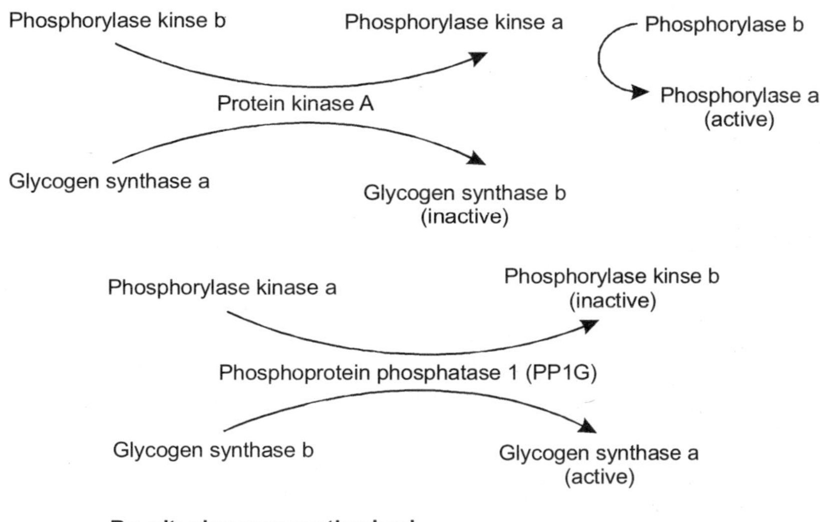

Fig. 5.7 : The pathway breakdown

Elevated G-6-P in both liver and muscle is associated with the fed state and increased availability or glucose (for storage).

Covalent Modification of Enzyme

Cyclic AMP dependant protein kinase Active	Glycogen Synthase Inactive, Phosphorylase Active
Phosphoprotein Phosphate Active	Glycogen Synthase Active, Phosphorylase Inactive

Hormonal Control Glucogen (liver) and adrenaline (muscle) action results in increased concentrated cyclic AMP within these cells and activation of PKA. This results in Activation of phosphorylase and inactivation of synthase.

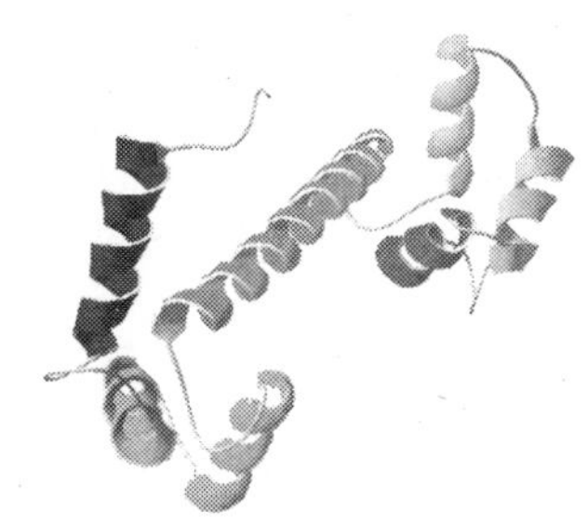

Fig. 5.8 : Calmodulin structure

Insulin release in the fed state leads to the dephosphorylation of both phosphorylase and glycogen synthase. This results in inactivation of phosphorylase and activation of glycogen synthase. Phosphorylase kinase is also activated by increased concentrations of calcium. This is explained by the fact that phosphorylase kinase has four subunits $\alpha\beta\gamma\delta$ and the δ-subunit is the calcium binding polypeptide calmodulin*. Binding of calcium activities the enzymes leading finally to increased

***Calmodulin (CaM)** (an abbreviation for Calcium Modulated protein) is a clcium-binding protein expressed in all eukaryotic cells. It can bind to and regulate a number of different protein targets, thereby affecting many differ cellular functions.

...(Contd.)

phosphorylation of the a and the β-subunits of phosphorylase kinase (and thus to activation). Calcium is an important signal for muscle contraction. Activation of glycogen breakdown will lead to the increased ATP formation required to support this. Adrenaline action on the liver is mediated via a_1 adrenoreceptors leading to increased levels of inositoltrisphosphate and calcium.

CORI CYCLE

The Cori cycle, named after its discoverers, Carl Cori and Gerty Cori, refers to the metabolic pathway in which lactate produced by anaerobic glycolysis in the muscles moves to the liver and is converted to glucose, which then returns to the muscles and is converted back to lactate. Muscular activity requires energy, which is provided by the breakdown of glycogen in the muscles. The breakdown of glycogen generates ATP which is used by muscles as an energy source. Since throughout muscular activity ATP is used to supply energy, it needs to be constantly replenished. Whilst the supply of oxygen is sufficient, glucose-6-phosphate produced by the breakdown of glycogen can be aerobically oxidized, feeding pyruvate into the TCA cycle. However, when oxygen is in short supply - i.e. during intense muscular activity - energy must be released from G-6-P through anaerobic respiration. Anaerobic respiration through glycolysis alone causes a build-up of pyruvate, which is converted to lactate. In the Cori cycle, glycolysis continues when pyruvic acid is converted back to lactic acid. Pyruvic acid is reduced to lactic acid by lactate dehydrogenase, which oxidises the NADH created during glycolysis back to NAD+, transferring NADH's two electrons to pyruvic acid, creating lactic acid. This occurs in order to maintain a concentration of NAD+ that can be reduced during glycolysis, so that glycolysis can continue. The lactic acid is then taken up by the liver in the next part of the cycle. The liver converts the

---

Function

CaM mediates processes such as inflammation, metabolism, apoptosis, muscle contraction, intracellular movement, short-term and long-term memory, nerve growth and the immune response. CaM is expressed in many cell types and can have different subcellular locations, including the cytoplasm, within organelles, or associated with the plasma or organelle membranes. Many of the proteins that CaM binds are unable to bind calcium themselves, and as such use CaM as a calcium sensor and signal transducer. CaM can also make use of the calcium stores in the endoplasmic reticulum, and the sarcoplasmic reticulum. CaM undergoes a conformational change upon binding to calcium, which enables it to bind to specific proteins for a specific response. CaM can bind up to four calcium ions, and can undergo post-translational modifications, such as phosphorylation, acetylation, methylation and proteolytic cleavage, each of which can potentially modulate its actions. Calmodulin can also bind to edema factor toxin from the anthrax bacteria.

Structure

Calmodulin is a small, acidic protein approximately 148 amino acids long (16706 Dalton) and, as such, is a favorite for testing protein simulation software. It contains four EF-hand "motifs", each of which binds a Ca^{2+} ion. The protein has two approximately symmetrical domains, separated by a flexible "hinge" region. Calcium participates in an intracellular signalling system by acting as a diffusible second messenger to the initial stimuli.

Mechanism

Calcium is bound via the use of the EF hand motif, which supplies an electronegative environment for ion coordination. After calcium binding, hydrophobic methyl groups from methionine residues become exposed on the protein via conformational change. This presents hydrophobic surfaces, which can in turn bind to Basic Amphiphilic Helices (BAA helices) on the target protein. These helices contain complementary hydrophobic regions. The flexibility of calmodulin's hinged region allows the molecule to "wrap around" its target. This property allows it to tightly bind to a wide range of different target proteins. The members are Calmodulin 1, 2 and 3.

lactic acid back to pyruvic acid and then to glucose through gluconeogenesis. The glucose then enters the blood and returns to the muscles to be used for energy if muscle activity has continued. If muscle activity has stopped by this time then the glucose is used to replenish the supplies of glycogen through glycogenesis. In the Cori cycle the gluconeogenic leg of the cycle is energy consuming. While there is a gain of 2 moles of ATP in the anaerobic glycolysis of glucose, there is a cost of 6 moles of ATP in the gluconeogenesis part of the cycle. The cost of the 4 moles of ATP means the cycle cannot be sustained continuously.

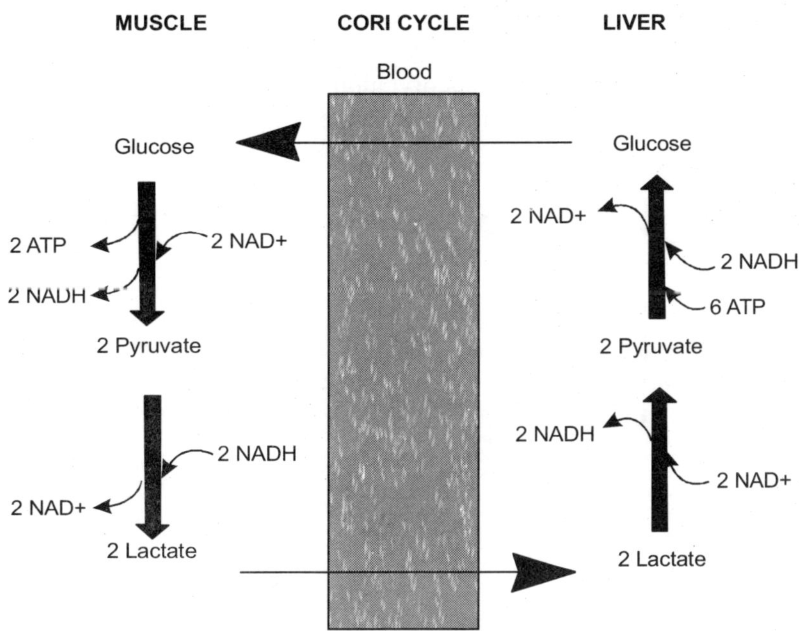

Fig. 5.9 : The Cori Cycle

Significance

The cycle's importance is based on the prevention of lactic acidosis in the muscle under anaerobic conditions. However, normally before this happens the lactic acid is moved out of the muscles into the liver. The cycle is also important in producing ATP, an energy source, during muscle activity. The Cori cycle functions more efficiently when muscle activity has ceased because the oxygen debt can be made up so that the citric acid cycle and electron transport chain also work.

GLYCOGEN STORAGE DISEASE

Defects in enzymes of glycogenolysis usually result in a glycogen storage disease. Subjects with defective branching system in enzyme in liver have a tendency to develop hypoglycemia but the most severe form is seen in those defective livers G-B phase. Subjects with defective muscle phosphorylase cannot support the high level of physical activity.

THE CITRIC ACID CYCLE/TRICARBOXYLIC ACID/KREB'S ACID

The citric acid cycle, also known as the tricarboxylic acid cycle (TCA cycle) or the Krebs cycle (or, rarely, the Szent-Györgyi–Krebs cycle), is a series of enzyme-catalysed chemical reactions of central importance in all living cells that use oxygen as part of cellular respiration. In eukaryotes, the citric acid cycle occurs in the matrix of the mitochondrion. The components and reactions of the citric acid cycle were established by seminal work from both Albert Szent-Györgyi and Hans Krebs.

In aerobic organisms, the citric acid cycle is part of a metabolic pathway involved in the chemical conversion of carbohydrates, fats and proteins into carbon dioxide and water to generate a form of usable energy. Other relevant reactions in the pathway include those in glycolysis and pyruvate oxidation before the citric acid cycle, and oxidative phosphorylation after it. In addition, it provides precursors for many compounds including some amino acids and is therefore functional even in cells performing fermentation.

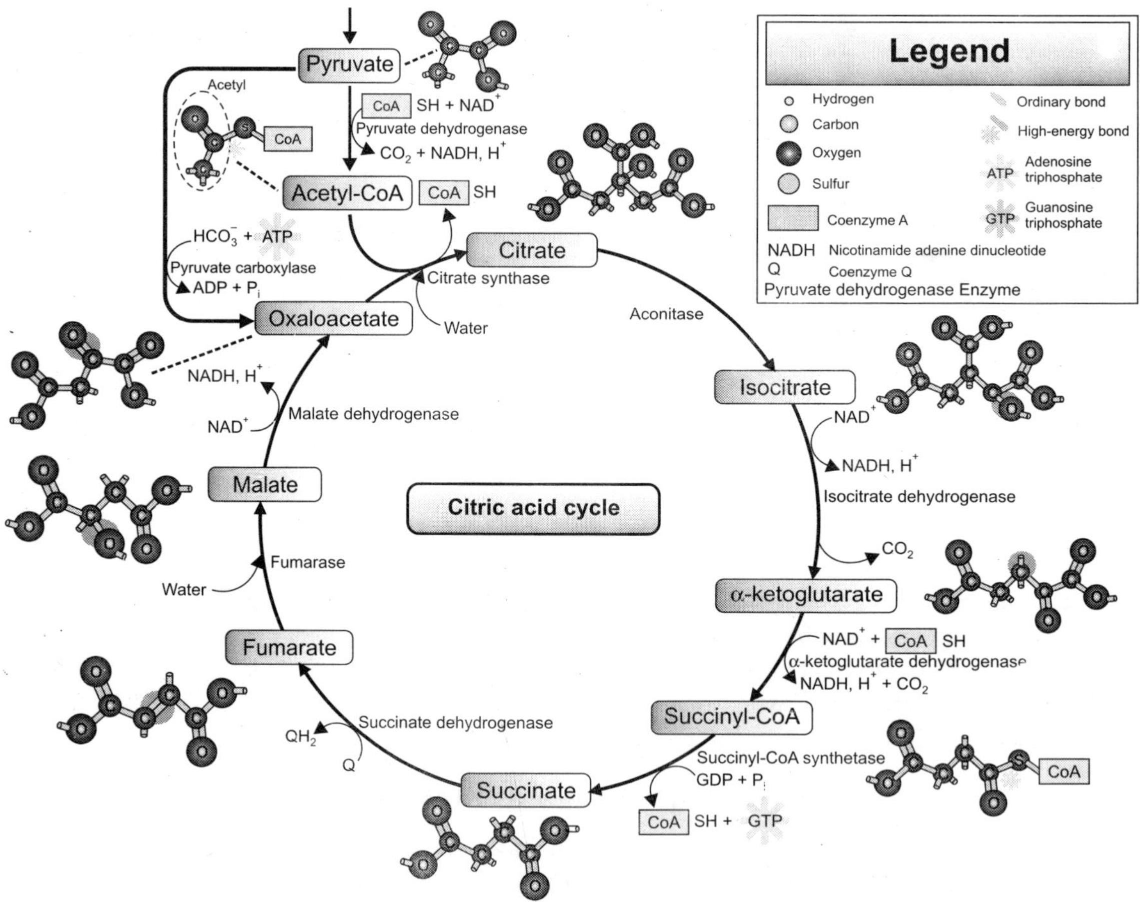

Fig. 5.10 : The Citric Acid Cycle

Two carbons are oxidized to CO_2, and the energy from these reactions is transferred to other metabolic processes by GTP (or ATP), and as electrons in NADH and QH_2. The NADH generated in the TCA cycle may later donate its electrons in oxidative phosphorylation to drive ATP synthesis; $FADH_2$ is covalently attached to succinate dehydrogenase, an enzyme functioning both in the TCA cycle and the mitochondrial electron transport chain in oxidative phosphorylation. $FADH_2$ thereby facilitates transfer of electrons to coenzyme Q, which is the final electron acceptor of the reaction catalyzed by the Succinate:ubiquinone oxidoreductase complex, also acting as an intermediate in the electron transport chain.

The citric acid cycle is continuously supplied with new carbons in the form of acetyl-CoA, entering at step 1 below:

Substrates	Products	Enzyme	Reaction type	Comment
1. Oxaloacetate + Acetyl CoA + H_2O	Citrate + CoA-SH	Citrate synthase	Aldol condensation	rate limiting stage, extends the 4C oxaloacetate to a 6C molecule
2. Citrate	*cis*-Aconitate + H_2O	Aconitase	Dehydration	reversible isomerisation
3. *cis*-Aconitate + H_2O	Isocitrate		Hydration	generates NADH (equivalent of 2.5 ATP)
4. Isocitrate + NAD+	Oxalosuccinate + NADH + H +		Oxidation	irreversible stage, generates a 5C molecule
5. Oxalosuccinate	α-Ketoglutarate + CO_2	Isocitrate dehydrogenase	Decarboxylation	generates NADH (equivalent of 2.5 ATP), regenerates the 4C chain (CoA excluded) or ADP->ATP,
6. Ketoglutarate + NAD+ + CoA-SH	Succinyl-CoA + NADH + H+ + CO_2	α-Ketoglutarate dehydrogenase	Oxidative decarboxylation	generates 1 ATP or equivalent
7. Succinyl-CoA + GDP + Pi	Succinate + CoA-SH + GTP	Succinyl-CoA synthetase	substrate level phosphorylation	uses FAD as a prosthetic group (FAD->$FADH_2$ in the first step of the reaction) in the enzyme,
8. Succinate + ubiquinone (Q)	Fumarate + ubiquinol (QH_2)	Succinate dehydrogenase	Oxidation	generates the equivalent of 1.5 ATP
9. Fumarate + H_2O	*L*-Malate	Fumarase	H_2O addition (hydration)	
10. *L*-Malate+ NAD+	Oxaloacetate + NADH + H^+	Malate dehydrogenase	Oxidation	generates NADH (equivalent of 2.5 ATP)

Mitochondria in animals including humans possess two succinyl-CoA synthetases, one that produces GTP from GDP, and another that produces ATP from ADP. Plants have the type that produces ATP (ADP-forming succinyl-CoA synthetase). Several of the enzymes in the cycle may be loosely-associated in a multienzyme protein complex within the mitochondrial matrix.

Products of the first turn of the cycle are: one GTP (or ATP), three NADH, one QH_2, two CO_2.

Because two acetyl-CoA molecules are produced from each glucose molecule, two cycles are required per glucose molecule. Therefore, at the end of all cycles, the products are: two GTP, six NADH, two QH_2, and four CO_2

Description	Reactants	Products
The sum of all reactions in the citric acid cycle is:	Acetyl-CoA + 3 NAD^+ + Q + GDP + P_1 + 2 H_2O	→ CoA-SH + 3 NADH + 3 H^+ + QH_2 + GTP + 2 CO_2
Combining the reactions occurring during the pyruvate oxidation with those occurring during the citric acid cycle, the following overall pyruvate oxidation reaction is obtained:	Pyruvic acid + 4 NAD^+ + Q + GDP + P_1 + 2 H_2O	→ 4 NADH + 4 H^+ + QH_2 + GTP + 3 CO_2
Combining the above reaction with the ones occurring in the course of glycolysis, the following overall glucose oxidation reaction (excluding reactions in the respiratory chain) is obtained:	Glucose + 10 NAD^+ + 2 Q + 2 ADP + 2 GDP + 4P_i + 2 H_2O	→ 10 NADH + 10 H^+ + 2 QH_2 + 2 ATP + 2 GTP + 6 CO_2

(the above reactions are equilibrated if P_i represents the $H_2PO_4^-$ ion, ADP and GDP the ADP_2^- and GDP_2^- ions, respectively, and ATP and GTP the ATP_3^- and GTP_3^- ions, respectively).

Estimates for the total number of ATP obtained after complete oxidation of one glucose in glycolysis, citric acid cycle, and oxidative phosphorylation given in the literature range from 30-38 molecules of ATP. A recent assessment of the total ATP yield obtained in these distinct reaction cycles, taking into account updated proton-to-ATP ratios, has arrived at an estimate of 29.85 ATP per glucose molecule.

Regulation

Although pyruvate dehydrogenase is not technically a part of the citric acid cycle, its regulation is included here.

The regulation of the TCA cycle is largely determined by substrate availability and product inhibition. NADH, a product of all dehydrogenases in the TCA cycle with the exception of succinate dehydrogenase, inhibits pyruvate dehydrogenase, isocitrate dehydrogenase and α-ketoglutarate dehydrogenase, and also citrate synthase. Acetyl-CoA inhibits pyruvate dehydrogenase, while succinyl-CoA inhibits succinyl-CoA synthetase and citrate synthase. When tested in vitro with TCA enzymes, ATP inhibits citrate synthase and α-ketoglutarate dehydrogenase; however, ATP levels do not change more than 10% in vivo between rest and vigorous exercise. There is no known allosteric mechanism that can account for large changes in reaction rate from an allosteric effector whose concentration changes less than 10%.

Calcium is used as a regulator. It activates pyruvate dehydrogenase, isocitrate dehydrogenase and α-ketoglutarate dehydrogenase. This increases the reaction rate of many of the steps in the cycle, and therefore increases flux throughout the pathway.

Citrate is used for feedback inhibition, as it inhibits phosphofructokinase, an enzyme involved in glycolysis that catalyses formation of fructose 1,6-bisphosphate, a precursor of pyruvate. This prevents a constant high rate of flux when there is an accumulation of citrate and a decrease in substrate for the enzyme.

Recent work has demonstrated an important link between intermediates of the citric acid cycle and the regulation of hypoxia inducible factors (HIF). HIF plays a role in the regulation of oxygen haemostasis, and is a transcription factor which targets angiogenesis, vascular remodeling, glucose utilization, iron transport and apoptosis. HIF is synthesized constitutively and hydroxylation of at least one of two critical proline residues mediates their interaction with the von Hippel Lindau E3 ubiquitin ligase complex which targets them for rapid degradation. This reaction is catalyzed by prolyl 4-hydroxylases. Fumarate and succinate have been identified as potent inhibitors of prolyl hydroxylases thus leading to the stabilization of HIF.

Major Metabolic Pathways Converging on the TCA Cycle

Several catabolic pathways converge on the TCA cycle. Reactions that form intermediates of the TCA cycle in order to replenish them (especially during the scarcity of the intermediates) are called anaplerotic reactions.

The citric acid cycle is the third step in carbohydrate catabolism (the breakdown of sugars). Glycolysis breaks glucose (a six-carbon-molecule) down into pyruvate (a three-carbon molecule). In eukaryotes, pyruvate moves into the mitochondria. It is converted into acetyl-CoA by decarboxylation and enters the citric acid cycle.

In protein catabolism, proteins are broken down by proteases into their constituent amino acids. The carbon backbone of these amino acids can become a source of energy by being converted to Acetyl-CoA and entering into the citric acid cycle.

In fat catabolism, triglycerides are hydrolyzed to break them into fatty acids and glycerol. In the liver the glycerol can be converted into glucose via dihydroxyacetone phosphate and glyceraldehyde-3-phosphate by way of gluconeogenesis. In many tissues, especially heart tissue, fatty acids are broken down through a process known as beta oxidation which results in acetyl-CoA which can be used in the citric acid cycle. Beta oxidation of fatty acids with an odd number of methylene groups produces propionyl CoA, which is then converted into succinyl-CoA and fed into the citric acid cycle.

The citric acid cycle is always followed by oxidative phosphorylation. This process extracts the energy (as electrons) from NADH and QH_2, oxidizing them to NAD^+ and Q, respectively, so that the cycle can continue. Whereas the citric acid cycle does not use oxygen, oxidative phosphorylation does.

The total energy gained from the complete breakdown of one molecule of glucose by glycolysis, the citric acid cycle and oxidative phosphorylation equals about 30 ATP molecules, in eukaryotes. The citric acid cycle is called an amphibolic pathway because it participates in both catabolism and anabolism.

A Simplified View of the Process

- The citric acid cycle begins with acetyl-CoA transferring its two-carbon acetyl group to the four-carbon acceptor compound (oxaloacetate) to form a six-carbon compound (citrate).
- The citrate then goes through a series of chemical transformations, losing first one, then a second carboxyl group as CO_2. The carbons lost as CO_2 originate from what was oxaloacetate, not directly from acetyl-CoA. The carbons donated by acetyl-CoA become part of the oxaloacetate carbon backbone after the first turn of the citric acid cycle. Loss of the acetyl-CoA-donated carbons as CO_2 requires several turns of the citric acid cycle. However, because of the role of the citric acid cycle in anabolism, they may not be lost since many TCA cycle intermediates are also used as precursors for the biosynthesis of other molecules.
- Most of the energy made available by the oxidative steps of the cycle is transferred as energy-rich electrons to NAD^+, forming NADH. For each acetyl group that enters the citric acid cycle, three molecules of NADH are produced.
- Electrons are also transferred to the electron acceptor Q, forming QH_2.
- At the end of each cycle, the four-carbon oxaloacetate has been regenerated, and the cycle continues.

PYRUVATE FORMATION FROM ACETYL CoA

The formation of Acetyl CoA from carbohydrates is less direct than from fat. Recall that carbohydrates, especially glucose are processed by glycolysis into pyruvate. Under anaerobic conditions, the pyruvate is converted into lactic acid or ethanol, depending on the organism. Under aerobic conditions, the pyruvate is transported into mitochondria in exchange for OH^- by the pyruvate carrier an antiporter. In the mitochondrial matrix pyruvate is oxidative dehydrogenase complex to form acetyl coenzyme.

$$\text{Pyruvate} + \text{CoA} + \text{NaD}^+ \rightarrow \text{AcetylCoA} + CO_2 + \text{NADH}$$

This irreversible reaction is the link between glycolysis and the citric acid cycle.*

The pyruvate dehydrogenase complex is a large, highly integrated complex of three kinds of enzyme. Pyruvate dehydrogenase is a member of a family of homologous complexes that includes the citric acid cycle enzyme α-ketoglutarate dehydrogenase.

The conversion of isocitrate into α-ketoglutarateis followed by a second oxidative decarboxylation reaction, the formation of succinyl CoA from α-ketoglutarate.

---

* *Note* that in the preparation of the glucose derivative pyruvate for the citric acid cycle an oxidative decarboxylation takes place and high – transfer potential electrons in form of NADH are captured. Thus the pyruvate dehydrogenese reaction has many of the key features of the reactions of the citric acid itself.

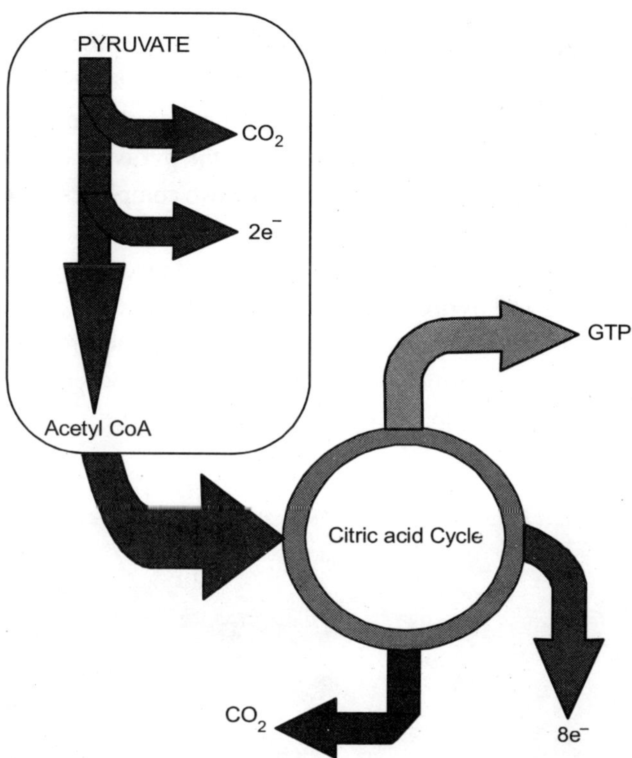

Fig. 5.11 : The Linkage between Glycolysis and the citric acid cycle

The oxidative decarboxylation of α-ketoglutarate closely resembles that of pyruvate also an α-ketoacid.

$$\text{Pyruvate} + \text{CoA} + \text{NAD}^+ \rightarrow \text{AcetylCoA} + \text{CO}_2 + \text{NADH}$$

$$\alpha\text{-ketoglutarate} + \text{NAD}^+ + \text{CoA} \longrightarrow \text{Succinyl Co enzyme} + \text{H}_2 + \text{NADH}$$

α - ketoglutarate

Succinyl Co enzyme

Both the reactions include the decarboxylation of an α-ketoacid and the subsequent formation of a high-transfer potential thio-ester linkage with CoA. The complex that catalyzes the oxidative decarboxylation of α-ketoglutarate is homologous to the pyruvate dehydrogenase complex and the reaction mechanism is entirely analogous. The α-ketoglutarate dehydrogenase component (E_2) and transsuccinase (E_1) are different form but homologous to the pyruvate dehydrogenase complex where as the dihydroliopyl dehydrogenase components (E_3) of the two complexes are identical. These complexes are giant, with molecular masses ranging from 4 to 10 million Daltons. As we will see their elaborate structures allow traveling from one active site to another, connected by thioethers to the core of the structure. The mechanism of the pyruvate dehydrogenase reaction is wonderful complex, more so than is suggested by its relatively simple stoichiometry. The reaction requires the participation of the three enzymes of the pyruvate dehydrogenase complex, each composed of several polypeptide chains and five enzymes. Thiamine pyrophosphate lipoic acid and FAD serve as catalytic co factors and CoA and NAD^+ are stoichiometric co factors.

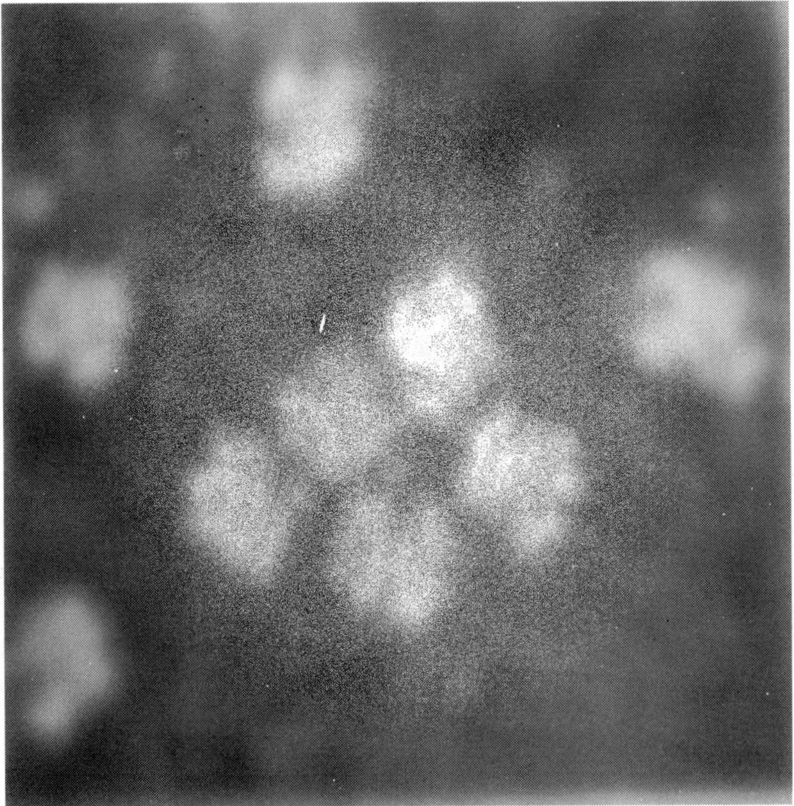

Fig. 5.12 : The Microscopic Photograph of Dehydrogenase Complex

The table of Pyruvate Dehydrogenase Complex of *E. Coli*

Enzyme	Abbre-viation	No of Chain	Prosthetic group	Reaction Catalysis
Pyruvate dehydrogenase component	E_1	24	TPP	Oxidative decarboxylation of Pyruvate
Dihydroliopyl transacetylase	E_2	24	Lipoamide	Transfer of the acetyl group to CoA
Dihydroliopyl dehydrogenase	E_3	12	FAD	Regeneration of the oxidized form of Lipoamide

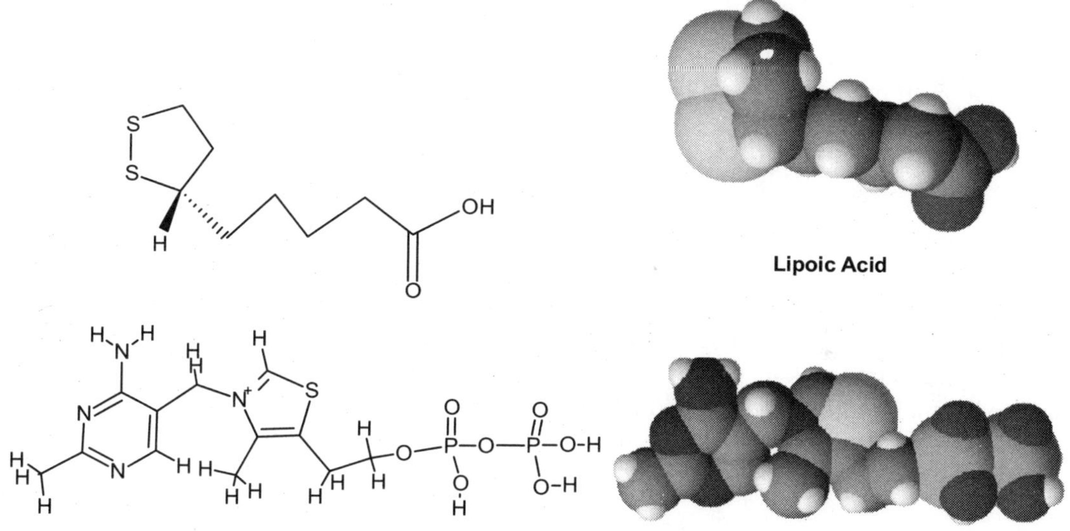

Lipoic Acid

Thyamine Pyrophopshate

At least two additional enzyme regulate the activity if the complex. The conversion iof pyruvate into acetyl CoA consists of these three steps decarboxylation, oxidation and transfer of the resultant acetyl group to CoA.

Pyruvate —(CO_2, decarboxylation)→ H_3C–COOH —($2e^-$, oxidation)→ H_3C–CO_2^- —(CoA, transfer of CoA)→ H_3C–CO–S–CoA (Acetyl CoA)

These steps must be coupled to preserve the free energy derived from the decarboxylation step to drive the formation of NADH and acetyl CoA. First pyruvate combines with TPP and is then decarboxylated.

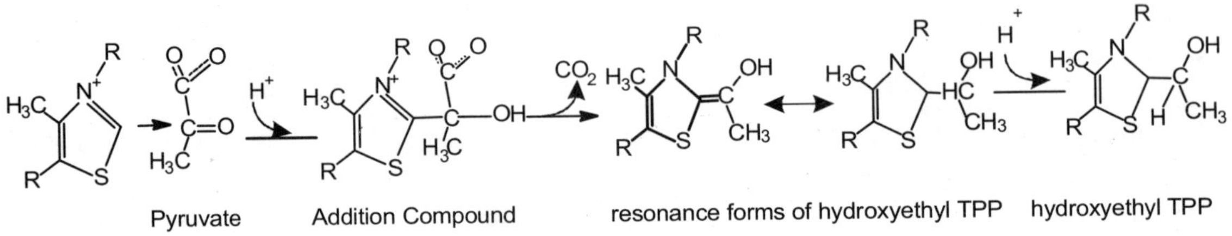

Pyruvate Addition Compound resonance forms of hydroxyethyl TPP hydroxyethyl TPP

Equation showing the decarboxylation of E_1 the pyruvate dehydrogenase component of pyruvate dehydrogenase complex.

This reaction is catalyzed by the pyruvate dehydrogenase component, is that the carbon atom between the nitrogen and sulphur atoms in the thiozole ring is much more acidic that most == CH group, with a pKa value of 10. This centre ionizes to form a carbanion, which readily absorbs to the carbonyl group of pyruvate.

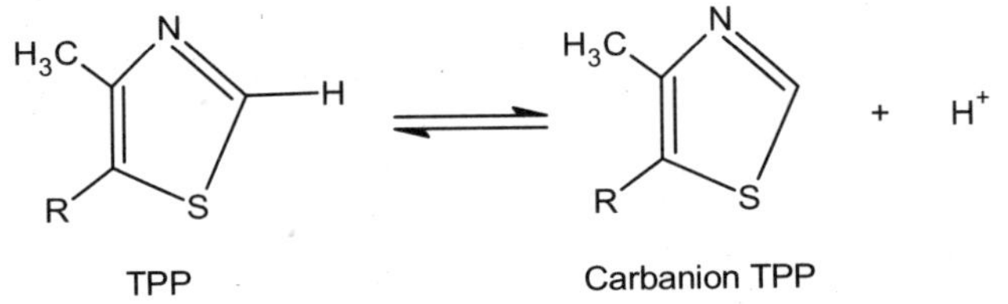

TPP Carbanion TPP

This addition is followed by the decarboxylation if pyruvate. The positively charged ring of the TPP acts as an electron sink that stabilizes the negative charge that is transferred to the ring as a part of decarboxylation. Protonation yields hydroxymethyl group attached to TPP is oxidized to form an acetyl group and concomitantly transferred to Lipoamide, a derivative of lipoic acid that is linked with the side chain of a residue by an amide linkage.

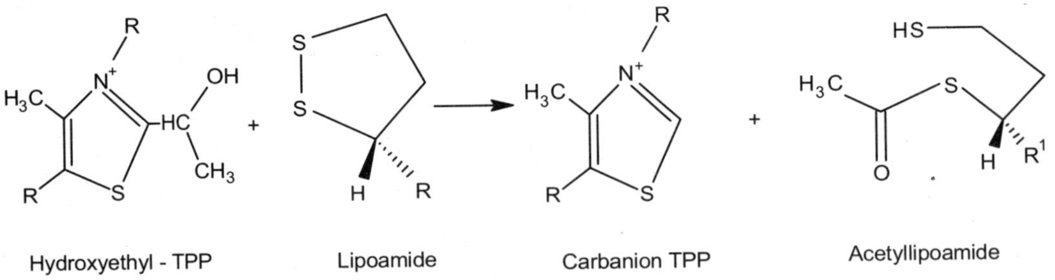

Hydroxyethyl - TPP Lipoamide Carbanion TPP Acetyllipoamide

The oxidant in this reaction is the disulphide group of Lipoamide, which is reduced to its disulphydryl form. This reaction also catalyzed by the pyruvate dehydrogenase component E_1 yields acetyllipoamide.

Third, the acetyl group is transferred from acetyllipoamide to CoA to form acetyl CoA. D. Dihydroliopyl transacetylase (E_2) catalyzes; the energy reach thioester bond is preserved as the acetyl group is transferred to CoA.

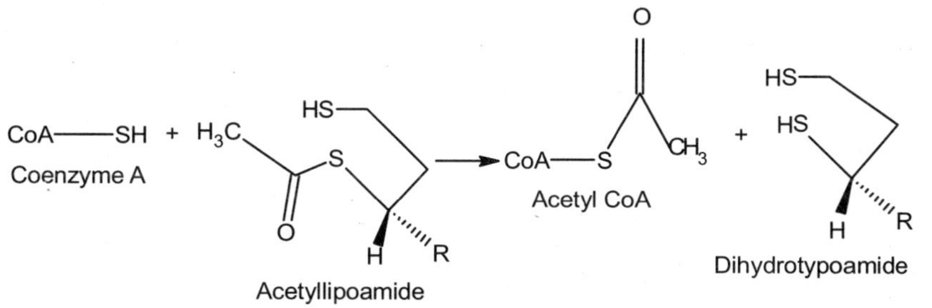

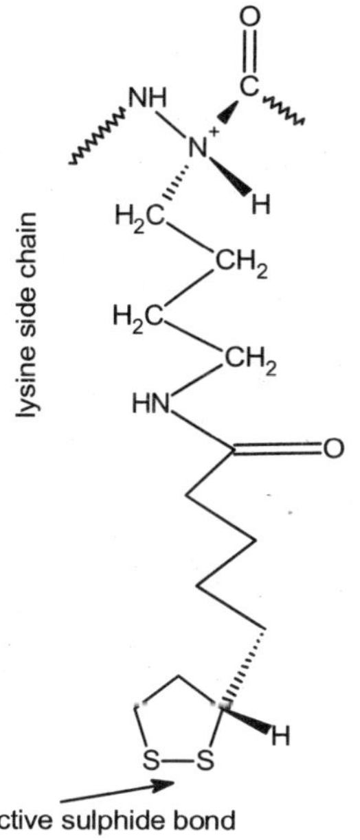

It is to be noted that CoA as carrier of many activated acetyl groups of which acetyl is the simplest acetyl CoA, the fuel for the citric acid cycle until has now been generated from pyruvate.

The pyruvate dehydrogenase complex cannot complex another catalytic cycle unit the dihydrolipoamide is oxide to Lipoamide. In a fourth step the oxidized from of Lipoamide is regenerated by dihydroliopyl dehydrogenase (E_3). Two electrons are transferred to an FAD prosthetic group of the enzyme and then to NAD.

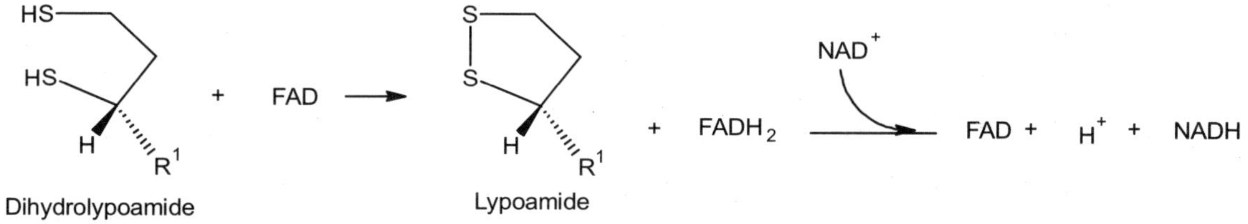

The electron transfer to FAD is unusual, because the common role for FAD is to receive electrons from NADH. The electron transfer potential of FAD is altered by its association with the enzyme and enables it to transfer electrons to NAD^+. Proteins tightly associated with FAD or flavin mono-nucleotide (FMN) are also called flavoproteins.

AS it has been previously stated that citric acid cycle begins with four carbon units, oxaloacetate, reacts with acetyl CoA and H_2O to yield citrate and CoA.

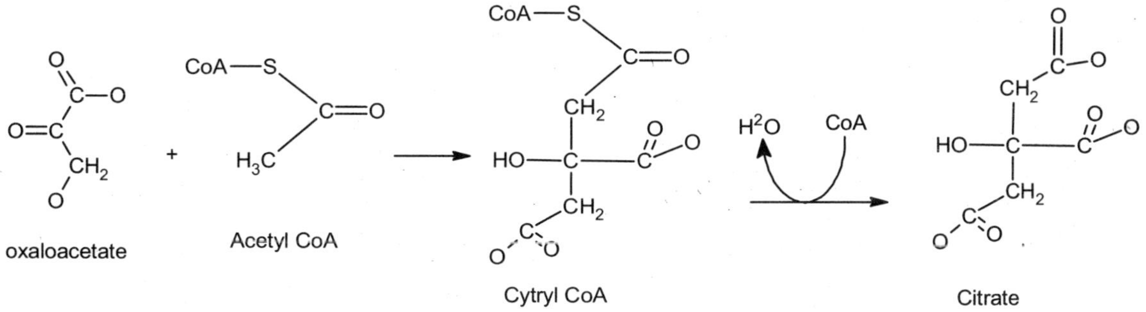

This reaction, which is aldol condensation followed by hydrolysis, is catalyzed by citrate synthase. Oxaloacetate first condenses with acetyl CoA to form citryl CoA, a high-energy thioester intermediate, drivers the overall reaction far in the direction of the synthesis of citrate. In essence the hydrolysis of the thioester powers the synthesis of the new molecule from two precursors. Because this reaction initiates the cycle, it is very important initiates the cycle, it is very important that side reactions are kept minimum.

Let us now briefly discuss how the citrate synthesis prevents wasteful processes such as the hydrolysis of acetyl CoA. Mammalian citrate synthase is a dimmer of identical 49-kd subunits. Each active site is located in a clef between large and small domains of a subunit adjacent to the subunit interface. The result of the X-Ray crystallographic studies of citrate synthase and its complexes with several substrates and inhibitors revealed that the enzyme undergoes large conformational changes in the course of catalysis. Citrate synthase exhibits sequential, ordered kinetics: oxaloacetate binds first, followed by acetyl CoA. The reason for the ordered binding is that oxaloacetate induces a major structural rearrangement leading to the creation of binding site for acetyl CoA. The open form of the enzyme observed in the absence of ligands is converted into a closed form by the binding of oxaloacetate See Fig. 5.13.

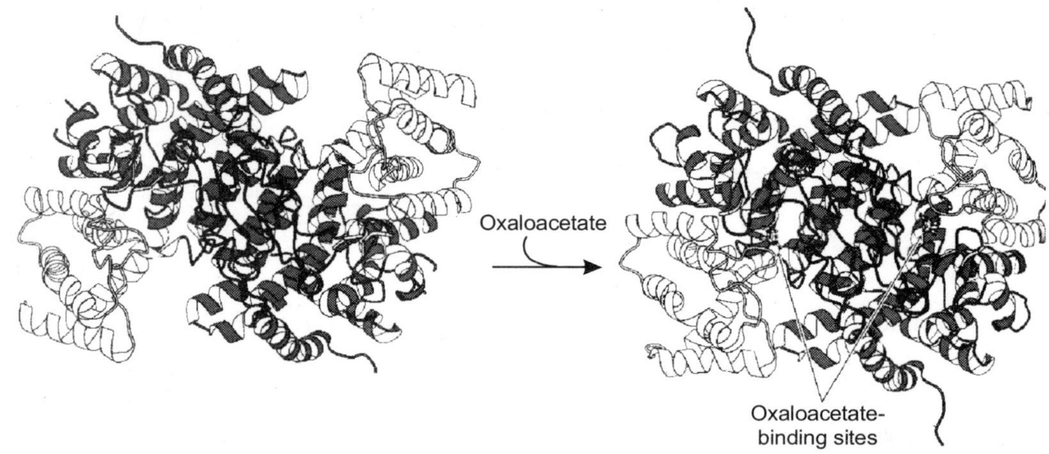

Fig. 5.13 : The Conformational Changes in the Citrate Synthase in Binding the Oxaloacetate

In each submit, the small domain rotates 19° relative to the large domain. Movements as large as 15Å are produced by the rotation of a helices elicited by quite small shift of side chains around bound oxaloacetate. This conformational transition is reminiscent of the cheft closure in hexokinase induced by the binding of glucose.

Citrate synthase catalyses the condensation reaction by bringing the substrates into close proximity orienting them and polarization of certain bonds. Two hystidine residues and an aspirate residue are important players.

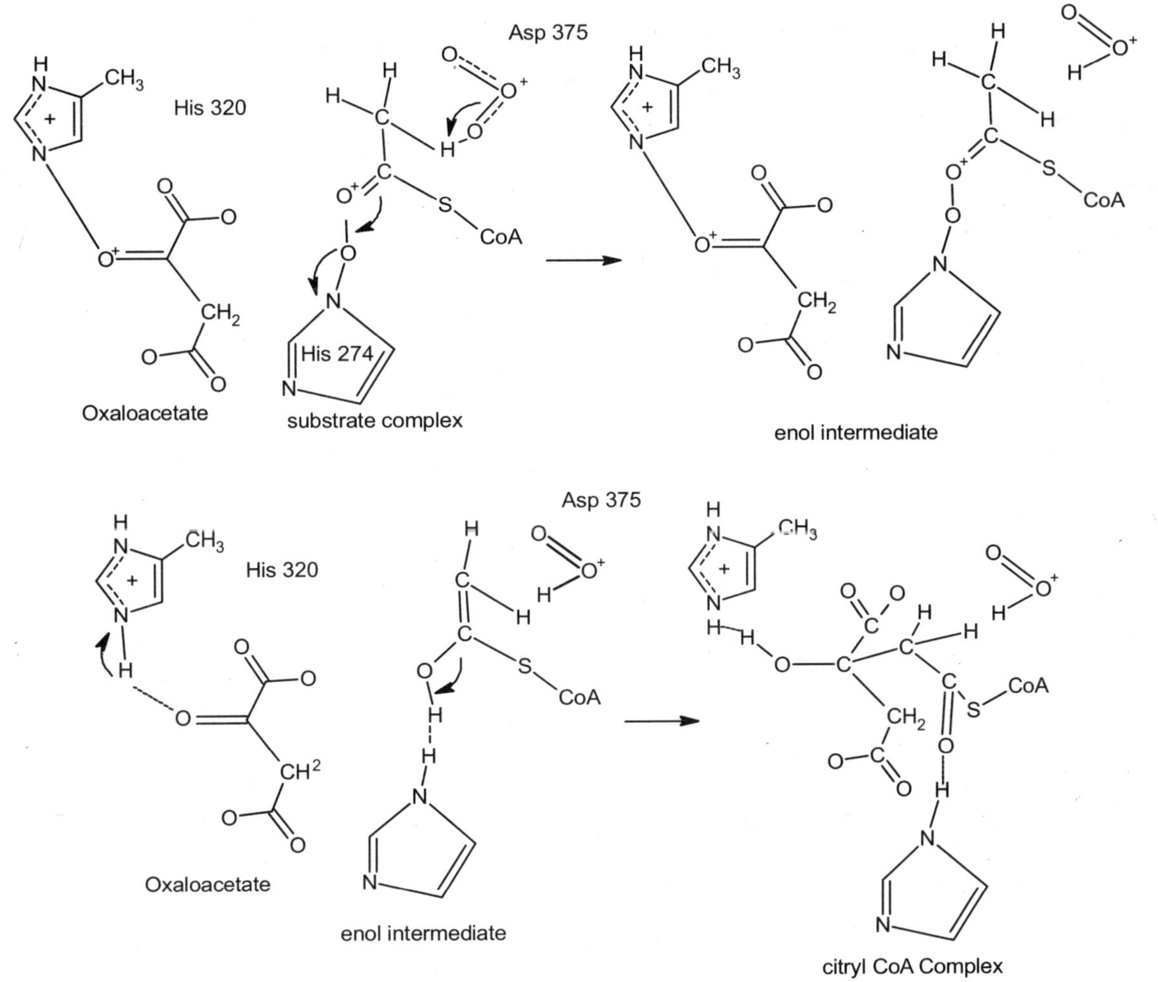

His 320 Asp 375

Oxaloacetate substrate complex enol intermediate

His 274

Asp 375 His 320

Oxaloacetate enol intermediate citryl CoA Complex

One of the hystidine residues (His 274) donates a proton to the carbonyl oxygen of acetyl CoA to promote the removal of a methyl proton by Asp 375. Oxaloacetate is activated by the transfer of a proton from His 320 to its carbonyl carbon atom. The concomitant attack of the enol of a carbon-carbon bond. The newly formed citryl CoA induces additional closed; His 274 participates again as a proton donor to hydrolyze the thioester, Coenzyme A leaves the enzyme then the citrate follows it and the enzyme returns to the initial open conformation.

We can now understand the wasteful hydrolysis of acetylation co-enzyme CoA is prevented. Citrate synthase is well suited to hydrolyze citryl CoA but not acetyl CoA. How is this discrimination accomplished? First acetyl CoA does not bind to the enzyme until oxaloacetate is bond and ready for condensation. Second, the catalytic residues crucial for the hydrolysis of the thioester linkage are not appropriately positioned until citryl CoA is formed. As with hexokinase and triose phosphate isomeriose induced fit prevents and undesirable side reaction.

The tertiary hydroxyl group is not properly located in the citrate molecule for the oxidative decarboxylation that follows. Thus, citrate isomerizes into isocitrate to enable the six carbon unit to undergo oxidative decarboxylation.

The isomerization of citrate is accomplished by a dehydration step followed by hydration step. The result is an inter change of a hydrogen atom and a hydroxyl group. The enzyme catalyzing both step is called aconitase because *cis*-aconitate is an intermediate.

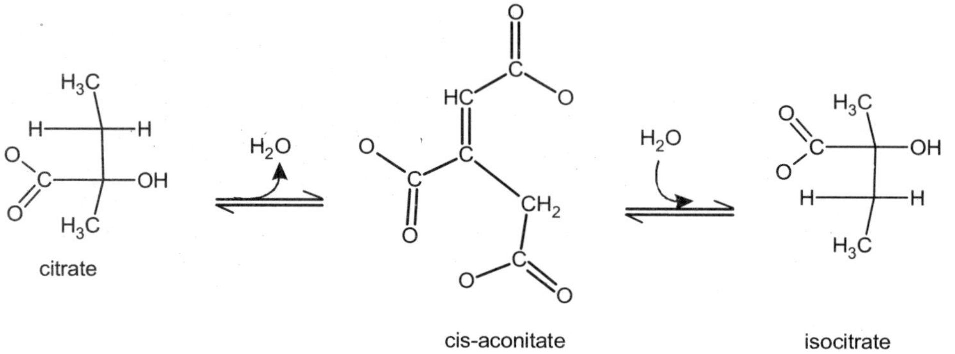

citrate

cis-aconitate

isocitrate

Aconitase is an iron-sulphur protein*, or non-hæme iron protein. It contains four iron atoms are complexed to four inorganic sulphides and three cystine sulphur atoms leaving one iron atoms available to bind citrate and then isocitrate through their carbohydrate and hydroxyl groups. This iron centre with in conjugation with other groups on the enzyme, facilitates the dehydration and rehydration reactions, we will consider the role of these iron sulphur cluster in the electron-transfer reactions of oxidative phosphorylation subsequently.

The oxidative decarboxylation of isocitrate is catalysed by isocitrate dehydrogenase.

$$\text{Isocitrate} + NAD^+ \rightarrow \alpha\text{-ketoglutarate} + CO_2 + NADH$$

Iron-sulphur proteins are proteins characterized by the presence of iron-sulphur clusters containing sulfide-linked di-, tri-, and tetrairon centers in variable oxidation states. Iron-sulphur clusters are found in a variety of metalloproteins, such as the ferredoxins, as well as NADH dehydrogenase, hydrogenases, Coenzyme Q - cytochrome c reductase, Succinate - coenzyme Q reductase and nitrogenase. Iron-sulphur clusters are best known for their role in the oxidation-reduction reactions of mitochondrial electron transport. Both Complex I and Complex II of oxidative phosphorylation have multiple Fe-S clusters. They have many other functions including catalysis as illustrated by aconitase, generation of radicals as illustrated by SAM-dependent enzymes, and as sulphur donors in the biosynthesis of lipoic

* Proteins in which non-haem iron is coordinated with cysteine sulphur and usually also with inorganic sulphur. Divided into three major categories: rubredoxins; "simple iron-sulphur proteins", containing only iron-sulphur clusters; and "complex iron-sulphur proteins", containing additional active redox centres such as flavin, molybdenum or haem.In most iron-sulphur proteins, the clusters function as electron-transfer groups, but in others they have other functions, such as catalysis of hydratase/dehydratase reactions, maintenance of protein structure, or regulation of activity 1997, 69, 1281 IUPAC Compendium of Chemical Terminology.

acid and biotin. Additionally some Fe-S proteins regulate gene expression. Fe-S proteins are vulnerable to attack by biogenic nitric oxide.

Structural Motifs

In almost all Fe-S proteins, the Fe centers is tetrahedral and the thiolato sulphur centers, from cysteinyl residues, are terminal ligands. The sulfide groups are either two- or three-coordinated. Three distinct kinds Fe-S clusters with these features are most common.

2Fe-2S Clusters

The simplest polymetallic system, $[Fe_2S_2]$ cluster, is constituted by two iron ions bridged by two sulfide ions and coordinated by four cysteinyl ligands (in Fe_2S_2 ferredoxins) or by two cysteines and two histidines (in Rieske proteins). The oxidized proteins contain two $Fe3^+$ ions, whereas the reduced proteins contain one Fe^{3+} and one Fe^{2+} ion. These species exist in two oxidation states, $(Fe^{III})_2$ and $Fe^{III}Fe^{II}$.

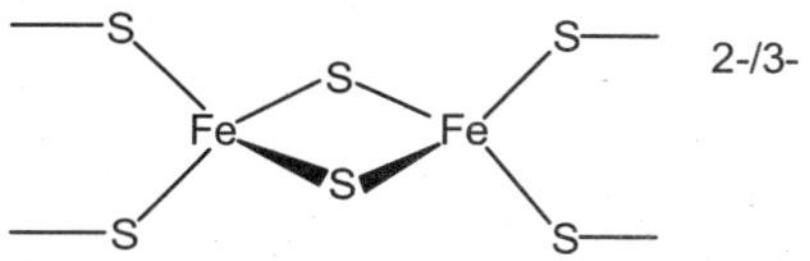

4Fe-4S Clusters

A common motif features a four iron ions and four sulfide ions placed at the vertices of a cubane-type structure. The Fe centers are typically further coordinated by cysteinyl ligands. The $[Fe_4S_4]$ electron-transfer proteins ($[Fe_4S_4]$ ferredoxins) may be further subdivided into low-potential (bacterial-type) and high-potential (HiPIP) ferredoxins. Low- and high-potential ferredoxins are related by the following redox scheme:

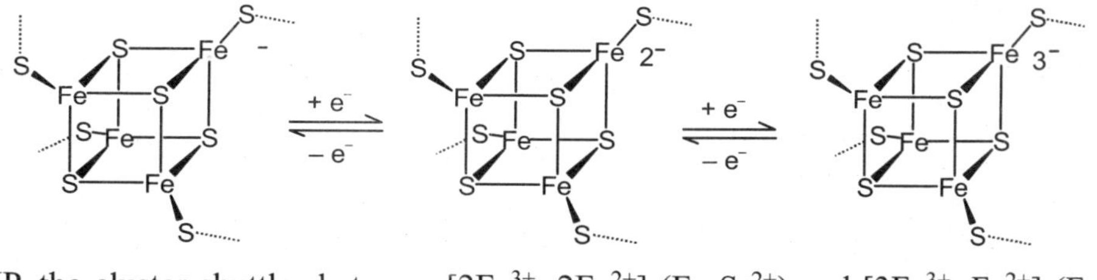

In HiPIP, the cluster shuttles between $[2Fe^{3+}, 2Fe^{2+}]$ $(Fe_4S_4^{2+})$ and $[3Fe^{3+}, Fe^{2+}]$ $(Fe_4S_4^{3+})$. The potentials for this redox couple range from 0.4 to 0.1 V. In the bacterial Fd's, the pair of oxidation states are $[Fe^{3+}, 3Fe^{2+}]$ $(Fe_4S_4^{+})$ and $[2Fe^{3+}, 2Fe^{2+}]$ $(Fe_4S_4^{2+})$. The potentials for this redox couple range from –0.3 to –0.7 V. The two families of 4Fe-4S clusters share the $Fe_4S_4^{2+}$ oxidation state. The difference in the redox couples is attributed to the degree of hydrogen bonding, which strongly modified the basicity of the cysteinyl thiolate ligands. A further redox couple, which is still more reducing than the bacterial Fd's is implicated in the nitrogenase.

Some 4Fe-4S clusters bind substrates and are thus classified as enzymes. In *aconitase*, the Fe-S cluster binds *aconitate* at the one Fe centre that lacks a thiolate ligand. The cluster does not undergo redox, but serves as a *Lewis acid* catalyst to convert aconitate to *isocitrate*. In the radical-SAM enzymes, the cluster binds and reduces *S-adenosylmethionine* to generate a radical, which is involved in many biosyntheses

3Fe-4S Clusters

Proteins are also known to contain $[Fe_3S_4]$ centres, which feature one iron less than the more common $[Fe_4S_4]$ cores. Three sulfide ions bridge two iron ions each, while the fourth sulfide bridges three iron ions. Their formal oxidation states may vary from $[Fe_3S_4]^+$ (all-Fe^{3+} form) to $[Fe_3S_4]^{2-}$ (all-Fe^{2+} form). In a number of iron-sulphur proteins, the $[Fe_4S_4]$ cluster can be reversibly converted by oxidation and loss of one iron ion to a $[Fe_3S_4]$ cluster, *e.g.,* the inactive form of aconitase possesses an $[Fe_3S_4]$ and is activated by addition of Fe^{2+} and reductant.

The intermediate in this reaction is oxaloacetate, an unstable β-ketoacid. While bound to the enzyme it loses CO_2 to form α-ketoglu**tarate.**

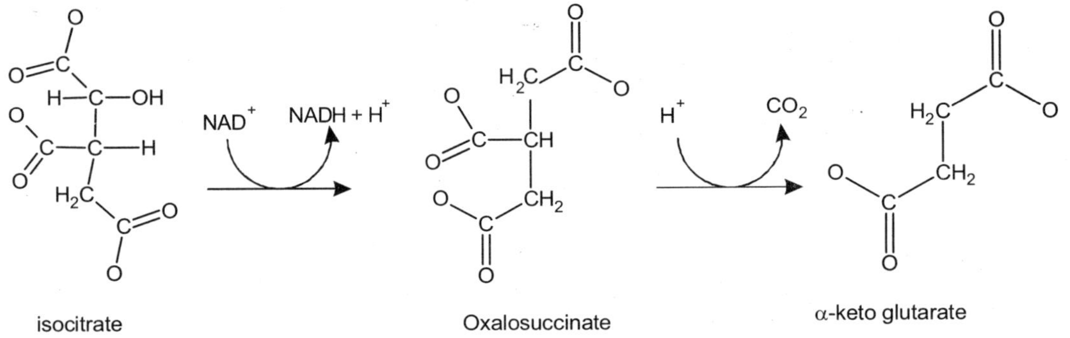

isocitrate —(NAD^+ → $NADH + H^+$)→ Oxalosuccinate —(H^+ → CO_2)→ α-keto glutarate

The rate of formation of α-ketoglutarate is important in determining the overall rate of the cycle. This oxidation generates the first high transfer potential electron carrier NADH in the cycle.

The net reaction of the citric acid cycle is

$$\text{Acetyl CoA} + 3NAD^+ + FAD + GDP + P_1 + 2H_2O \rightarrow 2CO_2 + 3NADH + FADH_2 + GDP + 2H^+ + CoA$$

GLUCONEOGENESIS

Gluconeogenesis (abbreviated GNG) is a metabolic pathway that results in the generation of glucose from non-carbohydrate carbon substrates such as pyruvate, lactate, glycerol, and glucogenic amino acids.

The vast majority of gluconeogenesis takes place in the liver and, to a smaller extent, in the cortex of kidneys. This process occurs during periods of fasting, starvation, or intense exercise and is highly endergonic. Gluconeogenesis is often associated with ketosis. Gluconeogenesis is also a target of therapy for type II diabetes, such as metformin, which inhibit glucose formation and stimulate glucose uptake by cells. This pathway is defined as the formation of glucose from non – carbohydrate sources. Gluconeogenesis is vital to normal brain function in the fasting state in that the brain receives its principal fuel, glucose, directly from the blood. Despite the high concentration of glycogen in the liver, its total content would be used up by about 16-24 hours of fasting. As a result glucose synthesis is vital. The liver is the principal site for gluconeogenesis although the kidney also has the pathway (and it is simulated in response to acidosis.

Most of the reactions of gluconeogenesis are catalyzed by the enzymes of the glycolytic sequence (Fig. 5.14).

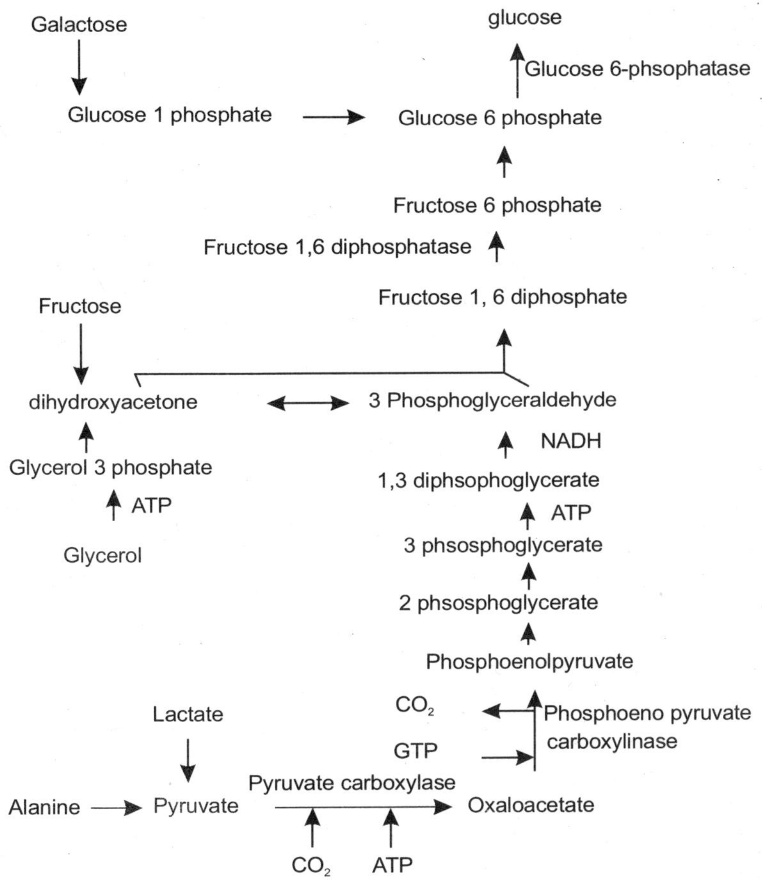

Fig. 5.14 : The Glycolytic sequence in the Gluconeogenesis Pathway

Because glycolysis and gluconeogenesis are opposing pathways, there has to be control so that glucose formation or breakdown will occur in a physiologically sound fashion. The flux through the respective pathways governed by:

(*a*) Allosteric effectors;

(*b*) Covalent modification of Enzymes; and

(*c*) Enzyme concentrations.

Covalent Modification of enzymes is brought about (in main) by fluctuations in the ratio of insulin to glucagon in blood. Insulin is the principal modulator in the fed state. When gluconeogenesis should be active. Three steps in glycolysis are virtually irreversible and have to be passed in gluconeogenesis; three substrate cycles are involved where control can be imposed. (See Fig. 5.15).

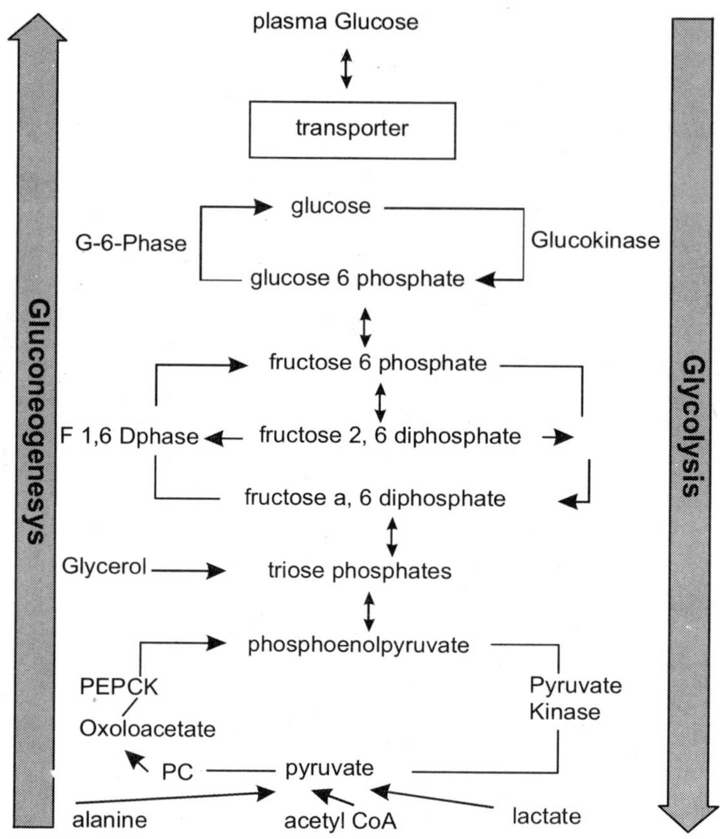

Fig. 5.15 : The Comparison of Gluconeogenesis and Glycolysis Pathways

GLUCONEOGENESIS FROM LACTATE

Lactate is a predominate source of carbon atoms for glucose synthesis by gluconeogenesis. During anaerobic glycolysis in skeletal muscle, pyruvate is reduced to lactate by lactate dehydrogenase (LDH). This reaction serves two critical functions during anaerobic glycolysis. First, in the direction of lactate formation the LDH reaction requires NADH and yields NAD^+ which is then available for use by the glyceraldehyde-3-phosphate dehydrogenase reaction of glycolysis. These two reactions are therefore, intimately coupled during anaerobic glycolysis. Secondly, the lactate produced by the LDH reaction is released to the blood stream and transported to the liver where it is converted to glucose. The glucose is then returned to the blood for use by muscle as an energy source and to replenish glycogen stores. This cycle is termed the Cori cycle. See Fig. 5.9.

GLUCOGENIC AMINO ACIDS

All 20 of the amino acids, excepting leucine and lysine, can be degraded to TCA cycle intermediates as discussed in the metabolism of amino acids. This allows the carbon skeletons of the amino acids to be converted to that in oxaloacetate and subsequently into pyruvate. The pyruvate thus formed can be utilized by the gluconeogenic pathway. When glycogen stores are depleted, in muscle during exertion and liver during fasting, catabolism of muscle proteins to amino acids contributes the major source of carbon for maintenance of blood glucose levels.

THE PENTOSE PHOPSHATE PATHWAY

The pentose phosphate pathway (also called hexose monophopshate shunt) is an alternative pathway for the oxidative metabolism of G-P leading to the production of pentose phosphate and NADPH. It is important to distinguish this pathway from glycolysis in that it is a significant pathway of glucose metabolism out does not result in ATP synthesis. The entire set of reactions can be summarized as follows:

Reactants	Products	Enzyme	Description
Glucose 6-phosphate + NADP+	→ 6-phosphoglucono α lactone + NADPH	glucose 6-phosphate dehydrogenase	Dehydrogenation. The hemiacetal hydroxyl group located on carbon 1 of glucose 6-phosphate is converted into a carbonyl group, generating a lactone, and, in the process, NADPH is generated.
6-phosphoglucono-α-lactone + H_2O	→ 6-phosphogluconate + H_+	6-phosphoglu-conolactonase	Hydrolysis
6-phosphogluconate + $NADP^+$	→ ribulose 5-phosphate + NADPH + CO_2	6-phosphogluconate dehydrogenase	Oxidative decarboxylation. $NADP^+$ is the electron acceptor, generating another molecule of NADPH, a CO_2, and ribulose 5-phosphate.

The overall reaction for this process is:

$$\text{Glucose 6-phosphate} + 2\ NADP^+ + H_2O \rightarrow \text{ribulose 5-phosphate} + 2\ NADPH + 2\ H^+ + CO_2$$

Glucose-6-phosphate dehydrogenase is the rate-controlling enzyme of this pathway. It is allosterically stimulated by $NADP^+$. The ratio of NADPH:$NADP^+$ is normally about 100:1 in liver cytosol. This makes the cytosol a highly-reducing environment. Formation of $NADP^+$ by a NADPH-utilizing pathway, thus, stimulating production of more NADPH.

The importance of the pentose phosphate pathway is as follows:

- It produces ribose-5-phosphate, which is required for the biosynthesis of purine and pyrimidine nucleotides and therefore for RNA and DNA the numerous nucleotides that play roles in the intermediatory metabolism including ATP, UTP, CTP GTP-S-Adenosyl-metionine FAD, NAD and NADP.

The pentose phosphate pathway's Nonoxidative phase:

Reactants	Products	Enzymes
ribulose 5-phosphate	→ ribose 5-phosphate	Ribulose 5-Phosphate Isomerase
ribulose 5-phosphate	→ xylulose 5-phosphate	Ribulose 5-Phosphate 3-Epimerase
xylulose 5-phosphate + ribose 5-phosphate	→ glyceraldehyde 3-phosphate + sedoheptulose 7-phosphate	transketolase
sedoheptulose 7-phosphate + glyceraldehyde 3-phosphate	→ erythrose 4-phosphate + fructose 6-phosphate	transaldolase
xylulose 5-phosphate + erythrose 4-phosphate	→ glyceraldehyde 3-phosphate + fructose 6-phosphate	transketolase

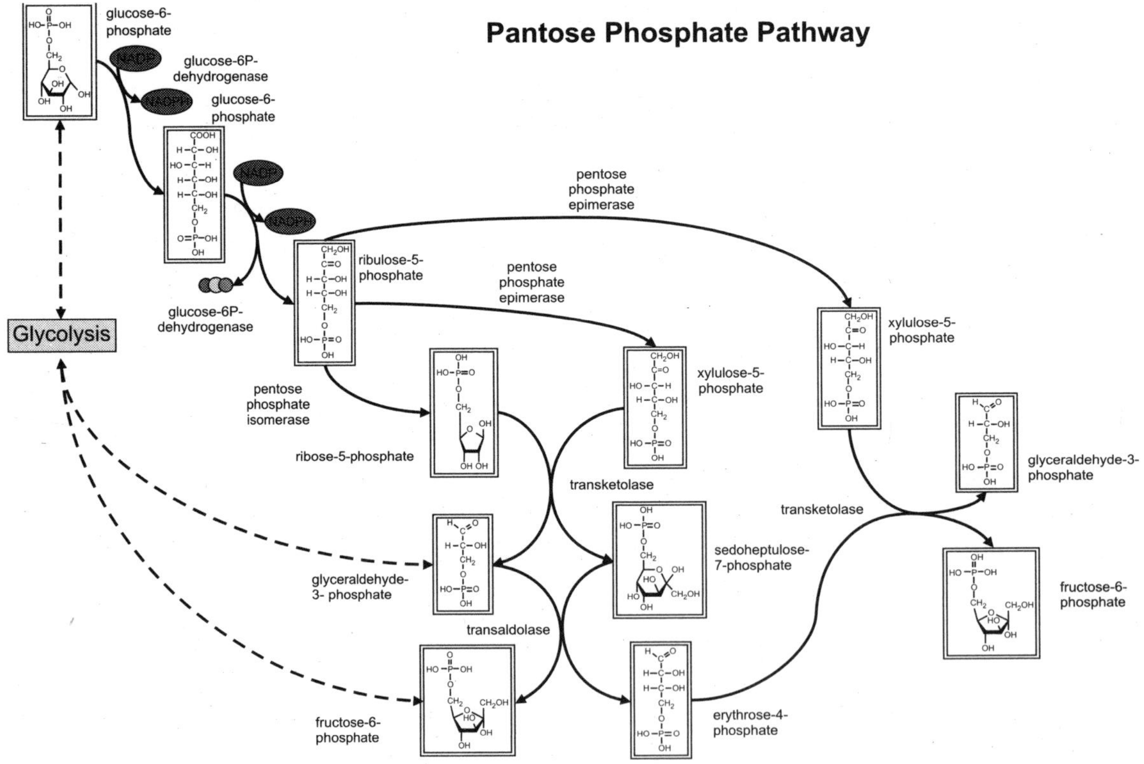

Fig. 5.16 : The Pentose Phosphate Pathway

- It produces reducing power in the form of NADPH which is required for the synthesis of fatty acids cholesterol and steroid hormones.
- In plants, the pathways are modified to participate in the formation of glucose from carbon dioxide in photosynthesis.

GLUCURONIC ACID

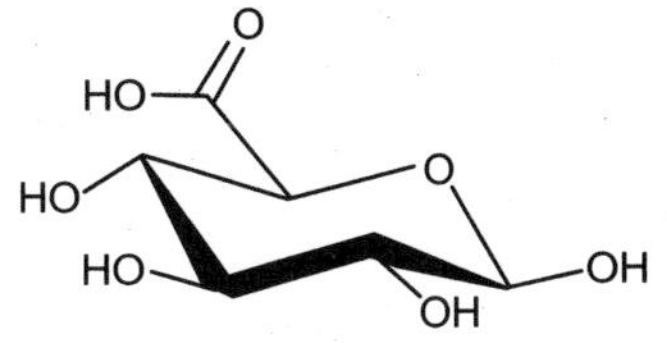

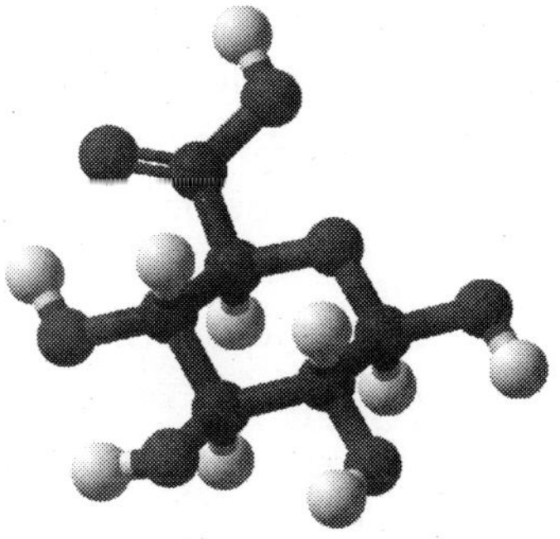

Glucuronic acid is a carboxylic acid. Its structure is similar to that of glucose. However, glucuronic acid's sixth carbon is oxidized to a carboxylic acid. Its formula is $C_6H_{10}O_7$.

The salts and esters of glucuronic acid are known as glucuronates; the anion $C_6H_9O_7-$ is the glucuronate ion.Glucuronic acid should not be confused with gluconic acid, a linear carboxylic acid resulting from the oxidation of a different carbon of glucose. Both glucuronic acid and gluconic acid are reported to be found in the fermented drink known as kombucha. Glucuronic acid has been attributed with the mildly-alcoholic effect that drinking the tea has for some.

FUNCTIONS GLUCURONIDATION

Glucuronic acid is highly soluble in water. In the animal body, glucuronic acid is often linked to the xenobiotic metabolism of substances such as drugs, pollutants, bilirubin, androgens, estrogens, mineralocorticoids, glucocorticoids, fatty acid derivatives, retinoids, and bile acids. These linkages involve glycosidic bonds, and this linkage process is known as glucuronidation. Glucuronidation occurs mainly in the liver, although the enzyme responsible for its catalysis, UDP-glucuronyltransferase, has been found in all major body organs, e.g., heart, kidneys, adrenal gland, spleen, and thymus. UDP-glucuronic acid (glucuronic acid linked via a glycosidic bond to uridine diphosphate) is an intermediate in the process and is formed in the liver. One example is this N-glucuronidation of an aromatic amine, 4-aminobiphenyl, by UGT1A4 or UGT1A9 from human, rat, or mouse liver.

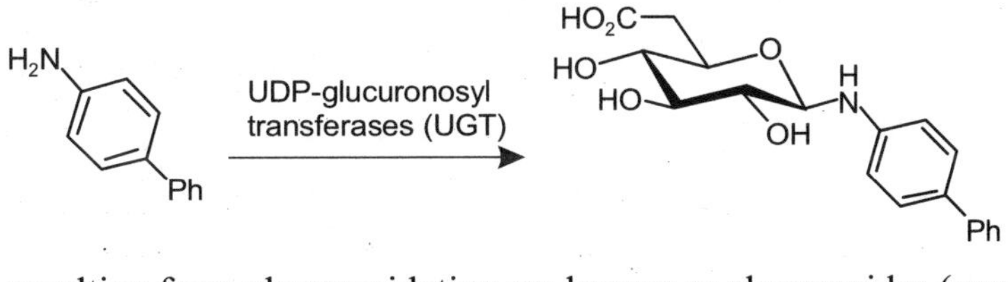

The substances resulting from glucuronidation are known as glucuronides (or glucuronosides) and are typically much more water-soluble than the non-glucuronic acid-containing substance from which they were originally synthesised. The human body uses glucuronidation to make a large variety of substances more water-soluble, and, in this way, allow for their subsequent elimination from the body upon urination. Hormones may also be glucuronidated to allow for easier transport around the body. Pharmacologists also commonly link drugs to glucuronic acid to allow for easier drug delivery.

The conjugation of xenobiotic molecules with hydrophilic molecular species such as glucuronic acid is known as phase II metabolism.

GLUCOSE TOLERANCE TEST

A glucose tolerance test in medical practice is the administration of glucose to determine how quickly it is cleared from the blood. The test is usually used to test for diabetes, insulin resistance, and sometimes reactive hypoglycemia. The glucose is most often given orally so the common test is technically an oral glucose tolerance test (OGTT). The test may be performed as part of a panel of tests, such as the comprehensive metabolic panel.

PROCEDURE FOR OGTT

The patient is instructed not to restrict carbohydrate intake in the days or weeks before the test. The test should not be done during an illness, as results may not reflect the patient's glucose metabolism when healthy. A full adult dose should not be given to a person weighing less than 43 kg (94 lb), or exaggerated glucoses may produce a false positive result. The patient should have been fasting for the previous 8-14 hours (water is allowed). Usually the OGTT is scheduled to begin in the morning (0700-0800) as glucose tolerance exhibits a diurnal rhythm with a significant decrease in the afternoon. A zero time (baseline) blood sample is drawn. The patient is then given a glucose solution to drink. The standard dose since the late 1970s has been 1.75 grams of glucose per kilogram of body weight, to a maximum dose of 75 g. It should be drunk within 5 minutes. Prior to 1975 a dose of 100 g was often used.

Blood is drawn at intervals for measurement of glucose (blood sugar), and sometimes insulin levels. The intervals and number of samples vary according to the purpose of the test. For simple diabetes screening, the most important sample is the 2 hour sample and the 0 and 2 hour samples may be the only ones collected. In research settings, samples may be taken on many different time schedules.

If renal glycosuria (sugar excreted in the urine despite normal levels in the blood) is suspected, urine samples may also be collected for testing along with the fasting and 2 hour blood tests.

INTERPRETATION OF OGTT RESULTS

Fasting plasma glucose should be below 6.1 mmol/l (110 mg/dl). Fasting levels between 6.1 and 7.0 mmol/l (110 and 126 mg/dl) are borderline ("impaired fasting glycaemia"), and fasting levels repeatedly at or above 7.0 mmol/l (126 mg/dl) are diagnostic of diabetes.

The 2 hour glucose level should be below 7.8 mmol/l (140 mg/dl). Levels between this and 11.1 mmol/l (200 mg/dl) indicate "impaired glucose tolerance." Glucose levels above 11.1 mmol/l (200 mg/dl) at 2 hours confirms a diagnosis of diabetes.

1999 WHO Diabetes criteria—Interpretation of Oral Glucose Tolerance Test

Glucose levels	NORMAL		impaired fasting glycaemia (IFG)		impaired glucose tolerance (IGT)		Diabetes Mellitus (DM)	
Venous Plasma	Fasting	2hrs	Fasting	2hrs	Fasting	2hrs	Fasting	2hrs
(mmol/l)	<6.1	<7.8	> 6.1 & <7.0	<7.8	<7.0	>7.8	>7.0	>11.1
(mg/dl)	<110	<140	>110 & <126	<140	<126	>140	>126	>200

A standard 2 hour OGTT is sufficient to diagnose or exclude all forms of diabetes mellitus at all but the earliest stages of development. Longer tests have been used for a variety of other purposes, such as detecting reactive hypoglycemia or defining subsets of hypothalamic obesity. Insulin levels are sometimes measured to detect insulin resistance or deficiency.

The OGTT is of limited value in the diagnosis of reactive hypoglycemia, since (1) normal levels do not preclude the diagnosis, (2) abnormal levels do not prove that the patient's other symptoms are related to a demonstrated atypical OGTT, and (3) many people without symptoms of reactive hypoglycemia may have the late low glucoses that are said to be characteristic. Using a glucose tolerance in this context resembles use of a Rorschach test in that it is often used to support a diagnosis that the patient and doctor are already reaching agreement on based on other evidence, but it is inadequate by itself to confirm or refute the diagnosis (unlike its use for diabetes).

When the glucose is given intravenously it is termed an intravenous glucose tolerance test (IVGTT). This has been used in the investigation of early insulin secretion abnormalities in prediabetic states.

GLYCOSURIA

Glycosuria or glucosuria is the excretion of glucose into the urine. Ordinarily, urine contains no glucose because the kidneys are able to reclaim all of the filtered glucose back into the bloodstream. Glycosuria is nearly always caused by elevated blood glucose levels, most commonly due to untreated diabetes mellitus. Rarely, glycosuria is due to an intrinsic problem with glucose reabsorption within the kidneys themselves, a condition termed renal glycosuria. Glycosuria leads to excessive water loss into the urine with resultant dehydration, a process called osmotic diuresis. Blood is filtered by millions of nephrons, the functional units that comprise the kidneys. In each nephron, blood flows from the arteriole into the glomerulus, a tuft of leaky capillaries. Bowman's capsule surrounds each glomerulus, and collects the filtrate that the glomerulus forms. The filtrate contains waste products (e.g. urea), electrolytes (e.g. sodium, potassium, chloride), amino acids, and glucose. The filtrate passes into the renal tubules of the kidney. In the first part of the renal tubule, the proximal tubule, glucose is reabsorbed from the filtrate, across the tubular epithelium and into the bloodstream. The proximal tubule can only reabsorb a limited amount of glucose. When the blood glucose level exceeds about 160-180 mg/dl (8.9-10 mmol/l), the proximal tubule becomes overwhelmed and begins to excrete glucose in the urine.

This point is called the renal threshold of glucose (RTG). Some people, especially children and pregnant women, may have a low RTG (less than ~7 mmol/L glucose in blood to have glucosuria).

If the RTG is so low that even normal blood glucose levels produce the condition, it is referred to as renal glycosuria. Glucose in urine can be identified by Benedict's qualitative test. Renal glycosuria, also known as renal glucosuria, is a rare condition in which the simple sugar glucose is excreted in the urine despite normal or low blood glucose levels. With normal kidney (renal) function, glucose is excreted in the urine only when there are abnormally elevated levels of glucose in the blood. However, in those with renal glycosuria, glucose is abnormally eliminated in the urine due to improper functioning of the renal tubules, which are primary components of nephrons, the filtering units of the kidneys.

GLYCOSYLATED HEAMOGLOBIN

Glycosylated Haemoglobin (GHb), the resultant of chronic hyperglycaemia, has hitherto been used as an index of long term glycaemic control in diabetics. Since it is a refiector of integrated long term glycaemic status, it can also be used as a screening test for the diagnosis of Diabetes Mellitus (DM). In this study of 1000 subjects, when GHb was compared to Oral Glucose Tolerance Test (OGTT) in the diagnosis of DM, GHb was found to have 93.98% sensitivity, 85.86% specificity and 93.14% efficiency in the diagnosis of DM. The predictive value of an elevated GHb is 98.2%. Hence it is concluded that GHb is a useful screening test for DM and that it may also be considered when criteria for the diagnosis of DM are discussed. Glycosylated Haemoglobin reflects retrospective time averaged glycaemic status of an individual and thus has became a useful tool in assessing the long term control of DM1, 2, 3. Being a reflector of chronic hyperglycaemia, an elevated GHb, can also be used as a screening test for the diagnosis of DM4,5. It has the added convenience that blood sample can be collected at any time of the day without any specific time interval in relation to the meal. Glycosylated Haemoglobin was determined in 1000 non-pregnant adults The OGTT was interpreted following the WHO criteria. Glycosylated Haemoglobin was estimated by modified Fluckinger and Winterhalter's method. The OGTT following WHO criteria, revealed that among the 1000 subjects studied, 797 had DM while 111 had Impaired Glucose Tolerance (IGT). The remaining 92 had normal glucose tolerance (Table 5.1). The mean GHb in those with normal glucose tolerance, IGT and Diabetes Mellitus was 6.9 ± 1.05%, 8.4 ± 2.50% and 12.6 ± 1.80% respectively. Table 5.2 shows the 95% confidence limits of GHb in each of these categories. The +2SD figure for those with those with normal glucose tolerance and the -2SD figure for those with DM are both 9%, indicating this to be a clear cut-off point between the normal- and diabetic population in this study. The mean GHb value in those with IGT lies between those of the other two two categories. The GHb was elevated in 749 diabetics, 60 subjects with IGT and 13 subjects with normal glucose tolerance. It was normal in 48 diabetics, 51 subjects with IGT and 79 normal individuals. Further analysis was restricted to those who had either normal glucose tolerance or DM. Table 5.2 shows the GHb results in relation to the OGTT results in subjects who had either diabetic curve or normal glucose tolerance. Oral Glucose Tolerance Test was taken as the reference "Gold Standard" against which the validity of GHb determination in the diagnosis of Diabetes Mellitus was assessed. Based on the OGTT and GHb results, subjects could be placed into one of the following 4 wells depicted in Fig. 5.17. Thus,

Table 5.1 : Results of OGTT

n = 1000

Diagnostic category (WHO)	Number	Percentage
DM	797	79.7
IGT	111	11.1
Normal	92	9.2

Table 5.2 : GHb in different categories of Glucose Tolerance

n = 1000

Categories of Glucose Tolerance	No. of subjects	GHb% Mean + S.D	GHb –2SD	GHb +2SD
Normal	92	6.9 ± 1.05	4.8	9.0*
IGT	111	8.4 ± 2.50	5.9	10.9
DM	797	12.6 ± 1.8	9.0*	16.2

*+2SD of GHb in those with normal glucose tolerance and -2SD of GHb in, those with DM are both 9%, including that a GHb level of 9% is the cut-off point between the, two categories.

		OGTT Abnormal	OGTT Normal
GHb	Elevated	a	b
	Normal	c	d

Well "a" represents diabetics with elevated GHb
Well "b" represents normal subjects with elevated GHb
Well "c" represents diabetics with normal GHb, and
Well "d" represents normal subjects with normal GHb

Fig. 5.17 : Classification of subjected based on OGTT and GHb

In other words,

Well "a" represents True positivity

Well "b" represents False positivity

Well "c" represents False negativity

Well "d" represents True Negativity of GHb estimations in the diagnosis of DM in relation to OGTT

After so placing the results of GHb determination in relation to OGTT results, we proceeded to analyse the various attributes of GHb in the diagnosis of DM. The sensitivity of GHb in the diagnosis of DM, i.e. its ability to correctly identify those with DM as compared to OGTT is given by the formula:

$$\frac{a}{a+c} \times 100$$

The specificity of GHb in the diagnosis of DM, i.e. its ability to correctly classify those who do not have Diabetes, as compared to OGTT is given by the formula :

$$\frac{d}{b+d} \times 100$$

The predictive value of raised GHb in the diagnosis of DM represents the percentage of diabetics among all those who have raised GHb and is given by the formula :

$$\frac{a}{a+c} \times 100$$

By the predictive value of normal GHb is meant the percentage of true normals by OGTT among all those who have a normal GHb and this is given by the formula :

$$\frac{d}{c+d} \times 100$$

Having analysed the sensitivity, specificity and predictive value of GHb in the diagnosis of DM, we proceeded to analyse its overall efficiency as a diagnostic test for DM. The efficiency of a test represents its ability to bisect the test subjects into either normals or abnormals-i.e: in this instance place them either into well 'a' or well 'd' and is given by the formula

$$\frac{a+d}{N} \times 100$$

were N represents the total number of subjects tested.

Various tests for the diagnosis of DM exist. Glycosuria depends on the renal threshold for glucose so much that it is less sensitive in diagnosing diabetes. Since DM can be present despite a normal fasting blood sugar, it is also less sensitive. Further even when it is elevated, National Diabetes Data Group (NDDG)a requires it to be elevated on more than one occasion to be diagnostic. Unequivocal elevation of random or postprandial blood sugars are also required to be demonstrated on more than one occasion. When all these tests are inconclusive and yet the diagnosis of DM is in doubt, an OGTT is recommended. All these blood sugar tests have the inconvenience of having to collect the blood sample at a specified time or preparing the patient. Glycosylated Haemoglobin determination which reflects integrated glycaemic status of about 2-3 months is a useful screening test for the diagnosis of

DM by the very nature of its evolution, viz. duration and degree of hyperglycaemia. In addition, it has the advantage that blood sample can be collected at any time without any specified time interval in relation to meal.

METABOLiSM OF LIPIDS

The major dilatory lipids are triglycerols (triglycerides as commonly called), which are esters of an alcohols the glycerol, and fatty acids. See Fig. 5.18 naturally occurring fats usually have different long chain fatty acids in all three ester positions. The fatty acid can be saturated, e.g. palminate or stearate or unsaturated e.g. oleate other fatty acids are poly unsaturated. Fats are hydrophobic molecules.

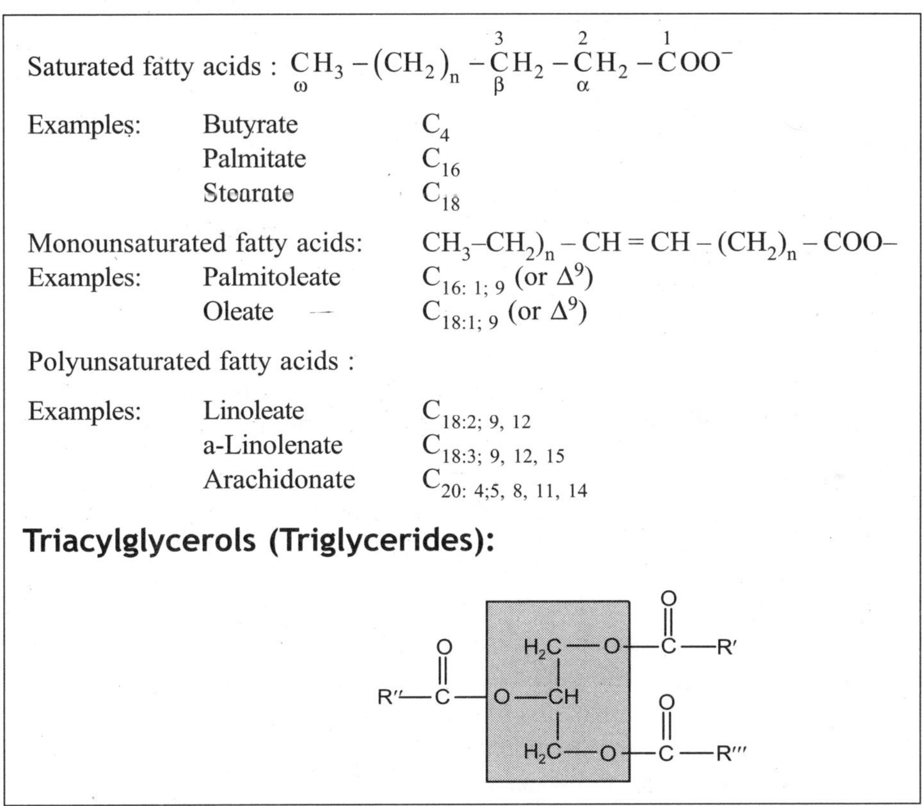

Fig. 5.18 : The fatty acid and triglycerols structures. Two different nomenclatures are shown for fatty acids. In one, the carbonyl carbon is carbon–1 and the rest of the carbons follow from that, which the terminal methyl group carbon having the number corresponding to the number of carbons in the fatty acid. The other system uses Greek alphabets starting from the carbon following the carbonyl carbon. In that scheme, the terminal methyl group is referred as the omega (ω) carbon. In the unsaturated fatty acids the number that follows colon (:) refers to the location of the double bonds. In most unsaturated fatty acids found in the body the bonds are *cis*.

DIGESTION AND ABSORPTION OF FATS

Dietary triglycerides are absorbed after undergoing partial hydrolysis with bile salts playing a key role. The main steps are listed:

1. Salivary and Lingual lipases preferentially hydrolyze triglycerols composed of short and medium chain fatty acids (as found in cow milk)
2. In the small intestine, bile salts emulsify fats
3. Pancreatic lipase and co-lipase are secreted from the pancreas. Co-Lipase and provides an anchor for the pancreatic lipase at the triglycerols-bile salt water interface. In the small intestine pancreatic lipase catalyses the hydrolysis of triglycerols, to monoacylglycerol by removing fatty acids from 1 and 3 positions.
4. Bile salt micelles are formed that contains triacylglecerols, mono-glycerides, fatty acids and fat soluble vitamins and these allows for the mucosal cells.
5. Fatty acids of less than 10 to 12 carbons go directly to the liver via the portal vein
6. Triacylglecerols are reformed intestinal mucosal cells form long chain fatty acids and mono – glycerides and are then incorporated into chlomicrons which are lipoproteins involved in the delivery of fatty acids.
7. Cholesterol esters in the diet are hydrolyzed by the action of cholesteryl ester hydrolase. Unesterified cholesterol and cholesteryl esters are included in the fat micelles.
8. Phospholipids in the diet are hydrolyzed by the action of pancreatic phospholipase A, which removes the fatty acid at the carbon 2-position leaving a lysophopsholipid, a powerful detergent. The fatty acids released and the lysophopsholipids are incorporated into micelles and transported into mucosal cells and appear chylomicrions.

MICHELLE FORMATION

A micelle is formed when a variety of molecules including soaps and detergents are added to water. The molecule may be a fatty acid, a salt of a fatty acid (soap), phospholipids, or other similar molecules. The molecule must have a strongly polar "head" and a non-polar hydrocarbon chain "tail". When this type of molecule is added to water, the non-polar tails of the molecules clump into the center of a ball like structure, called a micelle, because they are hydrophobic or "water hating". The polar head of the molecule presents itself for interaction with the water molecules on the outside of the micelle. The theoretical model shows 54 molecules of dodecylphosphocholine (DPC) and about 1200 H_2O molecules. Each lipid has a polar head group (phosphocholine) and a hydrophobic tail (dodecyl = C_{12}).

The Figure 5.19 represents a cross section of a micelle. The gray spheres on the interior represent the long hydrocarbon chains of the dodecyl groups which are massed together because they are non-polar. The polar head groups of the phosphate are shown as red and orange spheres. The amine nitrogen is shown in blue surrounded by the gray methyl groups. The water molecules are represented as red and white spheres surrounding the outside of the micelle and penetrates all of the spaces in the head group region.

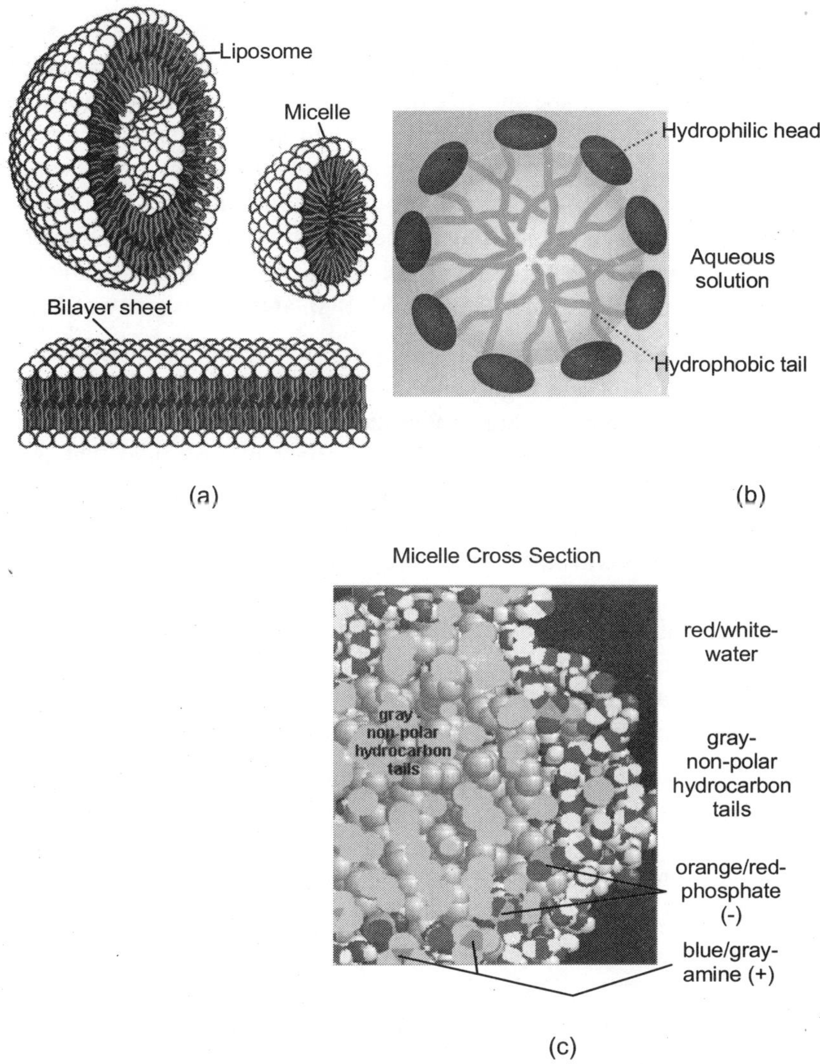

Fig. 5.19 : Scheme of an Inverse Micelle formed by Phospholipids in an Organic Solvent

The hydrophobic tails are shown Spacefill. H_2O is excluded from this entire interior volume. The hydrocarbon chains vary in their individual conformations (e.g. trans/gauche configuration at each carbon-carbon bond), but adapt so as to fill all of the interior space. A typical micelle in aqueous solution forms an aggregate with the hydrophilic "head" regions in contact with surrounding solvent, sequestering the hydrophobic single tail regions in the micelle centre. This phase is caused by the insufficient packing issues of single tailed lipids in a bilayer. The difficulty filling all the volume of the interior of a bilayer, while accommodating the area per head group forced on the molecule by the hydration of the lipid head group leads to the formation of the micelle. This type of micelle is known as a normal phase micelle (oil-in-water micelle). Inverse micelles have the headgroups at the centre with the tails extending out (water-in-oil micelle).

Micelles are approximately spherical in shape. Other phases, including shapes such as ellipsoids, cylinders, and bilayers are also possible. The shape and size of a micelle is a function of the molecular geometry of its surfactant molecules and solution conditions such as surfactant concentration, temperature, pH, and ionic strength. The process of forming micellae is known as micellisation and forms part of the phase behaviour of many lipids according to their polymorphism. The ability of a soapy solution to act as a detergent has been recognised for centuries. However, it was only at the beginning of the twentieth century that the constitution of such solutions was scientifically studied. Pioneering work in this area was carried out by James William McBain at the University of Bristol.

As early as 1913 he postulated the existence of "colloidal ions" to explain the good electrolytic conductivity of sodium palmitate solutions. These highly mobile, spontaneously formed clusters came to be called micelles, a term borrowed from biology and popularized by G.S. Hartley in his classic book "Paraffin Chain Salts, A Study in Micelle Formation". Individual surfactant molecules that are in the system but are not part of a micelle are called "monomers." Lipid micelles represent a molecular assembly in which the individual components are thermodynamically in equilibrium with monomers of the same species in the surrounding medium. In water, the hydrophilic "heads" of surfactant molecules are always in contact with the solvent, regardless of whether the surfactants exist as monomers or as part of a micelle. However, the lipophilic "tails" of surfactant molecules have less contact with water when they are part of a micelle – this being the basis for the energetic drive for micelle formation. In a micelle, the hydrophobic tails of several surfactant molecules assemble into an oil-like core the most stable form of which has no contact with water.

By contrast, surfactant monomers are surrounded by water molecules that create a "cage" of molecules connected by hydrogen bonds. This water cage is similar to a clathrate and has an ice-like crystal structure and can be characterized according to the hydrophobic effect. The extent of lipid solubility is determined by the unfavorable entropy contribution due to the ordering of the water structure according to the hydrophobic effect.Micelles composed of ionic surfactants have an electrostatic attraction to the ions that surround them in solution, the latter known as counterions. Although the closest counterions partially mask a charged micelle (by up to 90%), the effects of micelle charge affect the structure of the surrounding solvent at appreciable distances from the micelle. Ionic micelles influence many properties of the mixture, including its electrical conductivity.

Adding salts to a colloid containing micelles can decrease the strength of electrostatic interactions and lead to the formation of larger ionic micelles. This is more accurately seen from the point of view of an effective change in hydration of the system. Micelles only form when the concentration of surfactant is greater than the critical micelle concentration (CMC), and the temperature of the system is greater than the critical micelle temperature, or Krafft temperature. The formation of micelles can be understood using thermodynamics: micelles can form spontaneously because of a balance between entropy and enthalpy. In water, the hydrophobic effect is the driving force for micelle formation, despite the fact that assembling surfactant molecules together reduces their entropy. At very low concentrations of the lipid, only monomers are present in true solution. As the concentration of the lipid is increased, a point is reached at which the unfavorable entropy considerations, derived from the hydrophobic end of the molecule, become dominant. At this point, the lipid hydrocarbon chains of a portion of the lipids must be sequestered away from the water. Therefore, the lipid starts to form micelles. Broadly speaking, above the CMC, the entropic penalty of assembling the surfactant molecules is less than the entropic penalty of caging the surfactant monomers with water molecules. Also important are enthalpic considerations, such as the electrostatic interactions that occur between the charged parts surfactants.

In a non-polar solvent, it is the exposure of the hydrophilic head groups to the surrounding solvent that is energetically unfavorable, giving rise to a water-in-oil system. In this case the hydrophilic groups are sequestered in the micelle core and the hydrophobic groups extend away from the centre. These inverse micelles are proportionally less likely to form on increasing headgroup charge, since hydrophilic sequestration would create highly unfavourable electrostatic interactions. When surfactants are present above the CMC (Critical micelle concentration), they can act as emulsifiers that will allow a compound that is normally insoluble (in the solvent being used) to dissolve. This occurs because the insoluble species can be incorporated into the micelle core, which is itself solubilized in the bulk solvent by virtue of the head groups' favourable interactions with solvent species.

The most common example of this phenomenon is detergents, which clean poorly soluble lipophilic material (such as oils and waxes) that cannot be removed by water alone. Detergents also clean by lowering the surface tension of water, making it easier to remove material from a surface. The emulsifying property of surfactants is also the basis for emulsion polymerization. Micelle formation is essential for the absorption of fat-soluble vitamins and complicated lipids within the human body. Bile salts formed in the liver and secreted by the gall bladder allow micelles of fatty acids to form. This allows the absorption of complicated lipids (e.g., lecithin) and lipid soluble vitamins (A, D, E and K) within the micelle by the small intestine.

GASTRIC MUCOSA AND LIPID METABOLISM

The gastric mucosa is the mucous membrane layer of the stomach which contains the glands and the gastric pits. In men it is about 1 mm thick and its surface is smooth, soft, and velvety. It consists of epithelium, lamina propria, and the muscularis mucosae.

In its fresh state, it is of a pinkish tinge at the pyloric end and of a red or reddish-brown color over the rest of its surface. In infancy it is of a brighter hue, the vascular redness being more marked.

It is thin at the cardiac extremity, but thicker toward the pylorus. During the contracted state of the organ it is thrown into numerous plaits or rugae, which, for the most part, have a longitudinal direction, and are most marked toward the pyloric end of the stomach, and along the greater curvature. These folds are entirely obliterated when the organ becomes distended.

When examined with a lens, the inner surface of the mucous membrane presents a peculiar honeycomb appearance from being covered with funnel-like depressions or foveolae of a polygonal or hexagonal form, which vary from 0.12 to 0.25 mm. in diameter. These are the ducts of the gastric glands, and at the bottom of each may be seen one or more minute orifices, the openings of the gland tubes. Gastric glands arc simplc or branchcd tubular glands that emerge on the deeper part of the gastric foveola, inside the gastric areas and outlined by the folds of the mucosa.

There are three types of glands: cardiac glands (in the proximal part of the stomach), oxyntic glands (the dominating type of gland), and pyloric glands. The cardiac glands mainly contain mucus producing cells. The bottom part of the oxyntic glands is dominated by zymogen (chief) cells that produce pepsinogen (an inactive precursor of the pepsin enzyme). Parietal cells, which secrete hydrochloric acid are scattered in the glands, with most of them in the middle part. The upper part of the glands consist of mucous neck cells; in this part the dividing cells are seen. The pyloric glands contain mucus-secreting cells. Several types of endocrine cells are found in all regions of the gastric mucosa. In the pyloric glands contain gastrin producing cells (G cells); this hormone stimulates acid production from the parietal cells. ECL (enterochromaffine-like) cells, found in the oxyntic glands release histamine, which also is a powerful stimulant of the acid secretion. The A cells produce glucagon, which mobilizes the hepatic glycogen, and the enterochromaffin cells that produce serotonin, which stimulates the contraction of the smooth muscles.

The surface of the mucous membrane is covered by a single layer of columnar epithelium . This epithelium commences very abruptly at the cardiac orifice, where there is a sudden transition from the stratified epithelium of the esophagus. The epithelial lining of the gland ducts is of the same character and is continuous with the general epithelial lining of the stomach.

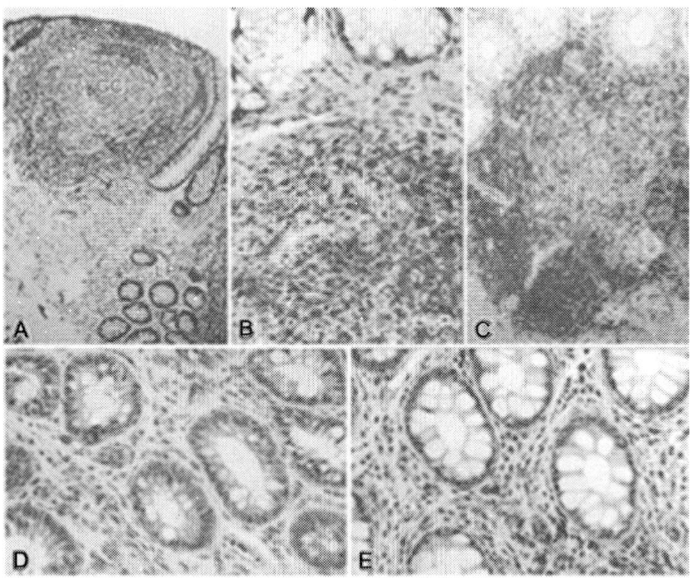

Fig. 5.20 :The Mucosal Cells viewed under the Phase Contrast Microscope in the Author's Lab with a Magnification of ´1200 under H & E stain

ROLE OF BILE AND PANCREATIC SECRETION

Bile is a complex fluid containing water, electrolytes and a battery of organic molecules including bile acids, cholesterol, phospholipids and bilirubin that flows through the biliary tract into the small intestine. There are two fundamentally important functions of bile in all species:

- Bile contains bile acids, which are critical for digestion and absorption of fats and fat-soluble vitamins in the small intestine.
- Many waste products, including bilirubin, are eliminated from the body by secretion into bile and elimination in feces.

Adult humans produce 400 to 800 ml of bile daily, and other animals proportionately similar amounts. The secretion of bile can be considered to occur in two stages:

- Initially, hepatocytes secrete bile into canaliculi, from which it flows into bile ducts. This hepatic bile contains large quantities of bile acids, cholesterol and other organic molecules.
- As bile flows through the bile ducts it is modified by addition of a watery, bicarbonate-rich secretion from ductal epithelial cells.

In species with a gallbladder (man and most domestic animals except horses and rats), further modification of bile occurs in that organ. The gall bladder stores and concentrates bile during the fasting state. Typically, bile is concentrated five-fold in the gall bladder by absorption of water and small electrolytes - virtually all of the the organic molecules are retained.

Secretion into bile is a major route for eliminating cholesterol. Free cholesterol is virtually insoluble in aqueous solutions, but in bile, it is made soluble by bile acids and lipids like lethicin. Gallstones, most of which are composed predominantly of cholesterol, result from processes that allow cholesterol to precipitate from solution in bile.

Role of Bile Acids in Fat Digestion and Absorption

Bile acids are derivatives of cholesterol synthesized in the hepatocyte. Cholesterol, ingested as part of the diet or derived from hepatic synthesis is converted into the bile acids cholic and chenodeoxycholic acids, which are then conjugated to an amino acid (glycine or taurine) to yield the conjugated form that is actively secreted into cannaliculi.

Bile acids are facial amphipathic, that is, they contain both hydrophobic (lipid soluble) and polar (hydrophilic) faces. The cholesterol-derived portion of a bile acid has one face that is hydrophobic (that with methyl groups) and one that is hydrophilic (that with the hydroxyl groups); the amino acid conjugate is polar and hydrophilic.

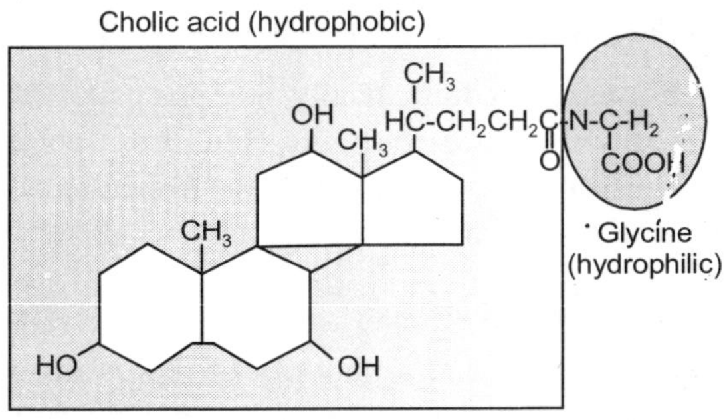

Fig. 5.21 : The Structure of Cholic Acid

Their amphipathic nature enables bile acids to carry out two important functions:

- **Emulsification of lipid aggregates:** Bile acids have detergent action on particles of dietary fat which causes fat globules to break down or be emulsified into minute, microscopic droplets. Emulsification is not digestion per se, but is of importance because it greatly increases the surface area of fat, making it available for digestion by lipases, which cannot access the inside of lipid droplets.
- **Solubilization and transport of lipids in an aqueous environment:** Bile acids are lipid carriers and are able to solubilize many lipids by forming micelles - aggregates of lipids such as fatty acids, cholesterol and monoglycerides - that remain suspended in water. Bile acids are also critical for transport and absorption of the fat-soluble vitamins.

Role of Bile Acids in Cholesterol Homeostasis

Hepatic synthesis of bile acids accounts for the majority of cholesterol breakdown in the body. In humans, roughly 500 mg of cholesterol are converted to bile acids and eliminated in bile every day. This route for elimination of excess cholesterol is probably important in all animals, but particularly in situations of massive cholesterol ingestion.

Interestingly, it has recently been demonstrated that bile acids participate in cholesterol metabolism by functioning as hormones that alter the transcription of the rate-limiting enzyme in cholesterol biosynthesis.

Enterohepatic Recirculation

Large amounts of bile acids are secreted into the intestine every day, but only relatively small quantities are lost from the body. This is because approximately 95% of the bile acids delivered to the duodenum are absorbed back into blood within the ileum.

Venous blood from the ileum goes straight into the portal vein, and hence through the sinusoids of the liver. Hepatocytes extract bile acids very efficiently from sinusoidal blood, and little escapes the healthy liver into systemic circulation. Bile acids are then transported across the hepatocytes to be resecreted into canaliculi. The net effect of this enterohepatic recirculation is that each bile salt molecule is reused about 20 times, often two or three times during a single digestive phase.

It should be noted that liver disease can dramatically alter this pattern of recirculation - for instance, sick hepatocytes have decreased ability to extract bile acids from portal blood and damage to the canalicular system can result in escape of bile acids into the systemic circulation. Assay of systemic levels of bile acids is used clinically as a sensitive indicator of hepatic disease.

Pattern and Control of Bile Secretion

The flow of bile is lowest during fasting, and a majority of that is diverted into the gallbladder for concentration. When chyme from an ingested meal enters the small intestine, acid and partially digested fats and proteins stimulate secretion of cholecystokinin and secretin. As discussed previously, these enteric hormones have important effects on pancreatic exocrine secretion. They are both also important for secretion and flow of bile:

- **Cholecystokinin**: The name of this hormone describes its effect on the biliary system - cholecysto = gallbladder and kinin = movement. The most potent stimulus for release of cholecystokinin is the presence of fat in the duodenum. Once released, it stimulates contractions of the gallbladder and common bile duct, resulting in delivery of bile into the gut.
- **Secretin**: This hormone is secreted in response to acid in the duodenum. Its effect on the biliary system is very similar to what was seen in the pancreas - it simulates biliary duct cells to secrete bicarbonate and water, which expands the volume of bile and increases its flow out into the intestine.

The processes of gallbladder filling and emptying described here can be visualized using an imaging technique called scintography. This procedure is utilized as a diagnostic aid in certain types of hepatobiliary disease.

Pancreatic juice is composed of two secretory products critical to proper digestion: digestive enzymes and bicarbonate. The enzymes are synthesized and secreted from the exocrine acinar cells, whereas bicarbonate is secreted from the epithelial cells lining small pancreatic ducts.

The pancreas secretes a magnificent battery of enzymes that collectively have the capacity to reduce virtually all digestible macromolecules into forms that are capable of, or nearly capable of being absorbed. Three major groups of enzymes are critical to efficient digestion:

Proteases

Digestion of proteins is initiated by pepsin in the stomach, but the bulk of protein digestion is due to the pancreatic proteases. Several proteases are synthesized in the pancreas and secreted into the lumen of the small intestine. The two major pancreatic proteases are trypsin and chymotrypsin, which are synthesized and packaged into secretory vesicles as an the inactive proenzymes trypsinogen and chymotrypsinogen. As you might anticipate, proteases are rather dangerous enzymes to have in cells, and packaging of an inactive precursor is a way for the cells to safely handle these enzymes. The

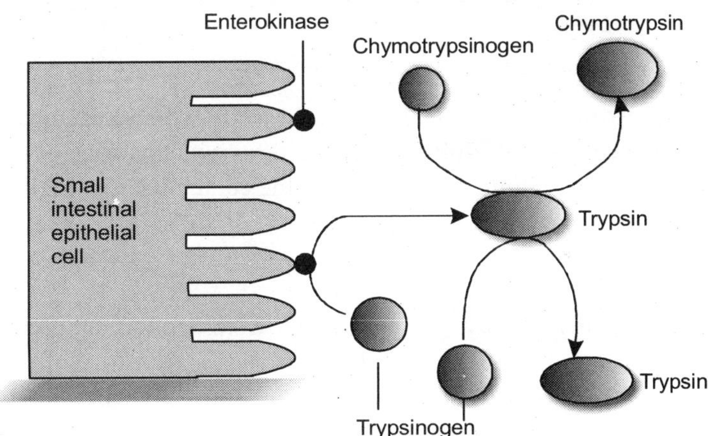

Fig. 5.22 : The action of Pancreas during Digestion

secretory vesicles also contain a trypsin inhibitor which serves as an additional safeguard should some of the trypsinogen be activated to trypsin; following exocytosis this inhibitor is diluted out and becomes ineffective - the pin is out of the grenade. Once trypsinogen and chymotrypsinogen are released into the lumen of the small intestine, they must be converted into their active forms in order to digest proteins. Trypsinogen is activated by the enzyme enterokinase, which is embedded in the intestinal mucosa.

Once trypsin is formed it activates chymotrypsinogen, as well as additional molecules of trypsinogen. The net result is a rather explosive appearance of active protease once the pancreatic secretions reach the small intestine.

Trypsin and chymotrypsin digest proteins into peptides and peptides into smaller peptides, but they cannot digest proteins and peptides to single amino acids. Some of the other proteases from the pancreas, for instance carboxypeptidase, have that ability, but the final digestion of peptides into amino acids is largely the effect of peptidases on the surface of small intestinal epithelial cells. More on this later.

Pancreatic Lipase

A major component of dietary fat is triglyceride, or neutral lipid. A triglyceride molecule cannot be directly absorbed across the intestinal mucosa. Rather, it must first be digested into a 2-monoglyceride and two free fatty acids. The enzyme that performs this hydrolysis is pancreatic lipase, which is delivered into the lumen of the gut as a constituent of pancreatic juice.

Sufficient quantities of bile salts must also be present in the lumen of the intestine in order for lipase to efficiently digest dietary triglyceride and for the resulting fatty acids and monoglyceride to be absorbed. This means that normal digestion and absorption of dietary fat is critically dependent on secretions from both the pancreas and liver.

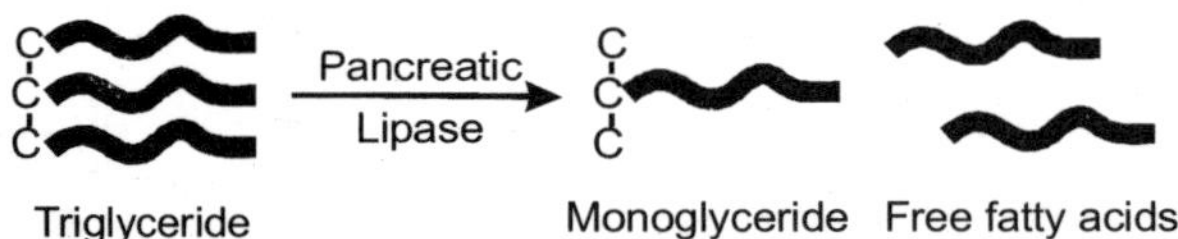

Pancreatic lipase has recently been in the limelight as a target for management of obesity. The drug orlistat (Xenical) is a pancreatic lipase inhibitor that interferes with digestion of triglyceride and thereby reduces absorption of dietary fat. Clinical trials support the contention that inhibiting lipase can lead to significant reductions in body weight in some patients.

Amylase

The major dietary carbohydrate for many species is starch, a storage form of glucose in plants. Amylase (technically alpha-amylase) is the enzyme that hydrolyses starch to maltose (a glucose-glucose disaccharide), as well as the trisaccharide maltotriose and small branchpoints fragments called limit dextrins. The major source of amylase in all species is pancreatic secretions, although amylase is also present in saliva of some animals, including humans.

Other Pancreatic Enzymes

In addition to the proteases, lipase and amylase, the pancreas produces a host of other digestive enzymes, including ribonuclease, deoxyribonuclease, gelatinase and elastase.

Bicarbonate and Water

Epithelial cells in pancreatic ducts are the source of the bicarbonate and water secreted by the pancreas. Bicarbonate is a base and critical to neutralizing the acid coming into the small intestine from the stomach. The mechanism underlying bicarbonate secretion is essentially the same as for acid secretion parietal cells and is dependent on the enzyme carbonic anhydrase. In pancreatic duct cells, the bicarbonate is secreted into the lumen of the duct and hence into pancreatic juice.

TRANSPORT OF LIPIDS

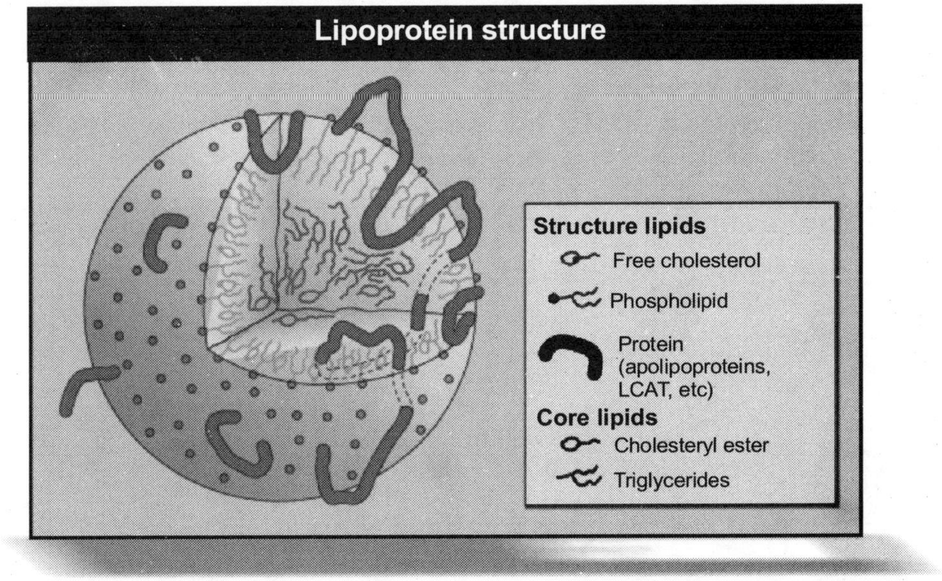

Fig. 5.23 :The Structure of lipoprotein

Once lipids are disassembled in the intestinal lumen and mucosal cell (enterocyte) they are reassembled in the mucosal cell as chylomicrons (CM's) and very low density lipoproteins (VLDL's). These vehicles contain primarily nonpolar cholesterol esters and triglycerides in the core and polar cholesterol, protein, and phospholipids in their membranes. They are transported via the lymph and blood circulation to the liver, fat depots, and muscles. There the endothelial enzyme lipoprotein lipase removes the lipid contents.

Lipid carrying vehicles are also made by the liver primarily as very low density lipoproteins (VLDL) and these function to move lipids made by the body itself into tissues. On the other hand, high density lipoproteins (HDL), which are made in the intestines and liver, function primarily to reverse this process and transport lipids from tissue to liver hepatocytes.1 HDL's are of two types HDL3 and HDL2. HDL3 is an empty package composed of a bilayer lipid membrane plus proteins. Lysolecithin cholesterol acyl transferase (LCAT) and apoprotein A associated with HDL3 remove free cholesterol from the blood, esterify it and fill the HDL3 package.The LCAT enzyme uses the fatty acid in the number two position of lecithin to esterify to cholesterol. If this fatty acid is saturated, the process is inhibited if it is unsaturated, the process is enhanced. Thus, cholesterol blood clearing by HDL3 is linked to dietary intake of saturated and unsaturated fatty acids. High saturated triglycerides are often clinically associated with high blood cholesterol levels.As HDL3 swells with cholesterol ester, it becomes HDL2, which in the liver releases its cholesterol through the action of hepatic lipase. Released cholesterol is conjugated with the amino acids glycine (predominantly in most species) and taurine (predominantly in cats) to form bile salts which are then excreted in the bile into the small intestine.2,3 Some cholesterol is then reabsorbed via the enterohepatic circulation and some passes with the feces. The less reabsorbed, the lower the blood levels of cholesterol. A variety of complex factors influences the reuptake of bile cholesterol. For example, some of the beneficial effects of fiber and certain bowel microorganisms can be related to decreasing cholesterol uptake.4,5Characterization of lipid transport vehicles is based on physical density, size and ratios of constituents. Chylomicrons are the largest particles, the very low density lipoprotein (VLDL) is the next largest, the intermediate density lipoprotein (IDL) is the next largest, the low density lipoprotein (IDL) is the next largest, and then high density lipoproteins (HDL) are the smallest. In terms of their constituents, as the particle becomes smaller as it is hydrolyzed by

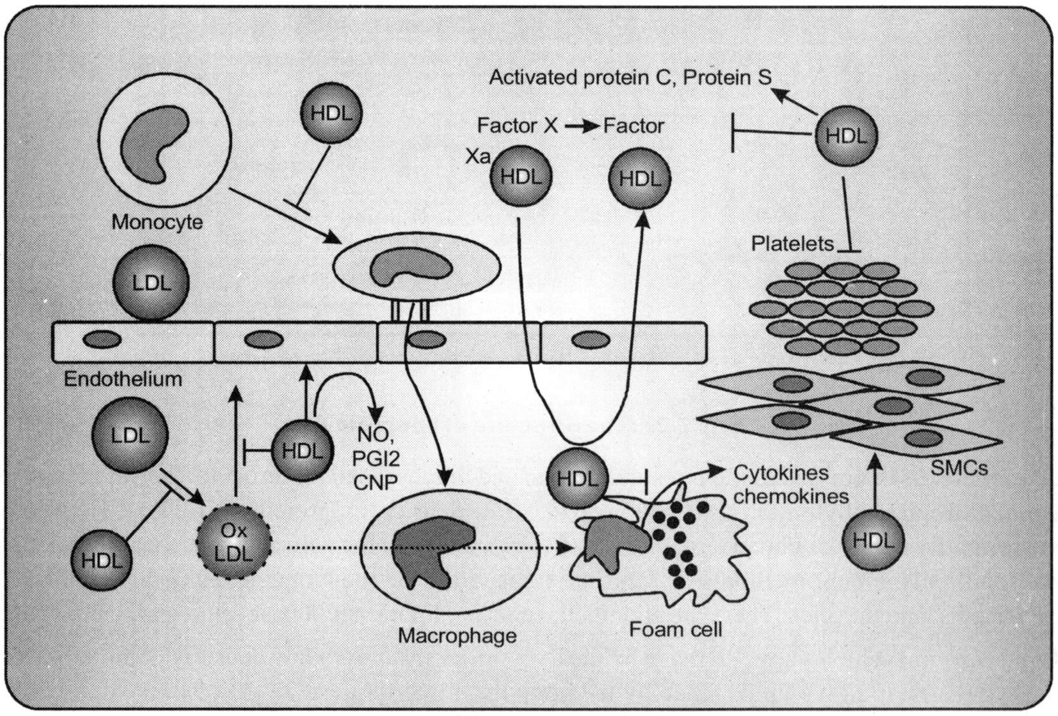

Fig. 5.24 : The Action of Lipoprotein

lipoprotein lipase on capillary endothelial cells, its protein and cholesterol content becomes greater, triglyceride content becomes smaller and its density increases.6 Thus chylomicrons are laden with lipid but lean of protein, whereas high density lipoproteins contain smaller amounts of lipid and larger measures of protein.

Diagnostically the measure of these lipid carriers in the blood is important as indicators of risk particularly to cardiovascular disease. If there are high levels of LDL's, this would be unfavorable whereas high levels of HDL's would be favorable. High levels of LDL's mean that there is a large amount of circulating cholesterol which may have atherogenic potential. On the other hand, a high level of HDL's would mean that lipid stores are being mobilized from tissue and metabolized in the liver to be excreted in the bile.

CLASSIFICATION OF LIPOPROTEINS

There are five major lipoproteins, each of which has a different function.

Chylomicrons — Chylomicrons are very large particles that carry dietary lipid. They are associated with a variety of apolipoproteins, including A-I, A-II, A-IV, B-48, C-I, C-II, C-III, and E.

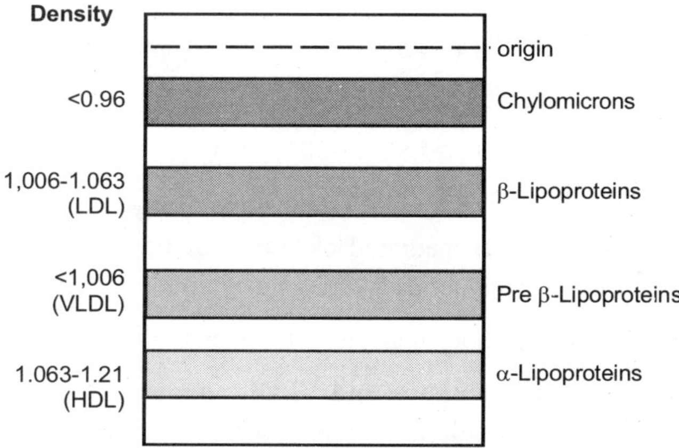

Fig. 5.25 : The Plasma Lipoproteins

Very low density lipoprotein — Very low density lipoprotein (VLDL) carries endogenous triglycerides and to a lesser degree cholesterol. The major apolipoproteins associated with VLDL are B-100, C-I, C-II, C-III,andE.

Intermediate density lipoprotein — Intermediate density lipoprotein (IDL) carries cholesterol esters and triglycerides.It is associated with apolipoproteins B-100, C-III, and E.

Low density lipoprotein — Low density lipoprotein (LDL) carries cholesterol esters and is associated with apolipoprotein B-100.

High density lipoprotein — High density lipoprotein (HDL) also carries cholesterol esters. It is associated with apolipoproteins A-I, A-II, C-I, C-II, C-III, D, and E.

Apolipoproteins — The major function of the different apolipoproteins can be summarized as follows: Understanding these functions is important clinically, because defects in apolipoprotein metabolism lead to abnormalities in lipid handling..The assembly and secretion of apolipoprotein B containing lipoproteins in the liver and intestines is dependent upon microsomal triglyceride transfer protein which transfers lipids to apolipoprotein B. In one study, apolipoprotein B and microsomal transfer protein genes were expressed in the human heart, strongly suggesting that the heart synthesizes and secretes apolipoprotein B containing lipoproteins. This may represent a pathway of "reverse triglyceride transport" by which the cardiac myocytes can unload surplus fatty acids not required for fuel.

- A-I — Structural protein for HDL; activator of lecithin-cholesterol acyltransferase (LCAT).
- A-II — Structural protein for HDL; activator of hepatic lipase.
- A-IV — Activator of lipoprotein lipase and LCAT.
- B-100 — Structural protein for VLDL, IDL, LDL, and Lp(a); ligand for the LDL receptor; required for assembly and secretion of VLDL.
- B-48 — Contains 48 percent of B-100; required for assembly and secretion of chylomicrons; does not bind to LDL receptor.
- C-I — Activator of LCAT.
- C-II — Essential cofactor for LPL.
- C-III — Interferes with apo-E mediated clearance of triglyceride-enriched lipoproteins by cellular receptors; inhibits triglyceride hydrolysis by lipoprotein lipase and hepatic lipase.
- D — May be a cofactor for cholesteryl ester transfer protein.
- E — Ligand for hepatic chylomicron and VLDL remnant receptor, leading to clearance of these lipoproteins from the circulation; ligand for LDL receptor. There are three different apo E alleles in humans: E2, which has cysteine residues at positions 112 and 158; E3, which occurs in 60 to 80 percent of Caucasians and has cysteine at position 112 and arginine at position 158; and E4, which has arginine residues at positions 112 and 158 . These alleles encode for a combination of apo E isoforms that are inherited in a codominant fashion. Compared to apo E3, apo E2 has reduced affinity and apo E4 has enhanced affinity for the LDL (apo B/E) receptor. These isoforms are important clinically because apo E2 is associated with familial dysbetalipoproteinemia (due to less efficient clearance of VLDL and chylomicrons) and apo E4 is associated with an increased risk of hypercholesterolemia and coronary heart disease. (See "Primary disorders of LDL-cholesterol metabolism", section on Polygenic hypercholesterolemia, and see "Approach to the patient with hypertriglyceridemia", section on Familial dysbetalipoproteinemia).
- Apo(a) — Structural protein for Lp(a); inhibitor of plasminogen activation on Lp(a).

CHEMICAL STRUCTURE OF LIPOPROTEIN

The general structure of a lipoprotein includes, as depicted at right: a core consisting of a droplet of triacylglycerols and/or cholesteryl esters a surface monolayer of phospholipid, unesterified cholesterol and specific proteins (apolipoproteins, e.g., apoprotein B-100 in low density lipoprotein).

Lipoproteins differ in the ratio of protein to lipids, and in the particular apoproteins and lipids that they contain. They are classified based on their density:

- chylomicron (largest; lowest in density due to high lipid/protein ratio; highest in triacylglycerols as % of weight)
- VLDL (very low density lipoprotein; 2nd highest in triacylglycerols as % of weight)
- IDL (intermediate density lipoprotein)
- LDL (low density lipoprotein, highest in cholesteryl esters as % of weight)
- HDL (high density lipoprotein, highest in density due to high protein/lipid ratio).

Apolipoprotein Structure: Amphipathic a-helices (polar along one surface of a helix and hydrophobic along the other side) are common structural motifs. One view is that these a-helices may float on the phospholipid surface of the lipoprotein. Some domains of apolipoproteins have roles in interaction of lipoproteins with cell surface receptors.

Apolipoprotein A-I (apoA-I) of human HDL, in the absence of lipid, is found consist of an N-terminal antiparallel 4-helix bundle and a C-terminal domain that is also a-helical, as depicted at right.

A truncated apoA-I, engineered to lack the first 43 amino acids, was earlier found to have a more open structure, with a horseshoe shape, shown below right. Lack of the first a-helix at the N-terminus may prevent stabilization of the 4-helix bundle. On interacting with lipid, the compact structure of the intact apolipoprotein A-I is assumed to open up into a structure resembling the horseshoe shape observed for the truncated protein.

In the open configuration, proline residues are observed to interrupt a-helical segments, providing curvature that would be appropriate for wrapping around a spherical or elliptical lipid micelle.

A strip of hydrophobic residues runs along one edge of the amphipathic a-helix. In the crystal, antiparallel dimers were found to be formed by association of these hydrophobic residues. At right is a view of such a dimer in cartoon display. At the far right the same view of the apoA-I dimer is displayed as spacefill, with hydrophobic residues colored magenta and polar residues cyan.

Apolipoprotein E (apoE), a constituent of several classes of lipoprotein also has an N-terminal domain that folds as a 4-helix bundle in the absence of lipid. Based in part on a low resolution structure determined in the presence of phospholipids, it has been proposed that interaction with lipids converts apoE to an a-helical hairpin that wraps around the lipid particles.

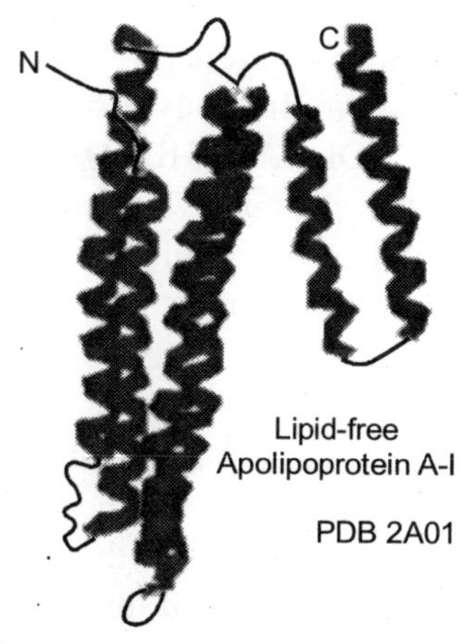

Fig. 5.26 : The Structure of Apo lipoprotein

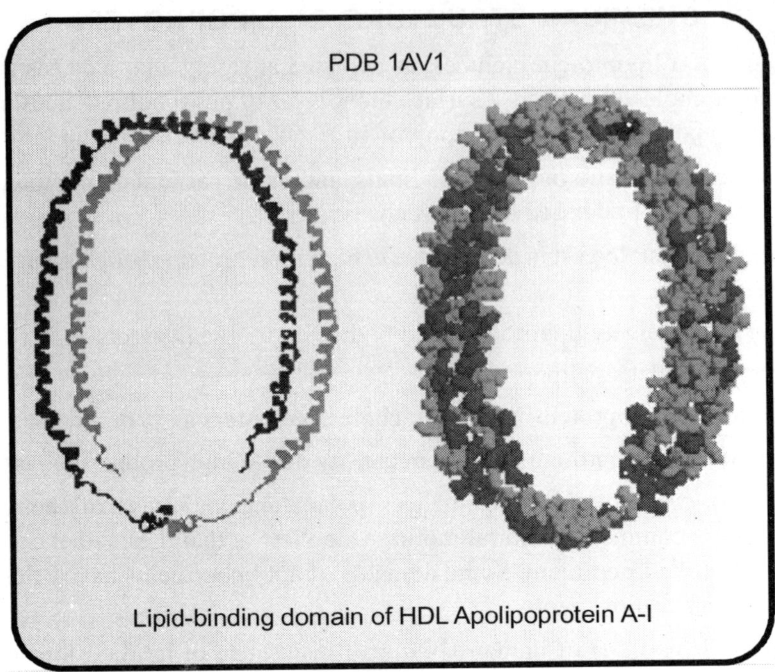

Fig. 5.27: Apolipoprotein A-I Lacking Residues 1-43, Structure

There is special interest in the structure and stability of apolipoprotein E. In addition to being a constituent of various lipoproteins, e.g. VLDL and HDL, a variant of apolipoprotein E, designated apoE4, is implicated in Alzheimer's disease and other neurological conditions.

METABOLISMS OF CHYLOMICRONS

Chylomicrons are large lipoprotein particles that transport dietary lipids from the intestines to other locations in the body. Chylomicrons are one of the 5 major groups of lipoproteins (chylomicrons, VLDL, IDL, LDL, HDL) which enable fats and cholesterol to move within the water based solution of the blood streamChylomicrons transport exogenous lipids to liver, adipose, cardiac and skeletal muscle tissue where their triglyceride components are unloaded by the activity of lipoprotein lipase. Consequently chylomicron remnants are left over which are taken up by the liver.

VLDL, LDL, HDL

Very low-density lipoprotein (VLDL) is a type of lipoprotein made by the liver. VLDL is one of the five major groups of lipoproteins (chylomicrons, VLDL, IDL, LDL, HDL) which enable fats and cholesterol to move within the water based solution of the blood stream. It is assembled in the liver from cholesterol and apolipoproteins. It is converted in the bloodstream to low-density lipoprotein (LDL). VLDL particles have a diameter of 30-80 nm. VLDL transports endogenous products where chylomicrons transport exogenous (dietary) products.VLDL transports endogenous triglycerides, phospholipids, cholesterol and cholesteryl esters. It functions as the body's internal transport mechanism for lipids.

Nascent VLDL circulates in blood and picks up apolipoprotein C-II and apolipoprotein E donated from High-Density Lipoprotein (HDL). At this point, the nascent VLDL becomes a mature VLDL. Once in circulation, the VLDL will come in contact with lipoprotein lipase (LPL) in the capillary beds in the body (adipose, cardiac, and skeletal muscle). The LPL will remove triglycerides from the VLDL for storage or energy production.

The VLDL now meets back up with HDL where apoC-II is transferred back to the HDL (but keeps apoE). In addition to this, the HDL transfers cholesteryl esters to the VLDL in exchange for phospholipids and triglycerides (via cholesteryl ester transfer protein).As more and more triglycerides are removed from the VLDL because of the action of the LPL enzyme, the composition of the molecule changes, and it becomes intermediate density lipoprotein (IDL). Fifty per cent of IDL are recognized by receptors in the liver cells (because of the apoB-100 and apoE they contain) and are endocytosed.

The other 50% of IDL lose their apoE. When their cholesterol content becomes greater than the triglyceride content, they become low-density lipoprotein (LDL), with the primary apolipoprotein being apoB-100. The LDL is taken into a cell via the LDL receptor (endocytosis) where the contents are either stored, used for cell membrane structure, or converted into other products (steroid hormones or bile acids) VLDL levels have been correlated with accelerated rates of atherosclerosis, and are elevated in a number of diseases and metabolic states.

Low-density lipoprotein (LDL) is a type of lipoprotein that transports cholesterol and triglycerides from the liver to peripheral tissues. The LDL is one of the five major groups of lipoproteins; these groups include chylomicrons, very low-density lipoprotein (VLDL), intermediate-density lipoprotein (IDL), low-density lipoprotein, and high-density lipoprotein (HDL). Like all lipoproteins, LDL enables fats and cholesterol to move within the water based solution of the blood stream. LDL also regulates cholesterol synthesis at these sites. It commonly appears in the medical setting as part of a cholesterol blood test, and since high levels of LDL cholesterol can signal medical problems like cardiovascular disease, it is sometimes called "bad cholesterol" (as opposed to high-density lipoprotein (HDL), which is frequently referred to as the "good cholesterol").

Each native LDL particle contains a single apolipoprotein B-100 molecule (Apo B-100, a protein with 4536 amino acid residues) that circulates the fatty acids, keeping them soluble in the aqueous environment. In addition, LDL has a highly-hydrophobic core consisting of polyunsaturated fatty acid known as linoleate and about 1500 esterified cholesterol molecules. This core is surrounded by a shell of phospholipids and unesterified cholesterol as well as a single copy of B-100 large protein (514 kD). LDL particles are approximately 22 nm in diameter and have a mass of about 3 million daltons, but since LDL particles contain a changing number of fatty acids, they actually have a mass and size distribution.

The LDL particles vary in size and density, and studies have shown that a pattern that has more small dense LDL particles—called "Pattern B"—equates to a higher risk factor for coronary heart disease (CHD) than does a pattern with more of the larger and less dense LDL particles ("Pattern A"). This is because the smaller particles are more easily able to penetrate the endothelium. "Pattern I," meaning "intermediate," indicates that most LDL particles are very close in size to the normal gaps in the endothelium (26 nm). The correspondence between Pattern B and CHD has been suggested by some in the medical community to be stronger than the correspondence between the LDL number measured in the standard lipid profile test.

Tests to measure these LDL subtype patterns have been more expensive and not widely available, so the common lipid profile test has been used more commonly. There has also been noted a correspondence between higher triglyceride levels and higher levels of smaller, denser LDL particles and alternately lower triglyceride levels and higher levels of the larger, less dense LDL. With continued research, decreasing cost, greater availability and wider acceptance of other "lipoprotein subclass analysis" assay methods, including NMR spectroscopy, research studies have continued to show a stronger correlation between human clinically obvious cardiovascular event and quantitatively-measured particle concentrations. When a cell requires cholesterol, it synthesizes the necessary LDL receptors, and inserts them into the plasma membrane. The LDL receptors diffuse freely until they associate with clathrin-coated pits. LDL particles in the blood stream bind to these extracellular LDL receptors.

The clathrin-coated pits then form vesicles that are endocytosed into the cell. After the clathrin coat is shed, the vesicles deliver the LDL and their receptors to early endosomes, onto late endosomes to lysosomes. Here the cholesterol esters in the LDL are hydrolysed. The LDL receptors are recycled back to the plasma membrane. Because LDLs transport cholesterol to the arteries and can be retained there by arterial proteoglycans starting the formation of plaques, increased levels are associated with atherosclerosis, and thus heart attack, stroke, and peripheral vascular disease. For this reason, cholesterol inside LDL lipoproteins is often called "bad" cholesterol. This is a misnomer.

The cholesterol transported on LDL is the same as cholesterol transported on other lipoprotein particles. The cholesterol itself is not "bad"; rather, it is how and where the cholesterol is being transported, and in what amounts over time, that causes adverse effects. Increasing evidence has revealed that the concentration and size of the LDL particles more powerfully relates to the degree of atherosclerosis progression than the concentration of cholesterol contained within all the LDL particles. The healthiest pattern, though relatively rare, is to have small numbers of large LDL particles and no small particles. Having small LDL particles, though common, is an unhealthy pattern; high concentrations of small LDL particles (even though potentially carrying the same total cholesterol content as a low concentration of large particles) correlates with much faster growth of atheroma, progression of atherosclerosis and earlier and more severe cardiovascular disease events and death.

The LDL is formed as VLDL lipoproteins lose triglyceride through the action of lipoprotein lipase (LPL) and become smaller and denser, containing a higher proportion of cholesterol. A hereditary form of high LDL is familial hypercholesterolemia (FH). Increased LDL is termed hyperlipoproteinemia type II (after the dated Fredrickson classification). LDL poses a risk for cardiovascular disease when it invades the endothelium and becomes oxidized, since the oxidized form is more easily retained by the proteoglycans. A complex set of biochemical reactions regulates the oxidation of LDL, chiefly stimulated by presence of free radicals in the endothelium. Nitric oxide down-regulates this oxidation process catalyzed by L-arginine.

In a corresponding manner, when there are high levels of asymmetric dimethylarginine in the endothelium, production of nitric oxide is inhibited and more LDL oxidation occurs. The mevalonate pathway serves as the basis for the biosynthesis of many molecules, including cholesterol. The enzyme 3-hydroxy-3-methylglutaryl coenzyme A reductase (HMG CoA reductase) is an essential component in the pathway. The use of statins (HMG-CoA reductase inhibitors) is effective against high levels of LDL cholesterol. Statins inhibit the enzyme HMG-CoA reductase in the liver, which stimulates LDL receptors, resulting in an increased clearance of LDL.

Clofibrate is effective at lowering cholesterol levels, but has been associated with significantly increased cancer and stroke mortality, despite lowered cholesterol levels.

Torcetrapib was a drug developed to treat high cholesterol levels, but its development was halted when studies showed a 60% increase in deaths when used in conjunction with atorvastatin versus the statin alone.

Niacin (B_3), lowers LDL by selectively inhibiting hepatic diacyglycerol acyltransferase 2, reducing triglyceride synthesis and VLDL secretion through a receptor HM74 and HM74A or GPR109A.

Insulin induces HMG-CoA reductase activity, whereas glucagon downregulates it. While glucagon production is stimulated by dietary protein ingestion, insulin production is stimulated by dietary carbohydrate. The rise of insulin is, in general, determined by the unfolding of carbohydrates into glucose during the process of digestion. Glucagon levels are very low when insulin levels are high.

A ketogenic diet may have similar response to taking niacin (lowered LDL and increased HDL) through beta-hydroxybutyrate, a ketone body, coupling the niacin receptor (HM74A).

Lowering the blood lipid concentration of triglycerides helps lower the amount of LDL, because VLDL gets converted in the bloodstream into LDL.

Fructose, a component of sucrose as well as high-fructose corn syrup, upregulates hepatic VLDL synthesis.

Because LDL appears to be harmless until oxidized by free radicals, it is postulated that ingesting antioxidants and minimizing free radical exposure may reduce LDL's contribution to atherosclerosis, though results are not conclusive.

Chemical measures of lipid concentration have long been the most-used clinical measurement, not because they have the best correlation with individual outcome, but because these lab methods are less expensive and more widely available. However, there is increasing evidence and recognition of the value of more sophisticated measurements. To be specific, LDL particle number (concentration), and to a lesser extent size, have shown much tighter correlation with atherosclerotic progression and cardiovascular events than is obtained using chemical measures of total LDL concentration contained within the particles. LDL cholesterol concentration can be low, yet LDL particle number high and cardiovascular events rates are high. Also, LDL cholesterol concentration can be relatively high, yet LDL particle number low and cardiovascular events are also low. If LDL particle concentration is tracked against event rates, many other statistical correlates of cardiovascular events, such as diabetes mellitus, obesity, and smoking, lose much of their additional predictive power.

The lipid profile does not measure LDL level directly but instead estimates it via the Friedewald equation using levels of other cholesterol such as HDL:

$$LDL - C \approx \text{Total cholesterol} - HDL - C - 0.20 * \text{Total triglycerides}$$

In mg/dl: LDL cholesterol = total cholesterol – HDL cholesterol – (0.20 × triglycerides)

In mmol/l: LDL cholesterol = total cholesterol – HDL cholesterol – (0.45 × triglycerides)

There are limitations to this method, most notably that samples must be obtained after a 12 to 14 h fast and that LDL-C cannot be calculated if plasma triglyceride is >4.52 mmol/L (400 mg/dL). Even at LDL-C levels 2.5 to 4.5 mmol/L, this formula is considered to be inaccurate. If both total cholesterol and triglyceride levels are elevated then a modified formula may be used

In mg/dl: LDL-C = Total-C – HDL-C – (0.16 × Trig)

This formula provides an approximation with fair accuracy for most people, assuming the blood was drawn after fasting for about 14 hours or longer. (However, the concentration of LDL particles, and to a lesser extent their size, has far tighter correlation with clinical outcome than the content of cholesterol with the LDL particles, even if the LDL-C estimation is about correct.)

In the USA, the American Heart Association, NIH, and NCEP provide a set of guidelines for fasting LDL-Cholesterol levels, estimated or measured, and risk for heart disease. As of 2003, these guidelines were:

Level mg/dL	Level mmol/L	Interpretation
<100	<2.6	Optimal LDL cholesterol, corresponding to reduced, but not zero, risk for heart disease
100 to 129	2.6 to 3.3	Near optimal LDL level
130 to 159	3.3 to 4.1	Borderline high LDL level
160 to 189	4.1 to 4.9	High LDL level
>190	>4.9	Very high LDL level, corresponding to highest increased risk of heart disease

These guidelines were based on a goal of presumably decreasing death rates from cardiovascular disease to less than 2% to 3% per year or less than 20% to 30% every 10 years. Note that 100 is not considered optimal; less than 100 is optimal, though it is unspecified how much less.

Over time, with more clinical research, these recommended levels keep being reduced because LDL reduction, including to abnormally low levels, has been the most effective strategy for reducing cardiovascular death rates in large double blind, randomized clinical trials; far more effective than coronary angioplasty/stenting or bypass surgery.

For instance, for people with known atherosclerosis diseases, the 2004 updated American Heart Association, NIH and NCEP recommendations are for LDL levels to be lowered to less than 70 mg/dL, unspecified how much lower. It has been estimated from the results of multiple human pharmacologic LDL lowering trials that LDL should be lowered to about 50 to reduce cardiovascular event rates to near zero. For reference, from longitudinal population studies following progression of atherosclerosis-related behaviours from early childhood into adult hood it has been discovered that the usual LDL in childhood, before the development of fatty streaks, is about 35 mg/dL. However, all the above values refer to chemical measures of lipid/cholesterol concentration within LDL, not LD Lipoprotein concentrations, probably not the better approach.

High-density lipoproteins (HDL) is one of the five major groups of lipoproteins (chylomicrons, VLDL, IDL, LDL, HDL) which enable lipids like cholesterol and triglycerides to be transported within the water based blood stream. In healthy individuals, about thirty percent of blood cholesterol is carried by HDL.

It is hypothesized that HDL can remove cholesterol from atheroma within arteries and transport it back to the liver for excretion or re-utilization—which is the main reason why HDL-bound cholesterol is sometimes called "**good cholesterol**", or HDL-C. A high level of HDL-C seems to protect against cardiovascular diseases, and low HDL cholesterol levels (less than 40 mg/dL) increase the risk for heart disease. When measuring cholesterol, any contained in HDL particles is considered as protection to the body's cardiovascular health, in contrast to "bad" LDL cholesterol. HDL are the smallest of the lipoprotein particles. They are the densest because they contain the highest proportion of protein. Their most abundant apolipoproteins are apo A-I and apo A-II. The liver synthesizes these lipoproteins as complexes of apolipoproteins and phospholipid, which resemble cholesterol-free flattened spherical lipoprotein particles. They are capable of picking up cholesterol, carried internally, from cells by interaction with the ATP Binding Cassette Transporter A1 (ABCA1). A plasma enzyme called lecithin-cholesterol acyltransferase (LCAT) converts the free cholesterol into cholesteryl ester (a more hydrophobic form of cholesterol) which is then sequestered into the core of the lipoprotein particle eventually making the newly synthesized HDL spherical. They increase in size as they circulate through the bloodstream and incorporate more cholesterol and phospholipid molecules from cells and other lipoproteins, for example by the interaction with the ABCG1 transporter and the phospholipid transport protein (PLTP).

HDL deliver their cholesterol mostly to the liver or steroidogenic organs such as adrenals, ovary and testes by direct and indirect pathways. The direct HDL removal pathways involve HDL receptors such as scavenger receptor BI (SR-BI) which mediate the selective uptake of cholesterol from HDL. In humans, the probably most relevant pathway is the indirect one, which is mediated by cholesteryl ester transfer protein (CETP). This protein exchanges triglycerides of VLDL against cholesteryl esters of HDL. As the result, VLDL are processed to LDL which are removed from the circulation by the LDL receptor pathway. The triglycerides are not stable in HDL, but degraded by hepatic lipase so that finally small HDL particles are left which restart the uptake of cholesterol from cells.

The cholesterol delivered to the liver is excreted into the bile and hence intestine either directly or indirectly after conversion into bile acids. Delivery of HDL cholesterol to adrenals, ovaries and testes are important for the synthesis of steroid hormones.

Several steps in the metabolism of HDL can contribute to the transport of cholesterol from lipid laden macrophages of atherosclerotic arteries, termed foam cells to the liver for secretion into the bile. This pathway has been termed reverse cholesterol transport and is considered as the classical protective function of HDL towards atherosclerosis. However, HDL carries many lipid and protein species, many of which have very low concentrations but are biologically very active. For example, HDL and their protein and lipid constituents help to inhibit oxidation, inflammation, activation of the endothelium, coagulation or platelet aggregation. All these properties may contribute to the ability of HDL to protect from atherosclerosis, and it is not yet known what is most important. In the stress response, serum amyloid A, which is one of the acute phase proteins and an apolipoprotein, is under the stimulation of cytokines (IL-1, IL-6) and cortisol produced in the adrenal cortex and carried to the damaged tissue incorporated into HDL particles. At the inflammation site, it attracts and activates leukocytes. In chronic inflammations, its deposition in the tissues manifests itself as amyloidosis. It has been postulated that the concentration of large HDL particles more accurately reflects protective action, as opposed to the concentration of total HDL particles. This ratio of large HDL to total HDL particles varies widely and is only measured by more sophisticated lipoprotein assays using either electrophoresis (the original method developed in the 1970s), or newer NMR spectroscopy methods, developed in the 1990s. Men tend to have noticeably lower HDL levels, with smaller size and lower cholesterol content, than women. Men also have an increased incidence of atherosclerotic heart disease. Epidemiological studies have shown that high concentrations of HDL (over 60 mg/dL) have protective value against cardiovascular diseases such as ischemic stroke and myocardial infarction. Low concentrations of HDL (below 40 mg/dL for men, below 50 mg/dL for women) increase the risk for atherosclerotic diseases.Data from the landmark Framingham Heart Study showed that for a given level of LDL, the risk of heart disease increases 10-fold as the HDL varies from high to low. Conversely, for a fixed level of HDL, the risk increases 3-fold as LDL varies from low to high.

Even people with very low LDL levels are exposed to some increased risk if their HDL levels are not high enough.

The American Heart Association, NIH and NCEP provides a set of guidelines for male fasting HDL levels and risk for heart disease.

Level mg/dL	Level mmol/L	Interpretation
<40	<1.03	Low HDL cholesterol, heightened risk for heart disease, <50 is the value for women
40-59	1.03-1.52	Medium HDL level
>60	>1.55	High HDL level, optimal condition considered protective against heart disease

Many laboratories used a two-step method : chemical precipitation of lipoproteins containing apoprotein B, then calculating HDL as cholesterol remaining in the supernate but there are also direct methods. Labs use the routine dextran sulfate-Mg2+ precipitation method with ultracentrifugation/ dextran sulfate-$Mg2^+$ precipitation as reference method. HPLC can be used Subfractions (HDL-2C, HDL-3C) can be measured and have clinical significance. A link has been shown between level of HDL and onset of dementia. Those with high HDL were less likely to have dementia. Low HDL-C in late-middle age has also been associated with memory loss.

DISORDERS OF LIPOPROTEIN METABOLISM

The most common lipoprotein disorders are hypercholesterolemia (type II hyperlipoproteinemia); hypertriglyceridemia (primarily types IV and V hyperlipoproteinemia); hypoalphalipoproteinemia; and high lipoprotein(a) (Lp[a]) levels. However, hypercholesterolemia and hypoalphalipoproteinemia may not be as prevalent among the elderly as among the general population because mortality risk is so high that patients with these disorders do not survive to old age.

The association between high serum cholesterol levels and coronary artery disease (CAD), whose prevalence and mortality rate increase with age, is well established. High levels of low-density lipoprotein cholesterol (LDL-C) and Lp(a) and a low level (< 35 mg/dL [< 0.91 mmol/L]) of high-density lipoprotein cholesterol (HDL-C) are significant independent positive risk factors for CAD and carotid artery atherosclerosis. A high level (>= 60 mg/dL [>= 1.55 mmol/L]) of HDL-C is a significant independent negative risk factor. However, the total cholesterol (TC)/HDL-C ratio is a more valuable measure of CAD risk than TC or LDL-C levels alone; the risk is higher in men when the ratio is > 6.4 and in women when the ratio is > 5.6.

For the elderly, the predictive value of high cholesterol levels in determining the risk of CAD is unclear, and the value of lowering high cholesterol levels—in terms of quality of life, morbidity, and mortality—is disputed. Some studies suggest that high cholesterol levels are an important risk factor for CAD in the elderly, others suggest that the risk decreases with age, and still others suggest a U-shaped relationship in which both high and low cholesterol levels are associated with an increased risk of morbidity and mortality. Some studies suggest sex differences: the mortality rate was lowest at a TC level of 215 mg/dL (5.55 mmol/L) for elderly men and 270 to 280 mg/dL (7.00 to 7.25 mmol/L) for elderly women.

There may also be a U-shaped relationship between cholesterol levels and death due to stroke. High cholesterol levels have been associated with increased risk of death due to nonhemorrhagic stroke, whereas low cholesterol levels have been associated with increased risk of death due to hemorrhagic stroke.

Studies of lipid-lowering treatments in the elderly, particularly primary prevention studies, have not consistently shown a significant reduction in the overall mortality rate. However, most studies have shown that for patients who have had a myocardial infarction (MI) or who have hemodynamically significant CAD, lowering lipid levels can stop or reverse disease progression and reduce the incidence of CAD events, CAD mortality, and all-cause mortality.

Lipoprotein disorders may be primary (usually familial and usually classified based on lipoprotein elevation pattern or secondary. However, genetic and secondary factors, such as various disorders and drugs, diet, obesity, physical activity, alcohol use, and cigarette smoking, are often interrelated.

The most common cause of secondary hypercholesterolemia is probably a diet high in saturated fat or cholesterol, whether or not a polygenic tendency for hypercholesterolemia exists. Covert hypothyroidism (with normal thyroxine and high thyroid-stimulating hormone levels) is a relatively common cause of high TC and LDL-C levels and is associated with an increased risk of premature MI.

The most common causes of secondary hypertriglyceridemia among the elderly are excessive alcohol intake, exogenous estrogen supplementation, poorly controlled diabetes, uremia, corticosteroid use, and β-blocker use. Isolated low HDL-C levels with normal triglyceride (TG) levels may result from smoking, androgenic steroid use, severe restriction of physical activity, or morbid obesity.

Generalized obesity may cause an increase in TG levels but not in cholesterol levels and, in persons >60, may not contribute to risk of CAD. However, abdominal/upper body (male-pattern) obesity appears to be correlated with increased risk.

Age and sex have a major influence on lipid levels. For persons living in most industrialized countries, cholesterol and TG levels increase through middle age. In men, mean levels of TC increase until about age 50, then plateau, and then decrease starting at about age 70. In women, they increase more gradually up to age 65 to 69, then decrease. Starting at about age 55 to 60, women have higher TC levels than men. The age-related increase in TC, particularly in women, results primarily from an increase in LDL-C levels and much less from a small increase in very-low-density lipoprotein cholesterol (VLDL-C) levels.

TG levels progressively increase from birth through adulthood. The rate of increase is greater in men than in women. TG levels increase until age 55 in men and until about age 70 in women, then decrease, gradually in men.

In men, mean levels of HDL-C decrease at puberty, increase at about age 45, and then level off at age 50 to 59. These changes may be an effect of testosterone; generally, levels of plasma testosterone and HDL-C are positively correlated in adult men. After puberty, women have higher HDL-C levels than men despite a decrease after age 65.

Premenopausal women have lower LDL-C and higher HDL-C levels than men, partly because of endogenous estrogens. This difference may contribute to the lower CAD rates in premenopausal women. At menopause (whether natural or surgical), women lose this protection against CAD: LDL-C and Lp(a) levels increase and HDL-C levels decrease. Patients who have had an MI, a stroke, or other manifestations of significant atherosclerosis (eg, peripheral arterial disease, carotid artery stenosis) before age 60 should be screened for familial lipoprotein abnormalities. The Lp(a) level should be measured in patients who have had an MI or a stroke or who have CAD and in patients who have other major CAD risk factors, because a high Lp(a) level probably acts synergistically with other risk factors. If a lipoprotein abnormality is detected, the physician should try to determine whether the lipoprotein disorder is primary or secondary and assess the patient for other CAD risk factors (e.g., smoking, hypertension, a high-fat diet, physical inactivity). If secondary causes can be ruled out, the disorder is usually one of the common familial hyperlipoproteinemias.

Screening and identification criteria for hypercholesterolemia are controversial. The National Cholesterol Education Program (NCEP) provides guidelines for identifying high TC and LDL-C levels and low and high HDL-C levels.

According to NCEP guidelines, no single cholesterol value should be used to classify a patient clinically, because values may vary from day to day. If the first screening test detects an abnormality, two subsequent tests are recommended. The NCEP does not give guidelines by age group; its guidelines are based on data from middle-aged persons and do not consider the increase in serum lipids that occurs with age. (As a result, 60% of persons > 65 would be classified as candidates for treatment.)

In contrast, the American College of Physicians guidelines recommend a single measurement of TC alone to identify patients who would benefit from lipid-lowering therapy. These guidelines note that evidence is insufficient to recommend or discourage the screening of men and women aged 65 to 75 for primary prevention, and screening is not recommended for persons > 75.

If elderly persons are screened, a full lipid profile should be obtained. For many elderly persons, the predominant reason for a high TC level is a high HDL-C level, not a high LDL-C level; therefore, their risk of CAD is decreased, not increased. Some persons have normal TC and TG levels but an HDL-C level below the 10th percentile and thus have a very high risk of CAD.

TG levels can be accurately measured only after a fast. If the TG level is < 400 mg/dL (< 4.52 mmol/L), the LDL-C component of the lipid profile can be estimated using the Friedewald equation: LDL-C = TC - [HDL-C + (TG/5)].

Basal lipoprotein measurements usually cannot be determined in the following situations: during a fever or major infection; within 4 weeks of an acute MI, a stroke, or major surgery; immediately after acute excessive alcohol intake; in diabetes mellitus that is severely out of control (fasting blood glucose level > 250 mg/dL (> 13.9 mmol/L), glycosylated hemoglobin > 9%); or during rapid weight loss.

Certain characteristic findings (e.g., tendinous, tuberous, or palmar—*Planar xanthomas*; *Arcus juvenilis corneae*) are diagnostically useful. Obesity (with or without essential hypertension), glucose intolerance, and hyperuricemia may indicate primary hypertriglyceridemia or hypoalphalipoproteinemia.

Secondary lipoprotein disorders are common, even among patients with a well-defined primary lipoprotein disorder, and may exacerbate the expression of the primary disorder, particularly severe hypertriglyceridemia. Thus, when a primary lipoprotein disorder is first diagnosed, a physical examination should be performed, and a drug, occupational, family, dietary, and alcohol-intake history should be obtained. Levels of thyroxine, thyroid-stimulating hormone, blood urea nitrogen or creatinine, and fasting blood glucose should be measured, and urinalysis and liver function tests should be performed.

The NCEP guidelines recommend treatment of persons in the top quintile for TC levels (although in the elderly, use of TC levels alone can be misleading) and of persons with an LDL-C level >= 160 mg/dL (>= 4.14 mmol/L) or lower depending on the number of CAD risk factors and the presence of CAD or other atheromatous disease. Patients are considered at high risk if they have two or more CAD risk factors (age > 45 for men or > 55 for women, premature menopause without estrogen replacement therapy, a family history of CAD, current cigarette smoking, hypertension, low HDL-C levels, and diabetes). One risk factor is subtracted if HDL-C levels are >= 60 mg/dL (>= 1.55 mmol/L).

Therefore, NCEP recommendations for initiation of dietary or drug treatment are based on LDL-C levels . The goals of treatment are to lower TC and LDL-C levels to the recommended levels, to lower TG levels to < 500 mg/dL (< 5.65 mmol/L) and thus avoid pancreatitis, and to increase HDL-C levels to > 35 mg/dL (> 0.91 mmol/L), preferably to >= 40 mg/dL (>= 1.03 mmol/L). For postmenopausal women, a high TG level is an important independent risk factor for CAD; thus, lowering the level to < 250 mg/dL (< 2.82 mmol/L) is probably valuable.

Because the NCEP guidelines are not based on data from nor designed specifically for the elderly, several factors should be considered when deciding whether elderly patients with a lipoprotein abnormality should be treated:

- Decisions should not be based primarily on the patient's age. Many elderly patients are physiologically and mentally much younger than their chronologic age.
- Some benefits of treatment may be realized almost immediately: Significant lowering of LDL-C levels can lead to increased endothelial cell production of nitrous oxide (a potent vasodilator) and can reduce platelet aggregation; significant lowering of TG levels sharply decreases plasminogen activator inhibitor activity and increases fibrinolysis. Often, treatment stops progression of atherosclerotic lesions or induces regression within 1 to 2 years.
- If patients have other life-limiting conditions, aggressive lipid-lowering treatment may be inappropriate or less useful.
- Elderly patients who take many drugs may be less likely to comply with drug regimens, and their risk of drug-drug interactions is increased. Although lipid-lowering treatment can reduce long-term costs for some patients, elderly patients with a fixed income may have difficulty affording the better tolerated and more effective drugs.
- Strict lipid-lowering diets may be effective in the elderly but may lead to or exacerbate malnutrition, as may voluntary dietary restriction. Unappetizing low-fat diets can cause anorexia and eventually lead to nutritional deficiencies. The potential morbidity of protein-energy malnutrition probably outweighs that of moderate hypercholesterolemia.

Dietary treatment: Because the elderly have difficulty maintaining adequate caloric and protein intake and are at risk of malnutrition, a moderate approach to diet is generally recommended. Patients can be advised to trim fat from meat; to increase intake of fish, soybean products, and beans; to avoid fried foods; and to use monounsaturated fats (e.g., olive oil). Consumption of foods rich in soluble fibre (e.g., oat bran) may also lower lipid levels. Phytosterols, present in soybean products, and stanol esters, available as a margarine product, decrease cholesterol absorption.

A strict diet may be preferred for some patients. Aerobic exercise should be included as part of treatment because the benefits of diet without exercise may be limited. A low-saturated-fat, low-cholesterol diet should be tried for 6 months before drug therapy is instituted.

For patients with a high LDL-C level, the American Heart Association and NCEP recommend a two-step diet. High-risk patients should start with the step 1 diet. Patients who are already following the equivalent of a step 1 diet should start with the step 2 diet. The full effects of dietary treatment at either step may not be achieved for 8 to 12 weeks; therefore, patients should advance to step 2 only if the therapeutic goal is not reached in 8 to 12 weeks. Weight loss helps lower TC, LDL-C, and TG levels but is at least as difficult for the elderly as for younger persons.

For patients with hypertriglyceridemia, a low HDL-C level, and an increased risk of CAD, dietary interventions are important. The major goal is to increase the HDL-C level, although lowering the TG level alone may be valuable for women > 50. When the TG level is > 1000 mg/dL (> 11.29 mmol/L), a sharp reduction in total fat intake (to < 30 g/day) helps prevent TG-induced pancreatitis. Other goals are to lose weight, because modest weight loss can markedly lower TG levels; to reduce alcohol intake to <= 3 drinks per week or, if TG levels are >= 500 mg/dL (>= 5.65 mmol/L), to discontinue alcohol; and to reduce the intake of total fat, saturated fat, and cholesterol using the step 2 diet.

For patients who have a TG level >= 1000 mg/dL (>= 11.29 mmol/L) with high levels of chylomicron and VLDL-TG or who have the much rarer primary hyperchylomicronemia, total fat intake should be restricted to 10 to 20% of total calories, primarily to prevent TG-induced pancreatitis. Also, the sharp decrease in VLDL-C levels and the increase in HDL-C levels that result may help prevent patients with high levels of chylomicron and VLDL-TG from developing severe premature CAD. Weight loss is crucial for obese patients. For patients with severe hypertriglyceridemia, estrogen is contraindicated.

For patients who have type III hyperlipoproteinemia and are overweight, the single most important strategy is weight loss. Patients should reduce intake not only of total fat, saturated fat, and cholesterol but also of dietary sugars.

For patients with hypoalphalipoproteinemia, the only consistently effective dietary strategy is weight loss, which may increase HDL-C levels. Other strategies to increase HDL-C levels include supplementation with fish oils rich in omega-3 fatty acids (4 to 12 g/day), smoking cessation (or at least reduction), and aerobic activity (three to five 30-minute periods per week). At least initially, exercise should be supervised by a physician. Intake of moderate amounts of alcohol can increase HDL-C levels, but only if hepatic synthetic function is normal.

Drug treatment: If hyperlipoproteinemia persists after secondary causes have been identified and treated when possible and after dietary treatment has been tried, drug treatment should be used. When a single drug is inadequate, two drugs may be required.

Bile acid-binding resins (e.g., cholestyramine, colestipol) interrupt the normal enterohepatic circulation of bile acids and indirectly increase the liver's catabolism of LDL-C by stimulating hepatocytes to synthesize more LDL receptors. These drugs can reduce CAD risk.

Resins are effective and safe as first-line drugs to lower the LDL-C level in patients whose primary abnormality is a high LDL-C level and whose TG level is < 250 mg/dL (< 2.82 mmol/L). However, they may increase the TG level. If the TG level increases to > 300 mg/dL (> 3.38 mmol/L) during resin therapy, nicotinic acid or gemfibrozil (1200 mg/day) may be added, particularly if the patient has high LDL-C and TG levels; the resin and nicotinic acid act synergistically. Alternatively, an HMG-CoA reductase inhibitor (statin) can be used instead of the resin or resin combination therapy.

Resins should be started in small doses (8 to 10 g/day), particularly for the elderly. Many patients respond to the lowest dose, but the dose should be adjusted according to effect on LDL-C levels. Constipation, the most common adverse effect, can usually be avoided by increasing dietary fiber or by using stool softeners. Because resins are not systemically absorbed, they have essentially no systemic adverse effects (except for rare, mild, reversible changes in liver enzymes).

Resins can augment warfarin's effects; however, if a resin and warfarin are taken within a short time, the resin can also bind warfarin. Thus, resins should be used cautiously, if at all, with warfarin-like anticoagulants. Resins should not be given concurrently with exogenous thyroid hormones, sex hormones, prednisone, or digoxin, all of which may be bound in the intestine by resins. These drugs should be given at least 2 hours before the first resin dose of the day.

Resins are probably contraindicated as single-drug therapy in patients with a high LDL-C level and a TG level > 300 mg/dL and in those with severe hemorrhoids or a history of bowel resection or severe constipation; a statin may be used.

Nicotinic acid (niacin) inhibits secretion of VLDL from the liver, lowering VLDL-C and LDL-C levels. In patients with CAD, it reduces the incidence of recurrent MI and all-cause mortality. Nicotinic acid can be used as a first-line drug, but it is usually added to resin therapy if the resin does not lower LDL-C levels sufficiently. The initial dose is 100 mg bid or tid. The frequency of administration and total daily dose should be increased slowly at about weekly intervals, as necessary. Generally, an initial maintenance dosage of 1.5 to 2 g/day is required. Every 6 to 8 weeks, liver function tests and stool tests for occult blood should be performed, and blood glucose, uric acid, and LDL-C levels should be measured.

Adverse effects of nicotinic acid are common, bothersome, and often severe. Flushing may be ameliorated by taking aspirin 80 to 325 mg 30 to 60 minutes before taking nicotinic acid, but aspirin may cause gastrointestinal adverse effects in the elderly. Fast-release formulations of nicotinic acid are preferable because hepatotoxicity appears to be more common and more severe with slow-release formulations. If possible, nicotinic acid should not be used concurrently with a statin, because risk of myositis and hepatotoxicity is increased.

HMG-CoA reductase inhibitors (statins) block intracellular cholesterol biosynthesis and force cells to synthesize more LDL receptors, thus increasing the catabolism of LDL cholesterol. Statins include atorvastatin, cerivastatin, fluvastatin, lovastatin, pravastatin, and simvastatin. Statins are becoming the cholesterol-lowering drugs of choice because they are generally safe, have a relatively favorable adverse effect profile, and effectively lower TC levels (by 15 to 45%) and LDL-C levels (by 20 to 40%) and increase HDL-C levels (by 5 to 15%). Statins also effectively lower TG levels (by 10 to 35%). Lovastatin should be taken with food. The others do not have this restriction and are usually taken at bedtime.

Adverse effects are relatively rare and usually transient. Because liver enzyme (particularly transaminase) levels may increase, liver function tests should be performed before treatment, at 6 and 12 weeks after initiation of treatment or after elevation in dose, and at 6-month intervals thereafter. Increases in transaminase levels should be monitored until levels return to normal. Usually, a statin is not discontinued unless the liver enzyme elevations exceed three times the upper limit of normal.

A statin may be the drug of choice for patients with a high LDL-C level and a TG level > 300 mg/dL and for those with severe hemorrhoids or a history of bowel resection or severe constipation. Gemfibrozil may be added if the TG level remains > 300 mg/dL and the patient is at high risk of or has had an atherosclerotic event. Statins can be used effectively with resins but probably should not be used with nicotinic acid because the risks of myopathy and hepatotoxicity are increased.

The concurrent use of a statin and cyclosporine commonly produces myopathy and may also produce rhabdomyolysis and myoglobinuria. This drug combination should thus be restricted to situations in which other lipid-lowering regimens are ineffective, and patients should be closely monitored.

Erythromycin and its derivatives should not be given concurrently with a statin because the risk of myopathy is increased. Gemfibrozil, a fibric acid derivative, increases the hydrolysis of VLDL-TG and the synthesis of HDL-C and apolipoprotein A-I. Gemfibrozil lowers LDL-C and TG levels and reduces CAD morbidity and mortality rates in appropriate patients. It is the drug of choice for elderly patients with hypertriglyceridemia or hypoalphalipoproteinemia. Gemfibrozil should be considered for patients with a high TG level (> 300 mg/dL), a low HDL-C level (< 35 mg/dL [< 0.91 mmol/L]), and a moderately high LDL-C level (< 190 mg/dL [< 4.92 mmol/L]). Well tolerated by most patients, gemfibrozil rarely causes gastrointestinal upset or myopathy, although myopathy is more common when the drug is given to patients with impaired renal function. For such patients, especially if they are also receiving cyclosporine, the dose should be reduced.

Combined gemfibrozil-statin therapy may be necessary for patients at highest risk of CAD who have evidence of atherosclerosis. Such patients often have combined hyperlipidemia, usually with high TC, high TG, and low HDL-C levels, which are maintained despite significant dietary, weight, and exercise modification. When these patients are treated with gemfibrozil alone, TG levels can usually be normalized and HDL-C levels can often be increased to > 35 mg/dL (> 0.91 mmol/L), but TC and LDL-C levels may not be adequately lowered and often increase. Conversely, when these patients are treated with a statin alone, LDL-C levels can usually be normalized, but TG levels often remain high, and HDL-C levels often remain low (< 35 mg/dL).

Because myopathy, rhabdomyolysis, myoglobinuria, and renal injury have been reported in patients taking combined gemfibrozil-statin therapy, the following guidelines are recommended:

- It should be reserved for secondary prevention.
- It should not be used if patients have a very low creatinine clearance rate, because the risk of myopathy is increased.
- It should not be used concurrently with cyclosporine or nicotinic acid, because the risk of myopathy is increased.
- It can be used if patients are reliable and are well informed about the possibility of myopathy.
- Creatine kinase levels should be measured and liver function tests should be performed at baseline and every 6 to 8 weeks thereafter.

- Patients should take gemfibrozil 1.2 g/day po and the lowest starting dose of the statin. The statin dose should be adjusted to the lowest dose that will lower LDL-C levels to the target range.

Probucol appears to increase the rate of LDL-C catabolism, probably through pathways not mediated by LDL receptors. One hypothesis is that the antioxidant effect of probucol may reduce atherosclerosis by reducing the atherogenic effect of oxidized LDL-C. The effect of probucol on CAD risk has not been established.

Probucol is considered a second-line drug in the NCEP guidelines. Probucol usually lowers LDL-C levels by about 8 to 15%, but because it also lowers HDL-C levels by as much as 25%, its role in treating patients with a high LDL-C level is uncertain. Xanthoma regression has been reported as HDL-C levels decrease.

Generally, probucol is well tolerated; diarrhea is the most common adverse effect. Because the drug can prolong the QT interval, it is probably contraindicated in patients with ventricular irritability and an initially prolonged QT interval and in those taking other drugs that prolong the QT interval. The combination of probucol and a resin is effective; the reduction in the HDL-C level is less than that with probucol alone, and constipation is much less common than with a resin alone.

Fish oils containing omega-3 fatty acids are available over the counter but can also be obtained in reasonable amounts by eating a 3-oz (cooked) portion of oily fish (e.g., herring, mackerel, salmon, shad, trout). Data about the long-term effectiveness and adverse effects of fish oil supplements are scarce; however, for short-term use, doses of <= 15 g/day appear to be safe and can lower TG levels, but they are not useful in lowering cholesterol levels. They may increase the hydrolysis of VLDL-TG. In patients with hypertriglyceridemia, fish oils may increase HDL-C levels by 10 to 15%. At much higher doses (usually > 50 g/day), they may be associated with thrombocytopenia and increased bleeding time; such doses are almost never used clinically. In rare cases when doses of > 20 g/day are used, platelet counts and bleeding time should be monitored. At such doses, fish oils may interfere with glucose control in patients with diabetes.

Estrogen replacement therapy: Postmenopausal women who receive unopposed estrogen replacement therapy have lower LDL-C levels (by 15 to 25%) and higher HDL-C levels (by 16 to 21%) than those who do not receive this therapy; as a result the LDL-C/HDL-C ratio is substantially decreased. Such therapy also lowers Lp(a) levels. Unopposed estrogen therapy appears to reduce the risk of cardiovascular death. The addition of a progestin to reduce the risk of endometrial hyperplasia and endometrial cancer may limit or eliminate the benefit observed with unopposed estrogen, and may even initially increase the risk of coronary artery disease.

Estrogen replacement therapy can be used alone or with other lipid-modifying treatments. Estrogen substantially elevates TG levels in women with preexisting hypertriglyceridemia, occasionally to levels that can cause lethal pancreatitis. Consequently, fasting TG levels should be measured before initiating estrogen replacement therapy (with or without a progestin). Estrogen replacement therapy is contraindicated in women with familial hypertriglyceridemia and a TG level > 300 mg/dL after modification of diet and alcohol intake.

Antioxidant therapy: The toxicity of LDL-C may be reduced by the use of antioxidants. According to one hypothesis, atherosclerosis progresses because "toxic LDL" triggers the development of fatty streaks. Toxic LDL is formed when the lipid component of the lipoprotein is oxidized by endothelial cells and smooth muscle cells. Oxidation occurs via a free radical mechanism involving superoxide anions and hydrogen peroxide. The oxidized LDL functions as a chemotactic agent for monocytes, transforming them to macrophages, which stimulate cholesterol esterification and the formation of foam cells.

α-Tocopherol (vitamin E) inhibits the oxidation of LDL in vitro. In several studies, vitamin E consumption appeared to be strongly and inversely correlated with CAD risk. Vitamin E supplementation for short periods produced no benefit, but supplementation for at least 2 years was associated with a lower risk of CAD in men and women.

Vitamin A, β-carotenoid, may affect atherosclerosis by scavenging oxidizing free radicals. In observational studies, the CAD mortality rate appeared to be strongly and inversely correlated with dietary carotene intake.

Ascorbic acid (vitamin C) is considered a secondary antioxidant; it works synergistically with vitamin E, regenerating vitamin E from the vitamin E radical. Vitamin C may also enhance the transformation of cholesterol into bile acids.

Treatment of special conditions: Treating elderly patients with an isolated low HDL-C level (usually well below 35 mg/dL [0.91 mmol/L]), a TC level < 200 mg/dL (< 2.26 mmol/L), and a TG level < 250 mg/dL (< 2.82 mmol/L) is a challenge. If lifestyle changes (weight loss, increased aerobic exercise, cessation of cigarette smoking) do not increase the HDL-C level and particularly if the patient has had an atherosclerotic event or is at high risk because of primary hypoalphalipoproteinemia or other CAD risk factors, drug treatment is warranted. However, the best approach has not been established. Nicotinic acid 1.5 to 6.0 g/day or gemfibrozil 1.2 g/day may be effective, particularly if the Lp(a) level is also high. If these drugs are ineffective or cannot be tolerated, a statin may be given to lower the TC level to < 160 mg/dL (< 4.14 mmol/L) or the LDL-C level to < 100 mg/dL (< 2.59 mmol/L); the HDL-C level usually does not change, but the TC/HDL-C ratio is usually substantially decreased.

High Lp(a) levels cannot be lowered by diet and can be lowered only modestly by nicotinic acid. Most other drugs that lower LDL-C levels do not substantially alter Lp(a) levels. Therefore, aggressive modification of other CAD risk factors, if present, is important.

OXIDATION OF FATTY ACIDS

Oxidation of fatty acids occurs in the mitochondria. The transport of fatty acyl-CoA into the mitochondria is accomplished via an acyl-carnitine intermediate, which itself is generated by the action of carnitine palmitoyltransferase I (CPT I, also called carnitine acyltransferase I, CA I) an enzyme that resides in the outer mitochondrial membrane. The acyl-carnitine molecule then is transported into the mitochondria where carnitine palmitoyltransferase II (CPT II, also called carnitine acyltransferase II, CA II) catalyzes the regeneration of the fatty acyl-CoA molecule.The process of fatty acid oxidation is termed β-oxidation since it occurs through the sequential removal of 2-carbon units by oxidation at the β-carbon position of the fatty acyl-CoA molecule.

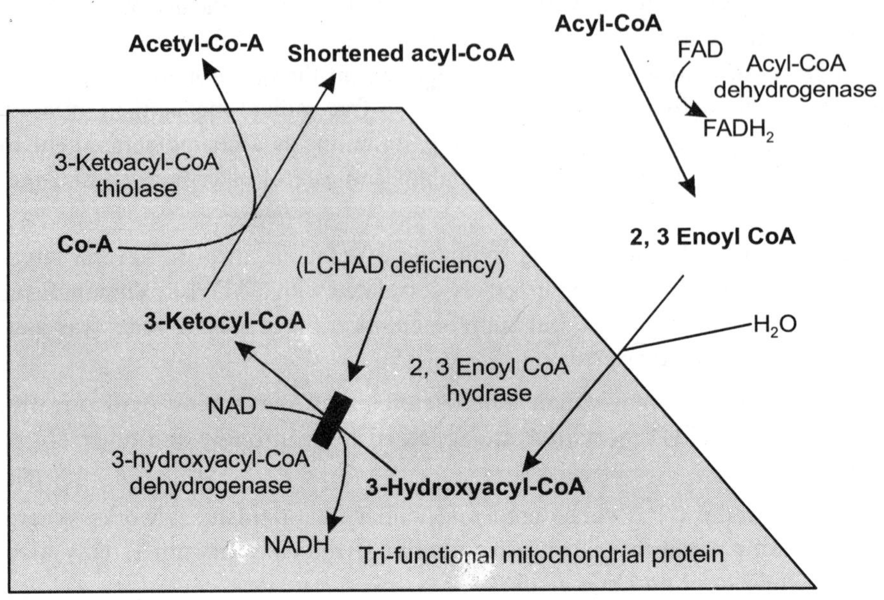

Fig. 5.28 : The Beta Oxidation Pathway

BETA OXIDATION

Beta oxidation is the process by which fatty acids, in the form of Acyl-CoA molecules, are broken down in mitochondria and/or in peroxisomes to generate Acetyl-CoA, the entry molecule for the Krebs cycle.

Free fatty acids can penetrate the plasma membrane due to their poor water solubility and high fat solubility. Once in the cytosol, a fatty acid reacts with ATP to give a fatty acyl adenylate, plus inorganic pyrophosphate. This reactive acyl adenylate then reacts with free coenzyme A to give a fatty acyl-CoA ester plus AMP.

Once inside the mitochondria, the β-oxidation of fatty acids occurs via four recurring steps:

Description	Equation of the reaction	Enzyme	End product
1	2	3	4
Oxidation by FAD: The first step is the oxidation of the fatty acid by Acyl-CoA-Dehydrogenase. The enzyme catalyzes the formation of a double bond between the C-2 and C-3.	Acyl-CoA → (FAD → $FADH_2$; Acyl-CoA-Dehydrogenase) → trans-Δ^2-Enoyl-CoA	acyl CoA dehydrogenase	trans-β2-enoyl-CoA

1	2	3	4
Hydration: The next step is the hydration of the bond between C-2 and C-3. The reaction is stereospecific, forming only the L isomer.	trans-Δ^2-Enoyl-CoA $\underset{-H_2O}{\overset{+H_2O}{\rightleftharpoons}}$ L-3-Hydroxyacyl-CoA (Enoyl-CoA-Hydralase)	enoyl CoA hydratase	L-β-hydroxyacyl CoA
Oxidation by NAD^+: The third step is the oxidation of L-β-hydroxyacyl CoA by NAD^+. This converts the hydroxyl group into a keto group.	L-3-Hydroxyacyl-CoA → 3-ketoacyl-CoA (NAD^+ → NADH $+H^+$; Hydroxyacyl-CoA-Dehydrogenase)	L-β-hydroxyacyl CoA dehy drogenase	β-ketoacyl CoA
Thiolysis: The final step is the cleavage of β-ketoacyl CoA by the thiol group of another molecule of CoA. The thiol is inserted between C-2 and C-3.	3-ketoacyl-CoA $\xrightarrow{+\text{CoA-SH}}$ Actyl-CoA + Acetyl-CoA (Thiolase)	β-ketothiolase	An acetyl CoA molecule, and an acyl CoA molecule, which is two carbons shorter

This process continues until the entire chain is cleaved into acetyl CoA units. The final cycle produces two separate acetyl CoA's, instead of one acyl CoA and one acetyl CoA. For every cycle, the Acyl CoA unit is shortened by two carbon atoms. Concomitantly, one molecule of FADH2, NADH and acetyl CoA are formed.

β-OXIDATION OF UNSATURATED FATTY ACIDS

β-oxidation of unsaturated fatty acids poses a problem since the location of a cis bond can prevent the formation of a trans-Δ2 bond. These situations are handled by an additional two enzymes.

Whatever the conformation of the hydrocarbon chain, β-oxidation occurs normally until the acyl CoA (because of the presence of a double bond) is not an appropriate substrate for *acyl CoA dehydrogenase*, or *enoyl CoA hydratase*:

- If the acyl CoA contains a cis-Δ3 bond, then cis-Δ3-Enoyl CoA isomerase will convert the bond to a trans-Δ2 bond, which is a regular substrate.
- If the acyl CoA contains a cis-Δ4 double bond, then its dehydrogenation yields a 2,4-dienoyl intermediate, which is not a substrate for enoyl CoA hydratase. However, the enzyme 2,4 Dienoyl CoA reductase reduces the intermediate, using NADPH, into trans-Δ3-enoyl CoA. As in the above case, this compound is converted into a suitable intermediate by 3,2-Enoyl CoA isomerase.

To summarize:

- odd numbered double bonds are handled by the isomerase.

- even numbered double bonds by the reductase (which creates an odd numbered double bond) and the isomerase.

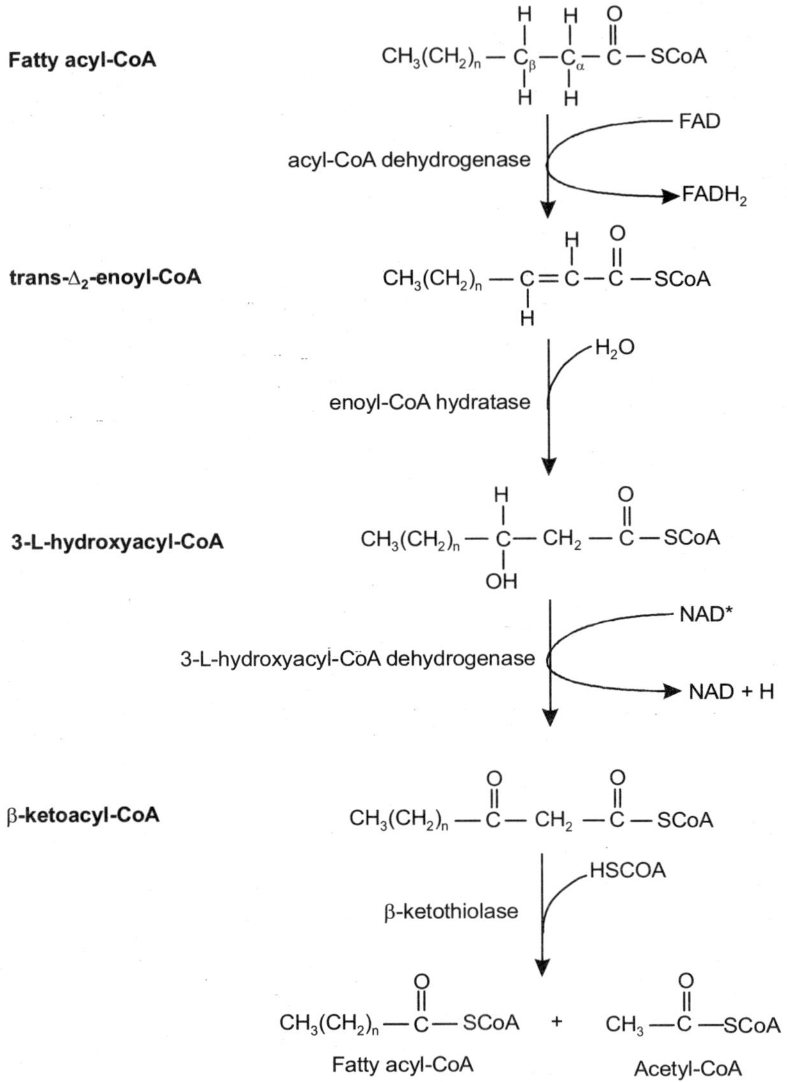

Fig. 5.29 : The Mechanism of Beta Oxidation

β-OXIDATION OF ODD-NUMBERED CHAINS

Fatty acids with an odd number of carbon are generally found in the lipids of plants and some marine organisms. Many ruminant animals form large amount of 3-carbon propionate during fermentation of carbohydrate in rumen.

Chains with an odd-number of carbons are oxidized in the same manner as even-numbered chains, but the final products are propionyl-CoA and acetyl-CoA.

Propionyl-CoA is first carboxylated using a bicarbonate ion into D-stereoisomer of methylmalonyl-CoA, in a reaction that involves a biotin co-factor, ATP, and the enzyme propionyl-CoA carboxylase. The bicarbonate ion's carbon is added to the middle carbon of propionyl-CoA, forming a D-methylmalonyl-CoA. However, the D conformation is enzymatically converted into the L conformation by methylmalonyl-CoA epimerase, then it undergoes intramolecular rearrangement which is catalyzed by methylmalonyl-CoA mutase(requires coenzyme-B_{12} as it's coenzyme) to form succinyl-CoA. The succinyl-CoA formed can then enter the citric acid cycle.

Because it cannot be completely metabolized in the citric acid cycle, the products of its partial reaction must be removed in a process called cataplerosis. This allows regeneration of the citric acid cycle intermediates, possibly an important process in certain metabolic diseases.

Fatty acid oxidation also occurs in peroxisomes, when the fatty acid chains are too long to be handled by the mitochondria. However, the oxidation ceases at octanyl CoA. It is believed that very long chain (greater than C-22) fatty acids undergo initial oxidation in peroxisomes which is followed by mitochondrial oxidation.

One significant difference is that oxidation in peroxisomes is not coupled to ATP synthesis. Instead, the high-potential electrons are transferred to O_2, which yields H_2O_2. The enzyme catalase, found exclusively in peroxisomes, converts the hydrogen peroxide into water and oxygen.

Peroxisomal β-oxidation also requires enzymes specific to the peroxisome and to very long fatty acids. There are three key differences between the enzymes used for mitochondrial and peroxisomal β-oxidation:

1. β-oxidation in the peroxisome requires the use of a peroxisomal carnitine acyltransferase (instead of carnitine acyltransferase I and II used by the mitochondria) for transport of the activated acyl group into the peroxisome.
2. The first oxidation step in the peroxisome is catalyzed by the enzyme acyl CoA oxidase.
3. The β-ketothiolase used in peroxisomal β-oxidation has an altered substrate specificity, different from the mitochondrial β-ketothiolase.

Peroxisomal oxidation is induced by high fat diet and administration of hypolipidemic drugs like clofibrate.

The ATP yield for every oxidation cycle is 14 ATP (according to the P/O ratio), broken down as follows:

Source	ATP	Total
1 $FADH_2$	x 1.5 ATP	= 1.5 ATP (some sources say 2 ATP)
1 NADH	x 2.5 ATP	= 2.5 ATP (some sources say 3 ATP)
1 acetyl CoA	x 10 ATP	= 10 ATP (some sources say 12 ATP)
Total		**= 14 ATP**

For an even-numbered saturated fat (C_{2n}), n – 1 oxidations are necessary and the final process yields an additional acetyl CoA. In addition, two equivalents of ATP are lost during the activation of the fatty acid. Therefore, the total ATP yield can be stated as:

(n - 1) * 14 + 10 – 2 = total ATP

For instance, the ATP yield of palmitate (C_{16}, $n = 8$) is:

(8 - 1) * 14 + 10 – 2 = 106 ATP

Represented in table form:

Source	ATP	Total
7 $FADH_2$	x 1.5 ATP	= 10.5 ATP
7 NADH	x 2.5 ATP	= 17.5 ATP
8 acetyl CoA	x 10 ATP	= 80 ATP
Activation		= –2 ATP
Total		**= 106 ATP**

For sources that use the larger ATP production numbers described above, the total would be 129 ATP equivalents per palmitate. Beta-oxidation of unsaturated fatty acids changes the ATP yield due to the requirement of two possible additional enzymes.

ROLE OF CARNITINE

Carnitine is a quaternary ammonium compound biosynthesized from the amino acids lysine and methionine. In living cells, it is required for the transport of fatty acids from the cytosol into the mitochondria during the breakdown of lipids (or fats) for the generation of metabolic energy. It is often sold as a nutritional supplement. Carnitine was originally found as a growth factor for mealworms and labeled vitamin Bt. Carnitine exists in two stereoisomers: its biologically active form is L-carnitine, while its enantiomer, D-carnitine, is biologically inactive. Carnitine transports long-chain acyl groups from fatty acids into the mitochondrial matrix, so that they can be broken down through β-oxidation to acetate to obtain usable energy via the citric acid cycle. In some organisms such as fungi, the acetate is used in the glyoxylate cycle for gluconeogenesis and formation of carbohydrates. Fatty acids must be activated before binding to the carnitine molecule to form acyl-carnitine. The free fatty acid in the cytosol is attached with a thioester bond to coenzyme A (CoA). This reaction is catalyzed by the enzyme fatty acyl-CoA synthetase and driven to completion by inorganic pyrophosphatase.

The acyl group on CoA can now be transferred to carnitine and the resulting acyl-carnitine transported into the mitochondrial matrix. This occurs via a series of similar steps:

1. Acyl-CoA is conjugated to carnitine by carnitine acyltransferase I (palmitoyltransferase) located on the outer mitochondrial membrane
2. Acyl-carnitine is shuttled inside by a carnitine-acylcarnitine translocase

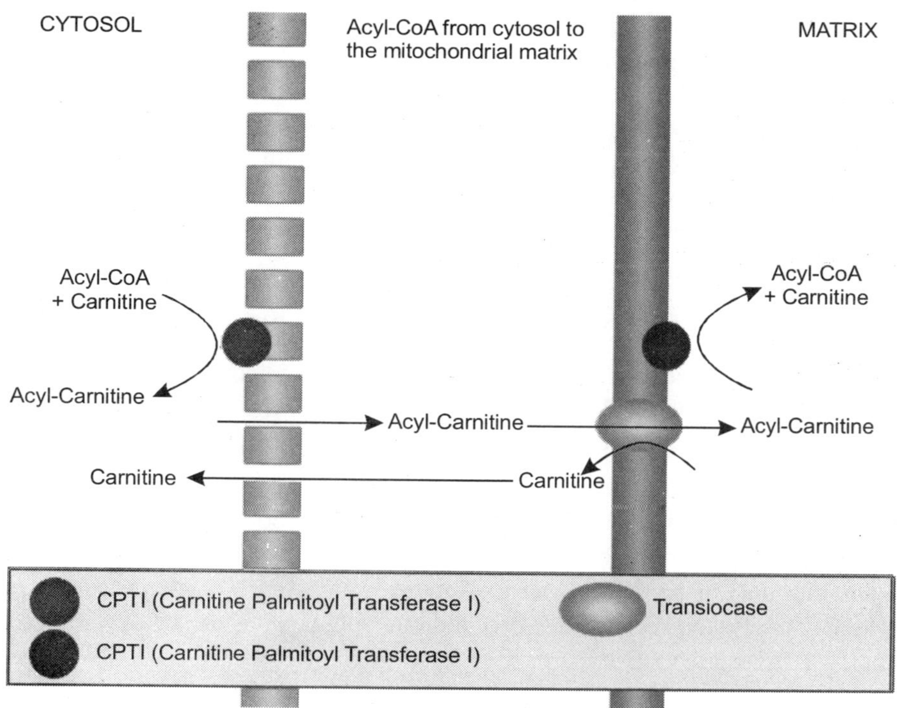

Fig. 5.30 : The Carnitine Involvement in Fatty Acid Metabolism

3. Acyl-carnitine is converted to acyl-CoA by carnitine acyltransferase II (palmitoyltransferase) located on the inner mitochondrial membrane. The liberated carnitine returns to the cytosol.

Human genetic disorders such as primary carnitine deficiency, carnitine palmitoyltransferase I deficiency, carnitine palmitoyltransferase II deficiency and carnitine-acylcarnitine translocase deficiency affect different steps of this process.

Carnitine acyltransferase I undergoes allosteric inhibition as a result of malonyl-CoA, an intermediate in fatty acid biosynthesis, in order to prevent futile cycling between β-oxidation and fatty acid synthesis.

FATTY ACID OXIDATION DISORDERS

Fatty Oxidation Disorders (FODs) are genetic metabolic deficiencies in which the body is unable to oxidize (breakdown) fatty acids to make energy because an enzyme is either missing or not working correctly. The main source of energy for the body is a sugar called glucose. Normally when the glucose runs out, fat is broken down into energy. However, that energy is not readily available to children and adults with an FOD.

Fatty acid transport and mitochondrial oxidation is a complex pathway that plays a major role in energy production during times of fasting and metabolic stress. Once free acids are released into the blood they are taken up by the liver and muscle cells and activated to coenzyme A esters. Then they are transported into the mitochondria and oxidized in a cyclic fashion by four sequential reactions that are each catalyzed by one of multiple enzymes. The acyl-CoA dehydrogenases are chain-length specific enzymes. Deficiencies or abnormalities in these result in very long chain acyl-CoA dehydrogenase deficiency (VLCAD), long chain acyl CoA dehydrogenase deficiency (LCAD), medium chain acyl-CoA dehydrogenase deficiency (MCAD)*, and short chain acyl-CoA dehydrogenase deficiency (SCAD). Any illness may lead to a fasting state that can then lead to the depletion of glucose stores. Once this occurs, fatty acid metabolism becomes the dominant energy source. If there is an abnormality in fatty acid metabolism, then life-threatening episodes of metabolic decompensation can ensue. Relatively simple dietary management may avoid symptoms.

KETOSIS AND KETOGENESIS

During high rates of fatty acid oxidation, primarily in the liver, large amounts of acetyl-CoA are generated. These exceed the capacity of the TCA cycle, and one result is the synthesis of ketone bodies, or ketogenesis. The ketone bodies are **acetoacetate, β-hydroxybutyrate**, and **acetone**.

The formation of acetoacetyl-CoA occurs by condensation of two moles of acetyl-CoA through a reversal of the thiolase catalyzed reaction of fat oxidation. Acetoacetyl-CoA and an additional acetyl-CoA are converted to β-hydroxy-β-methylglutaryl-CoA (HMG-CoA) by HMG-CoA synthase, an enzyme found in large amounts only in the liver. HMG-CoA in the mitochondria is converted to acetoacetate by the action of HMG-CoA lyase. Acetoacetate can undergo spontaneous decarboxylation to acetone, or be enzymatically converted to β-hydroxybutyrate through the action of β-hydroxybutyrate dehydrogenase.

When the level of glycogen in the liver is high the production of β-hydroxybutyrate increases. When carbohydrate utilization is low or deficient, the level of oxaloacetate will also be low, resulting in a reduced flux through the TCA cycle. This in turn leads to increased release of ketone bodies from the liver for use as fuel by other tissues. In early stages of starvation, when the last remnants of fat are oxidized, heart and skeletal muscle will consume primarily ketone bodies to preserve glucose for use by the brain. Acetoacetate and β-hydroxybutyrate, in particular, also serve as major substrates for the biosynthesis of neonatal cerebral lipids.

Ketone bodies are utilized by extrahepatic tissues through the conversion of β-hydroxybutyrate to acetoacetate and of acetoacetate to acetoacetyl-CoA. The first step involves the reversal of the β-hydroxybutyrate dehydrogenase reaction, and the second involves the action (shown below) of acetoacetate:succinyl-CoA transferase, also called β-ketoacyl-CoA-transferase.

The fate of the products of fatty acid metabolism is determined by an individual's physiological status. Ketogenesis takes place primarily in the liver and may by affected by several factors:

1. Control in the release of free fatty acids from adipose tissue directly affects the level of ketogenesis in the liver. This is, of course, substrate-level regulation.

---

* MCAD deficiency is the most common of the disorders with an incidence of 1/60000 to 1/10,000. Deficiency of VLCAD, LCAD, and SCAD is rare compared to MCAD.

2. Once fats enter the liver, they have two distinct fates. They may be activated to acyl-CoAs and oxidized, or esterified to glycerol in the production of triacylglycerols. If the liver has sufficient supplies of glycerol-3-phosphate, most of the fats will be turned to the production of triacylglycerols.
3. The generation of acetyl-CoA by oxidation of fats can be completely oxidized in the TCA cycle. Therefore, if the demand for ATP is high the fate of acetyl-CoA is likely to be further oxidation to CO_2.
4. The level of fat oxidation is regulated hormonally through phosphorylation of ACC, which may activate it (in response to glucagon) or inhibit it (in the case of insulin).

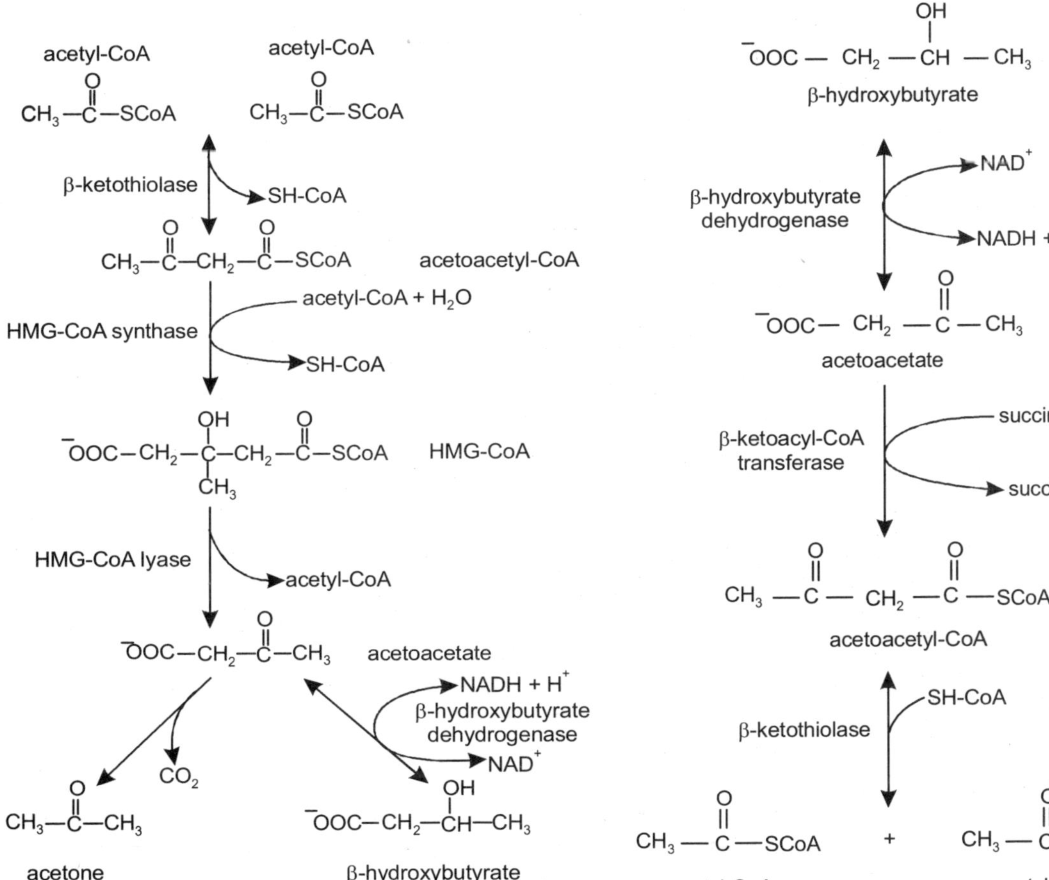

Fig. 5.31 : The Process of Ketogenesis

Fig. 5.32: The Utilization of Ketones

FATTY LIVER

Fatty liver, also known as **fatty liver disease (FLD), steatorrhoeic hepatosis,** or **steatosis hepatitis,** is a reversible condition where large vacuoles of triglyceride fat accumulate in liver cells via the process of steatosis. Despite having multiple causes, fatty liver can be considered a single disease that occurs worldwide in those with excessive alcohol intake and those who are obese (with or without effects of insulin resistance). The condition is also associated with other diseases that influence fat metabolism. Morphologically it is difficult to distinguish alcoholic FLD from non alcoholic FLD and both show micro-vesicular and macrovesicular fatty changes at different stages.

Fatty liver is commonly associated with alcohol or metabolic syndrome (diabetes, hypertension and dyslipidemia) but can also be due to any one of many causes:

Metabolic

Abetalipoproteinemia, glycogen storage diseases, Weber-Christian disease, Wolman disease, acute fatty liver of pregnancy, lipodystrophy.

Nutritional

Malnutrition, total parenteral nutrition, severe weight loss, refeeding syndrome, jejuno-ileal bypass, gastric bypass, jejunal diverticulosis with bacterial overgrowth.

Drugs and Toxins

Amiodarone, methotrexate, diltiazem, highly active antiretroviral therapy, glucocorticoids, tamoxifen, environmental hepatotoxins (e.g. phosphorus, toxic mushroom).

Other

Inflammatory bowel disease, HIV, Hepatitis C especially genotype 3.

The treatment of fatty liver is related to the cause. It is important to remember that simple fatty liver may not require treatment. The benefit of weight loss, dietary fat restriction, and exercise in obese patients is inconsistent.

Reducing or eliminating alcohol use can improve fatty liver due to alcohol toxicity. Controlling blood sugar may reduce the severity of fatty liver in patients with diabetes. Ursodeoxycholic acid may improve liver function test results, but its effect on improving the underlying liver abnormality is unclear.

LIPOTROPIC FATORS

Our blood contains certain constituents known as lipotropic factors. In addition to playing important roles in the mobilization and utilization of dietary fats, lipotropic factors act as the body's natural emulsifiers, holding blood lipids in solution and resisting lipid deposition within the cardiovascular system. Sufficient lipotropic factors can also keep homocystein levels in check to support cardiovascular health, including the health of the arterial walls and blood lipids. The body is able to synthesize its own lipotropic factors when given all the ingredients: choline, inositol, betaine, folic acid and B vitamins. These elements are key to the proper metabolism and elimination of homocysteine, a potent oxidant (something you don't want).

METABOLISM OF LIPIDS IN ADIPOSE TISSUE AND ITS INFLUENCE OF HORMONES

Adipose tissue is not just a static lump of fats it is in "Dynamic state" breakdown of fats and synthesis takes place all the time. Triglycerides stores in the body is continually undergoing lypolysis and esterification. These two processes are not forward and reverse process of the same reaction. They are entirely different pathways involving different reactants and enzymes.

Many of the nutritional metabolic and hormonal factors regulate either of these two mechanisms i.e. esterification and lypolysis. Resultant of these two processes determine the magnitude of free fatty acids (FFA) circulating in the blood. Triglycerides in the adipose tissue undergoes hydrolysis by a hormone TG – lipase enzyme to form free fatty acids and glycerol. Adipolytic lipase is triacyl glycerol (the main regulating enzyme) and the Diacyl glycerol lipase.

These lipases are distinct from lipoprotein lipase that catalyzes lipoprotein TG (present in chylomicrions and VLDL. The free fatty acids formed by lypolysis can be reconverted in the tissue to acyl CoA by Acyl – CoA synthase and re – esterified with a glycerol – P from TG.

Thus there is an continuous cycle of lypolysis and resterification within the tissue.

When the rate of re-esterification is less than rate < of lypolysis FFA accumulates and diffuses into the plasma where it raises the level of FFA ↑ in plasma

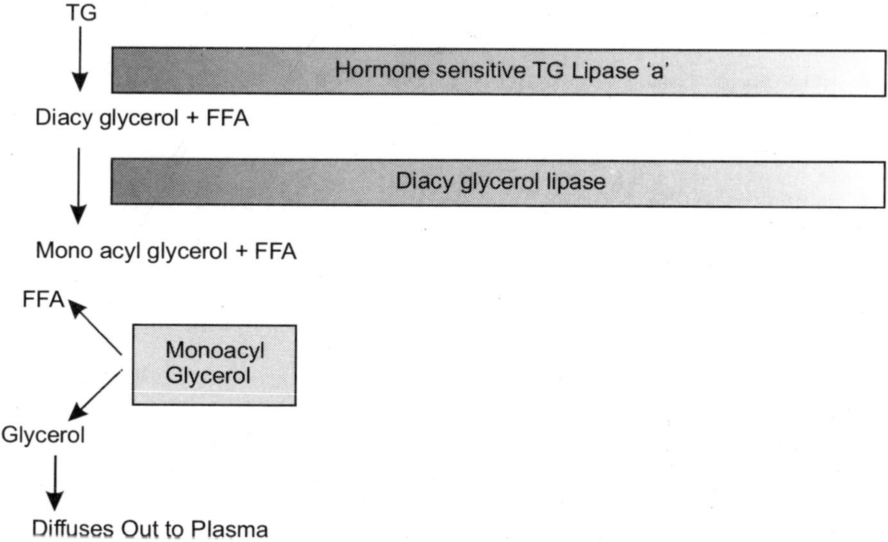

Fig. 5.33 : The Conversion of TG to FFA

Under conditions of adequate nutritional intake or when utilization of glucose by the adipose tissue is increased then more α-glycerol-P will be available. Re-esterification will be greater than lypolysis as a result FFA level outflow decreases and plasma FFA is not mediated by decreasing the rate of lypolysis.

It proves that the effect is due to provision of α-glycerol-P from glycolysis, which enhances esterification.

In diabetes mellitus and in an starvation, availability of glucose in adipose tissue is grossly reduced, resulting to lack of α-glycerol-P. Thus rate of re-esterification is decreased. Lypolysis is greater than re-esterification, resulting to accumulation of FFA and increasing in plasma FFA level.

INFLUENCE OF HORMONES ON ADIPOSE TISSUE

Rate of release of FFA from adipose tissue, is affected by many hormones which influence either the (*a*) rate of esterification (*b*) rate of lypolysis.

List of hormones that increase the rate of esterification:

1. Insulin is the principal hormone and
2. Prolactin – effective in large doses.

Net result of insulin on adipose tissue, which results in fall of circulating FAA.

(*a*) The above is brought about by decreasing the level of cyclic AMP level in the cells. This achieved by

(*i*) Inhibiting adenyl cyclase and

(*ii*) Increasing the phopshodiesterase activity.

Lowered cyclic AMP level in the cell inhibits the activity of hormone sensitive-TG lipase (conversion from 'b' → to 'a' does not occur. The action is mediated through C-AMP-dependant protein kinase, which is not activated. Thus it is not only decreases the release of free FA but also glycerol.

(*b*) Insulin also enhances the update of glucose in the adipose cells. Glucose oxidation provides α-glycerol-P through dihydroxyacetone-P enhancing esterification.

Increase uptake of glucose is activated by the insulin thereby causing the translocation of glucose 'transporters' from the Golgi apparatus to the plasma membrane.

There is also increase F.A synthesis as glucose is oxidized by HMP shunt pathway and provides NADPH

(*c*) Insulin also affects the enzyme by increasing the activity of Pyruvate dehydrogenase, acetyl CoA carboxylase and glycerol-P-acyl transferase, which reinforce the effects arising from increased glucose uptake on the enhancement of FA synthesis and also TG synthesis.

(*d*) Insulin also been shown to inhibit Hormone sensitive TG lipase 'a' independent c-AMP pathway.

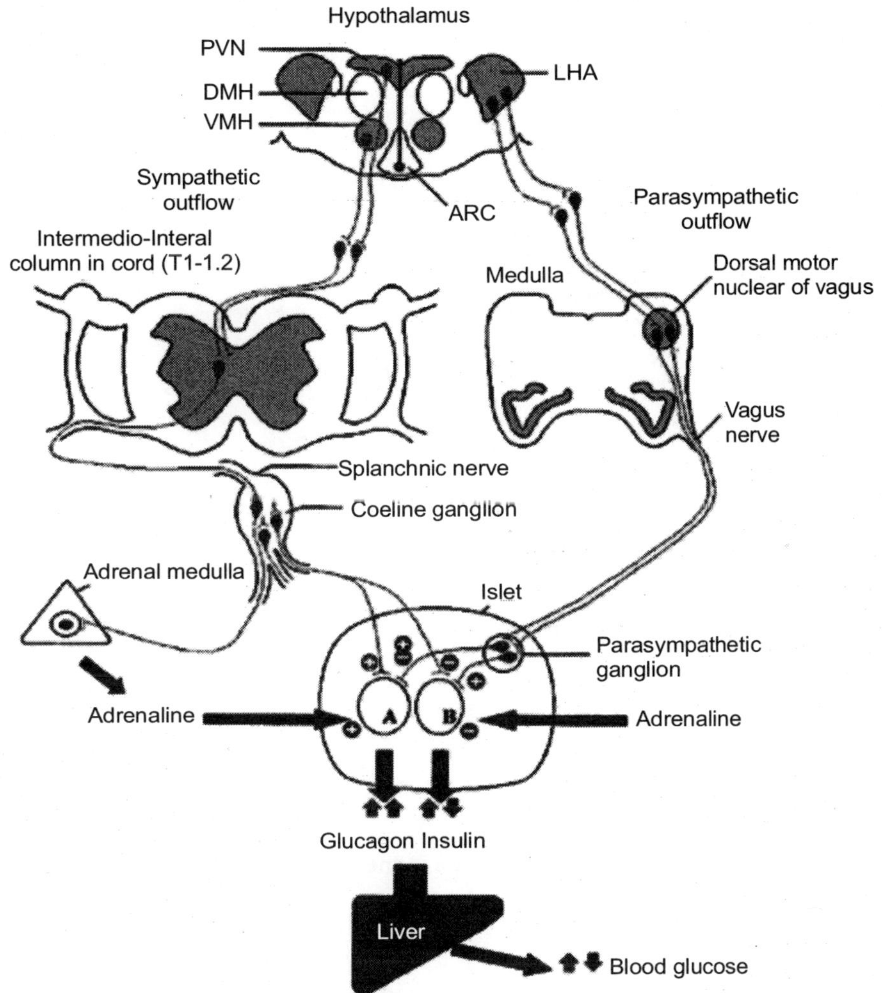

Fig. 5.34 : The Total Fat Metabolism including the Thyroid Hormonal Influence

Effect of prolactorin is similar to insulin provided it is given in larger doses.

List of hormones that increase the rate of Lypolysis are given below:

1. Caticolamines – epinephrine and nor – epinephrine are principal hormones.
2. Glucagon
3. Growth hormone
4. Glucocorticoids
5. ACTH a and b MSH TSH and vasopressin.

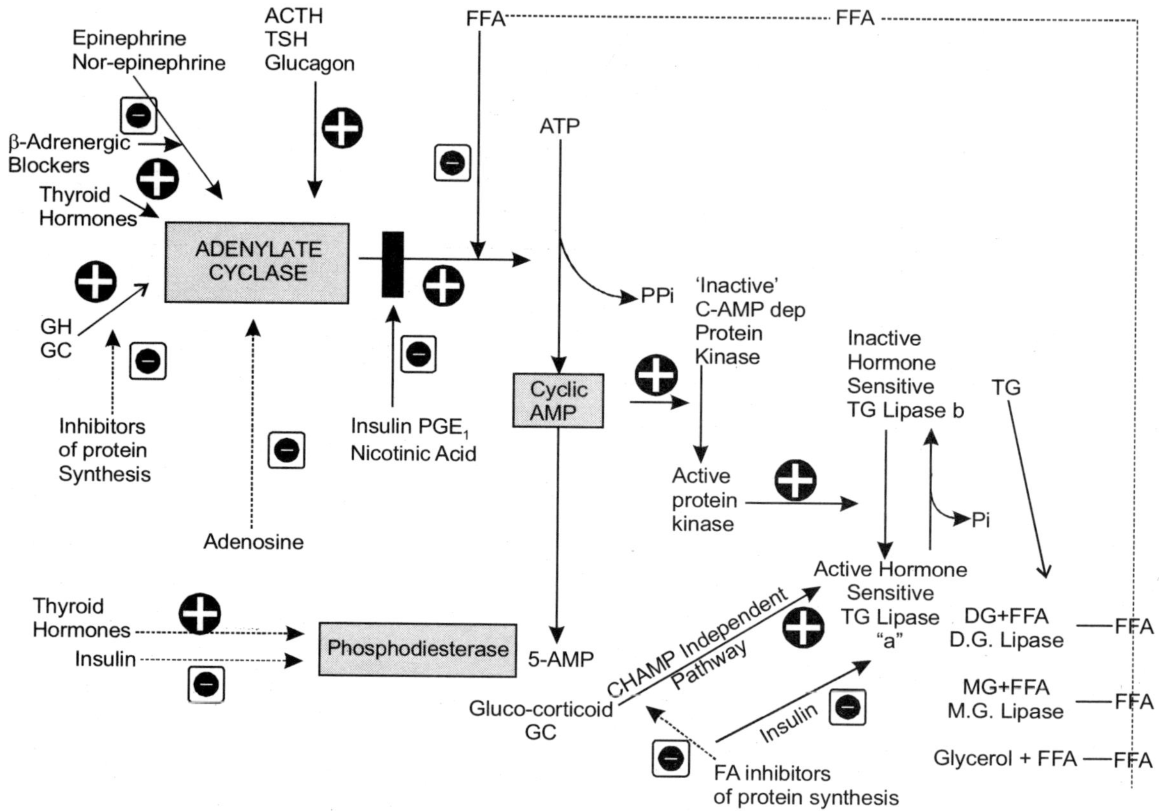

Fig. 5.35 : The Influence of Hormones in the Adipose Tissue Metabolism

These hormones accelerate the release of FFA form adipose tissue and raise the plasma FFA level by increasing lyposis of TG.

Most of them act by activating adenyl cyclase, thus increasing the cyclic AMP level in cells. Cyclic AMP in turn converts 'inactive' cyclic AMP dep protein kinase to the 'active' from, which phosphorylates inactive hormones sensitive TG lipase 'b' to active from 'a' and brings about lypolysis.

Growth hormone in promoting lypolysis is slow. It is dependent on synthesis of proteins involved in the formation of C-AMP i.e. adenylate cyclase

Glucocorticoids are responsible for the plasma FFA increment. This achieved by

(*i*) Glucocorticoids depress uptake of glucose and there is less α-glycerol-P available. Thus decreases rate of esterification.

(*ii*) Stimulates synthesis of adenylate cyclase thus increasing C-AMP level in the cells.

(*iii*) Facilitates adipokinetic property of growth hormones and

(*iv*) Increases synthesis of new lipase protein by a c-AMP independent pathway.

BIOSYNTHESIS OF FATTY ACIDS

Fatty acid biosynthesis is different from the β-oxidation degradation pathway:

1. Fatty acid biosynthesis occurs in the cytosol, β-oxidation occurs in the mitochondrial matrix.
2. All of the carbon atoms of fatty acids come from acetyl CoA.
3. Intermediates of fatty acid biosynthesis are covalently linked to an acyl carrier protein. In β-oxidation the fatty acids are attached to Coenzyme A.
4. In animals, the enzymes involved in fatty acid biosynthesis are contained in one long polypeptide chain, whereas the enzymes of β-oxidation are independent enzymes found in the matrix.
5. The oxidation/reduction reagents of fatty acid biosynthesis are NADP+/NADPH whereas the redox reagents of β-oxidation are NAD^+/NADH and FAD/FADH2.

The Design Strategy for Fatty Acid Biosynthesis.

- Fatty acids are constructed by the addition of two carbon units derived from acetyl-CoA.

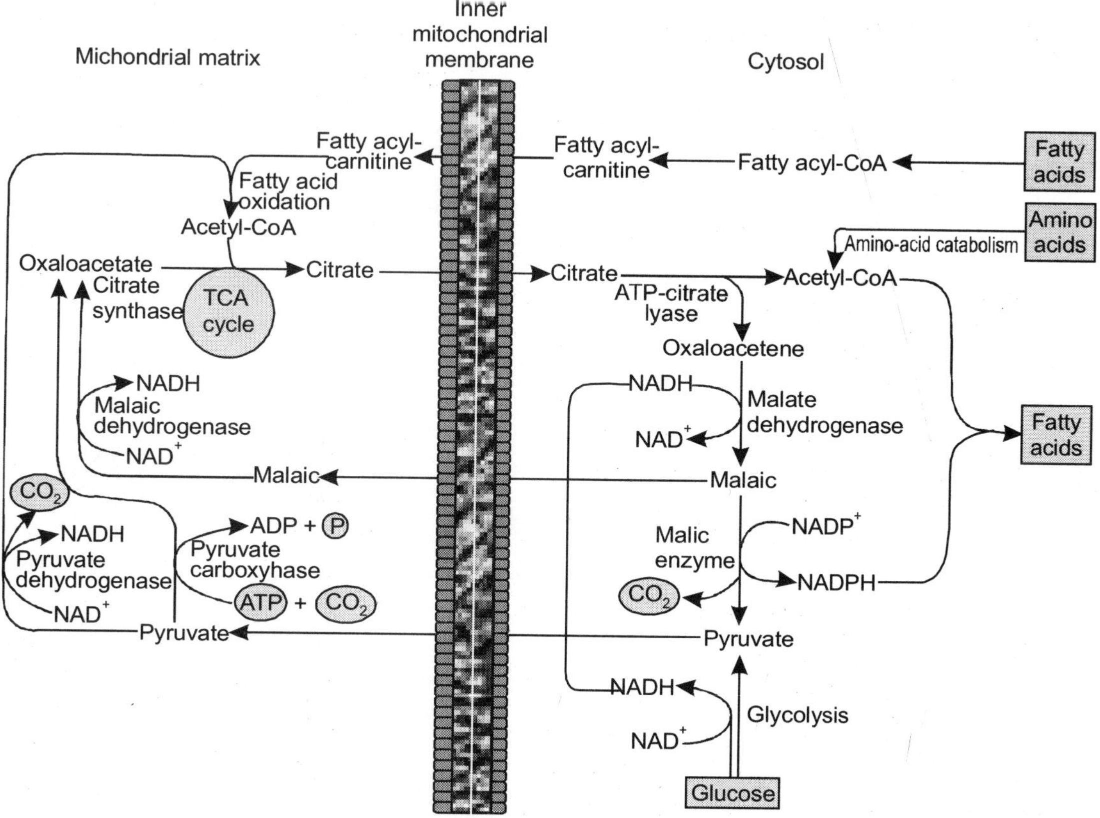

Fig. 5.36 : The Biosynthesis of Fatty Acids

- The acetate units are activated by the formation of malonyl-CoA at the expense of ATP.
- The driving force for the addition of two carbon units to the growing chain is the decarboxylation of malonyl-CoA.
- The chain elongation stops at palmitoyl-CoA.
- Other enzymes add double bonds or additional carbon atom to the carbon chain.

Fatty acid biosynthesis in the cytosol requires a sufficient concentration of NADPH and acetyl-CoA.

NADPH is generated in the cytosol by the pentose phosphate pathway, and by the malic enzyme which oxidizes malate into pyruvate and CO_2, generating NADPH. There are 3 principle ways of producing acetyl-CoA in the cytosol of the cell.

1. Amino acid degradation produces acetyl-CoA.
2. Fatty acid oxidation in the matrix of the mitochondria produces acetyl-CoA which is converted into citrate which is transported into the cytosol by the tricarboxylate transporter. ATP-citrate lyase convertes citrate in the cytosol into acetyl-CoA.
3. Glycolysis generates pyruvate which can be carboxylated in the mitochondria into oxaloacetate and then converted into citrate which is transported into the cytosol by the translocase. ATP-citrate lyase converts citrate in the cytosol into acetyl-CoA.

The acetyl-CoA formed by amino acid degradation is insufficient for fatty acid biosynthesis.

Acetyl-CoA Carboxylase (ACC)

Acetyl CoA molecules are the building blocks of fatty acid synthesis.

The acetyl CoA molecules need to be activated for fatty acid biosynthesis.

HO—C(=O)—O⁻
ATP
ADP
HO—C(=O)—O—P(=O)(O⁻)—O
HN
NH
H
H
H
H_2C
S
R
Pi
⁻O—C(=O)—N
NH
H
H
H
H_2C
S
R

CoA—S—C(=O)—C—H :B
H_2
CoA—S—C(O⁻)=CH_2
CoA—S—C(=O)—CH_2
CoA—S—C(=O)—CH
B
H
O
⁻O—C(=O)—N
NH
H
H
H
H_2C
S
R
CoA—C(=O)—C—C(=O)—O⁻
H_2
Malonyl CoA

Acetyl CoA is carboxylated to form malonyl CoA by acetyl CoA carboxylase which is a biotin containing enzyme.

The carboxylation reaction is irreversible and is the first committed step of fatty acid biosyntehsis.

The mechanism of this carboxylase is the same as pyruvate carboxylase and propionyl CoA carboxylase.

ATP is used to activate bicrbonate in the form of carboxyphosphate which leads to the carboxylation of biotin.

The activated CO_2 group is transferred to acetyl-CoA to form malonyl CoA.

Acetyl CoA carboxylase has three domains:

1. A biotin carboxyl group carrier protein.
2. Biotin carboxylase which adds CO_2 to biotin.
3. A transcarboxylase which transfers the CO_2 group from biotin to acetyl CoA to form malonyl CoA.

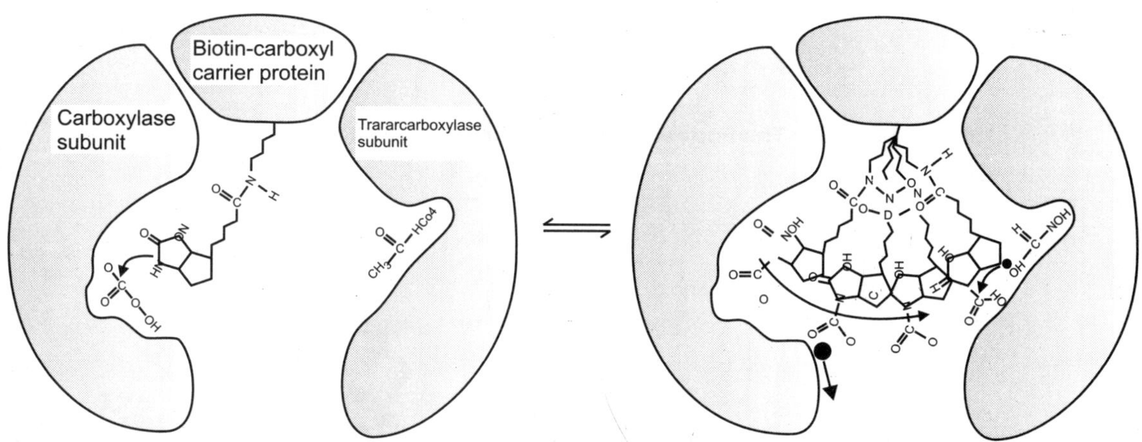

Fig. 5.37 : The Biotin Carboxylase Transferase

Because this is the first committed step of fatty acid biosynthesis, this acetyl CoA carboxylase (ACC) is allosterically regulated. In animals, ACC is a filamentous polymer composed of 230 kD protomers. Each protomer contains the biotin carboxyl carrier protein, the carboxylase and the transcarboxylase domains as well as allosteric regulatory sites. The polymeric form of this enzyme is active, the individual protomers are not. The activity of ACC is dependent of the equilibrium between the two forms of this enzyme.

$$\text{Inactive protaimers} \rightleftharpoons \text{Active polymers}$$

The final product of fatty acid biosynthesis is palmitoyl CoA. There is an allosteric binding site for palmitoyl-CoA which shifts the equilibrium toward the inactive protamers.

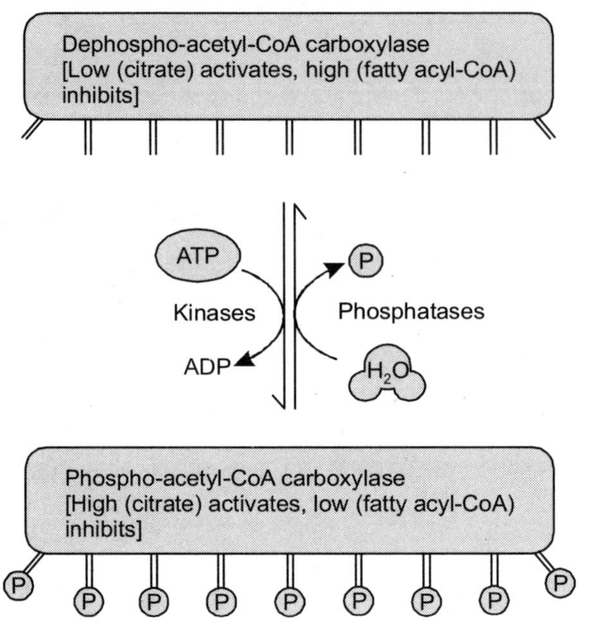

Fig. 5.38 : The Formation of Phospho-acetyl carboxylase

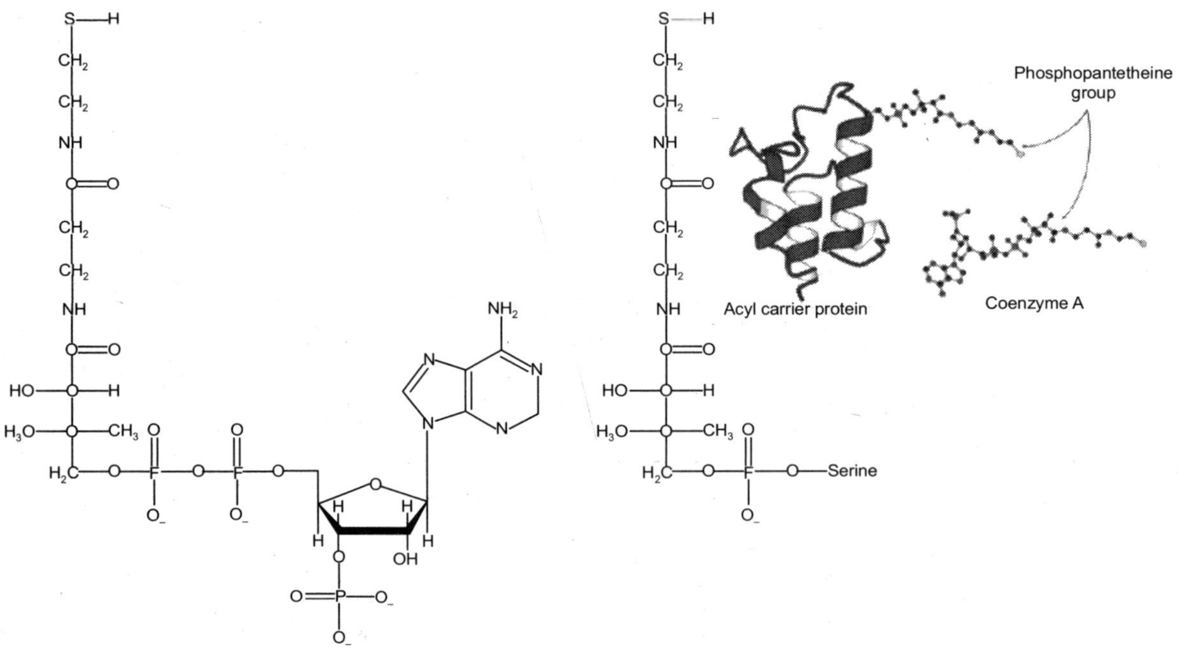

Fig. 5.39 : The Acyl Carrier Protein

Citrate which is a precursor for acetyl CoA formation in the cytosol is an allosteric activator of this enzyme. It binds to an allosteric binding site shifting the equilibrium towards the active polymers. The regulatory effects of citrate and palmitoyl CoA are modulated by the phosphorylation state of acetyl-CoA carboxylase. The animal ACC enzyme has 8-10 potential phosphorylation sites on each protomer. Some of these phosphorylation sites are regulatory others are silent and have no effect on enzyme activity. Acetyl CoA carboxylase is phosphorylated by protein kinases and dephosphorylated by protein phosphatases.

The unphosphorylated ACC binds citrate with high affinity and thus becomes fully active a very low citrate concentration. The phosphorylated ACC enzyme has a greatly reduced affinity for citrate, so high concentrations of citrate are required to activate the enzyme. Palmitoyl CoA binds preferably to the phosphorylated form of the enzyme, so when the ACC enzyme is phosphorylated it takes a small concentration of palmitoyl CoA to inactivate it. When the enzyme is dephosphorylated, the dephosphorylated enzyme has a low affinity for palmitoyl CoA and thus it takes a large concentration of palmitoyl CoA to inactivate it. The intermediates of fatty acid biosynthesis are not linked to Coenzyme A. Rather the intermediates of fatty acid biosynthesis are linked to an acyl carrier protein. The acyl carrier protein is very similar to Coenzyme A in that it contains a phosphopantetheine prosthetic group. Coenzyme contains the phosphopantetheine group attached to an adenosine nucleotide. In the acyl carrier protein, the phosphopantetheine group is attached to a serine residue.

ELONGATION OF FATTY ACIDS

Before fatty acid biosynthesis begins, fatty acid synthetase must be primed with acetyl CoA. The elongation phase of fatty acid biosynthesis begins with the formation of acetyl ACP and malonyl ACP.

Acetyl transacylase and malonyl transacylase catalyzes the reactions.

$$\text{Acetyl CoA} + \text{ACP} \longrightarrow \text{Acetyl ACP} + \text{CoA}$$

$$\text{Malonyl CoA} + \text{ACP} \longrightarrow \text{Malonyl ACP} + \text{CoA}$$

Acetyl transacylase will also catalyze the following reaction at a much slower rate:

$$\text{Propionyl CoA} + \text{ACP} \longrightarrow \text{Propionyl ACP} + \text{CoA}$$

This reaction allows for the synthesis of fatty acid chains of odd length. Malonyl transacylase is specific for malonyl CoA.

Once acetyl KSase and malonyl ACP have been formed, elongation can begin. First the acetyl group of acetyl ACP is transferred to a sulfhydryl residue of ketoacyl-ACP synthase also known as acyl-malonyl ACP condensing enzyme. The decarboxylation of malonyl ACP generates an enolate anion which is a good nucleophile that attacks the carbonyl of thioester of acetyl-S-KSase to form acetoacetyl ACP. The exergonic decarboxylation reaction drives the condensation reaction. The CO2 group added to acetyl CoA by acetyl CoA carboxylase is given up in this reaction. In effect this reaction is driven by ATP. ATP was used to activate bicarbonate in the acetyl CoA carboxylase reaction. The free energy was conserved in the malonyl CoA and released upon decarboxylation to drive the synthesis of acetoacetyl

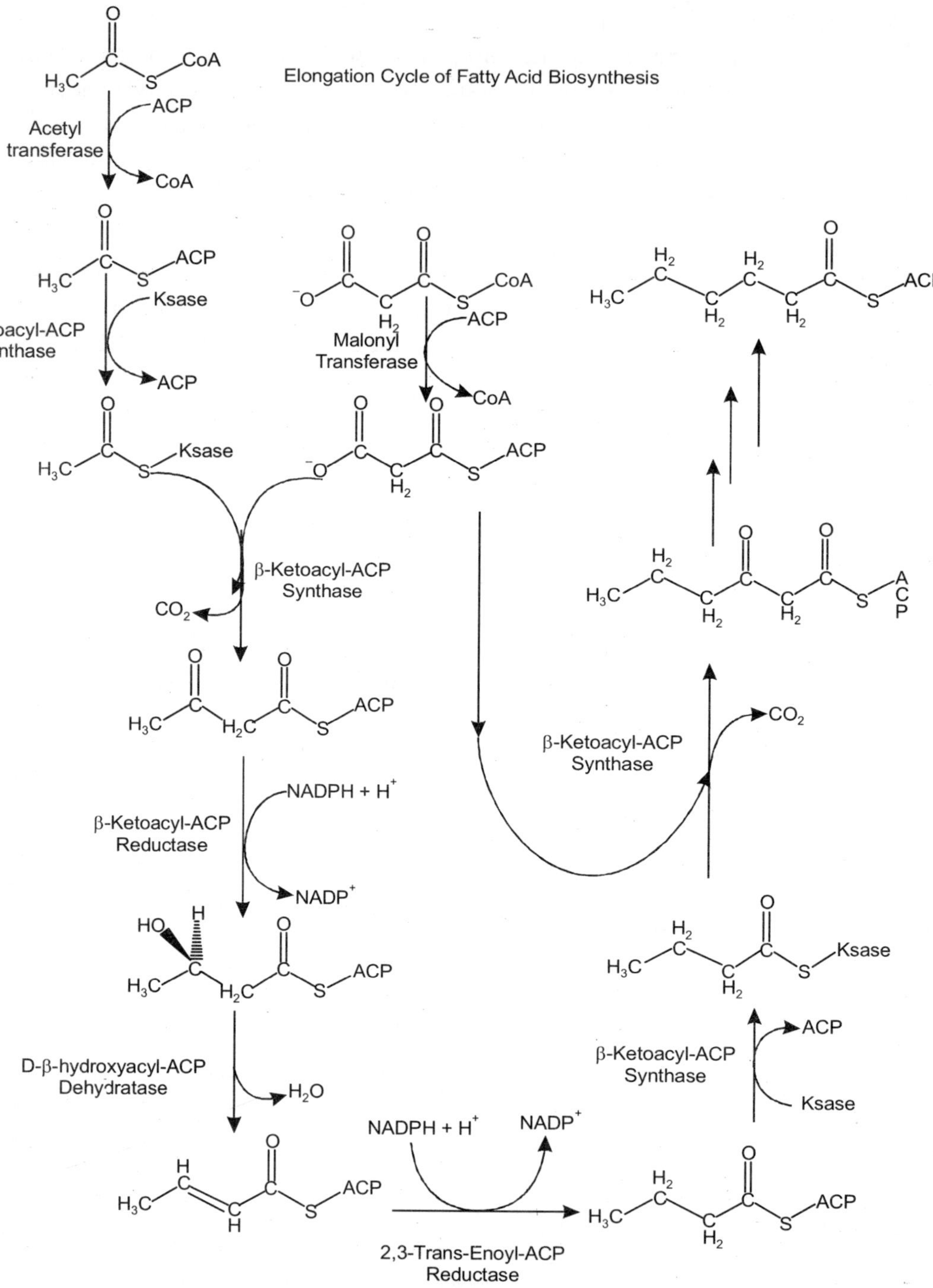

Fig. 5.40 : The Elongation of Fatty Acid Biosynthesis

CoA. The next three reactions are the reverse of β-oxidation. First the ketone is reduced to the alcohol by the enzyme β-ketoacyl ACP reductase. This reaction differs from the reverse of the β-oxidation reaction in that the D-isomer rather than the L-isomer is formed and NADPH is the reducing agent rather than NADH. The alcohol is dehydrated D-β-hydroxyacyl ACP dehydratase to form crotonyl ACP which is reduced by enoyl-ACP reductase to form butyryl ACP. This step is also different than the mere reversal of β - oxidation in that NADPH is the reducing agent instead of $FADH_2$.

In the next round of fatty acid biosynthesis, butyryl ACP it transferred to the sulfhydryl group of β-ketoacyl-ACP synthase. The decarboxylation of a second molecule of malonyl CoA generates another nucleophile that attacks the carbonyl of the thioester of butyryl-S-KSase to form a C6-β-ketoacyl ACP which is then reduced the β-hydoxyacyl ACP, dehydrated to from the C6-.2-enoyl-ACP which is then reduced to form the C6-acyl-ACP which is then transferred to the sulfhydryl group of β-ketoacyl-ACP synthase, condensed with a third maloyl CoA ect. Ect. Ect., unitil palmitoyl ACP is formed. A thioesterase hydrolyzes the thioester bond of palmitoyl ACP to yield palmitate and ACP. The overall reaction is shown below:

Acetyl CoA + 7 Malonyl CoA + 14 NADPH + $14H^+ \rightarrow$ Palmitate + $7CO_2$ + 14 NADP+ + 8CoA + $6H_2O$

The formation of 7 malonyl CoA requires:

7 Acetyl CoA + 7 CO_2 + 7ATP $\rightarrow$ 7 Malonyl CoA + 7 ADP + 7 Pi + $7H^+$

Combining these two steps:

8 Acetyl CoA + 7 ATP + 14 NADPH + $7H^+ \rightarrow$ Palmitate + $14NADP^+$ + 7ADP + 7Pi + 8CoA + $6H_2O$

DETAIL ACTION OF BIOTIN

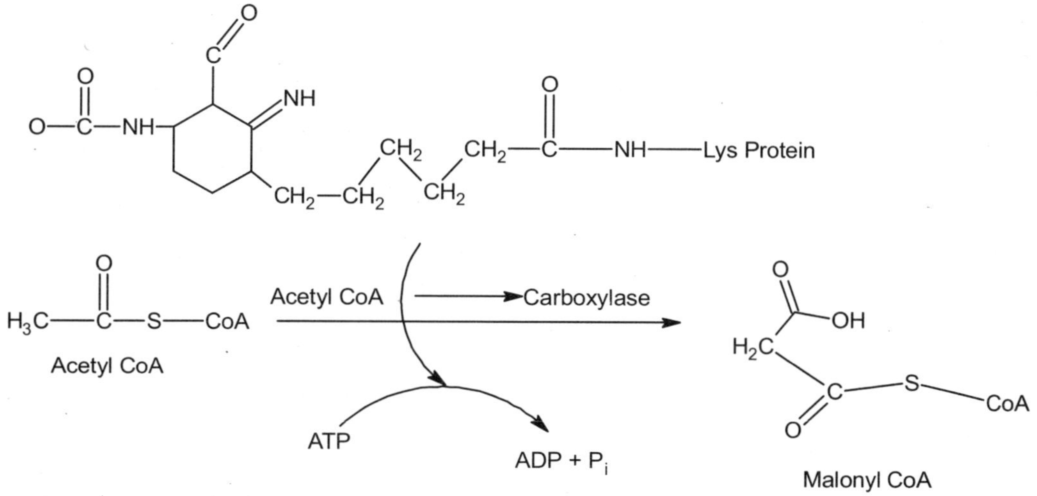

Biotin is the prosthetic group of certain enzymes that catalyze CO_2 transfer reaction (CO_2 fixation reaction). In biological system, biotin functions as the co enzyme for the enzyme called Carboxylase which catalyzes the CO_2 fixation (carboxylation).

In this process biotin is first converted to carboxy biotin complex by reaction with HCO_3- and ATP "CO_2 biotin complex is the source of" active "CO_2 which transferred to the substrate, CO_2 becomes attached to the biotin co enzyme as above.

Example of carboxylation or CO_2 fixation reactions in biologic systems are given ahead and above.

1. *Conversion of acetyl CoA to Matonyl CoA*: In the first step of extra de novo fatty acid synthesis, the acetyl CoA is converted to malonyl CoA, the reaction is catalyzed by the enzyme acetyl CoA carboxylase.

CH_3
CH_2
C
O S CoA
Propionyl CoA
CO_2
BIOTIN
Propionyl CoA
Carboxylase
ATP ADP + Pi
CoA S
CH_3
HC C O OH
C O
Methyl Matenyl CoA

2. *Conversion of Propionyl CoA to Methyl Malonyl CoA*: The enzyme catalyzing the reaction is propionyl – CoA
3. *Conversion of Pyruvic Acid to Oxaloacetate*: The enzyme that catalyzes the reaction is pyruvate carboxylase.
4. *Purine Synthesis:* Biotin has been made involved in the fixation of CO_2 for the formation of Carbon 6 of Purine nucleus. The above reaction is impaired in biotin deficient yeast cells, which suggests that Biotin plays an important role in Purine synthesis.
5. *Conversion of β-methyl crotonyl CoA to β-methyl gluconyl CoA*: In their conversion in leucine metabolism the reaction is catalyzed by the enzyme β- methyl crotonyl – CoA carboxylase.
6. *Other enzyme systems*: A number of other enzyme systems are reportedly influenced by Biotin. These include succinic acid dehydrogenase and decarboxylase and the deaminases of the amino acids aspartic acid, serine and threonine.

MULTIENZYME COMPLEX CONCEPT

In contrast to bacterial fatty acid biosynthesis, Eukaryotes fatty acid synthase is a multienzyme complexcontained in 2 different polypeptide chains. The α subunit is 213 kD, the β subunit is 203 kD. Animal fatty acid synthase complexes are dimers of αβ subunits. The separate activities of each dimer of αβ subunits are shown in the figure below. The α subunits contain the β-ketoacyl-ACP synthase (KSase) domain and the β-ketoacyl reductase domain. The β subunit contains the acetyl transferase domain, the malonyl transferase domain, the β-hydroxyacyl dehydrogenase domain and the enoyl reductase domain. The subunits are arranged in a head to tail fashion that allows the first domain of one subunit of fatty acid synthase to interact with the second and third domains of the other subunits.

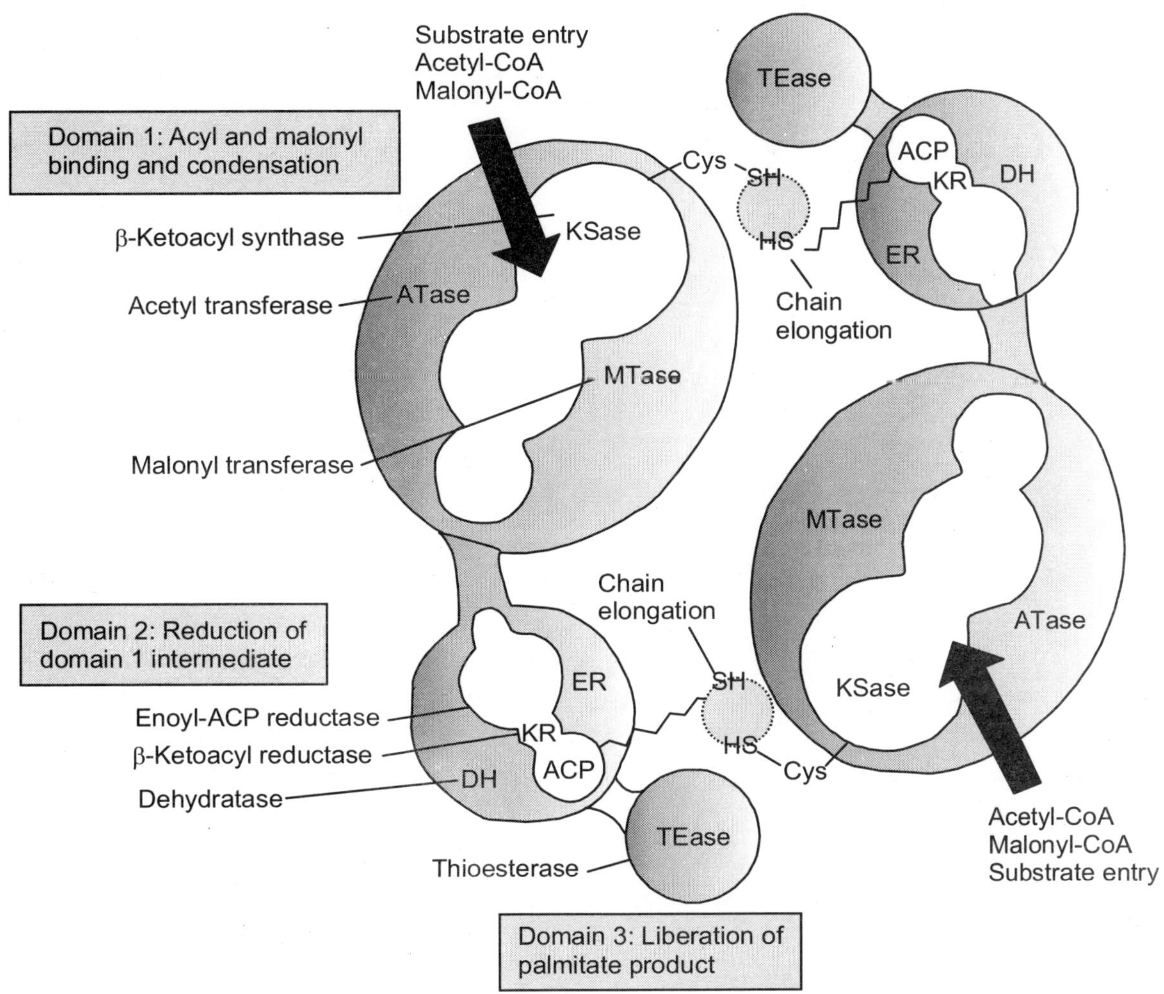

Fig. 5.41 : The Multienzyme Complex Concept

The first step of the fatty acid synthase reaction is the formation of acetyl-O-enzyme intermediate between an acetyl group of acetyl CoA and an active site serine residue of the acetyl transferase domain as shown above on the left. In a similar manner, a malonyl-O-enzyme intermediate is formed between malonyl CoA and an active site serine residue of the malonyl transferase domain as shown above on the

Acetyl Transferase

$H_3C-C(=O)-S-CoA$ + O—H (SER–ENZ) + :B → $H_3C-C(O^-)(-O-SER-ENZ)-S-CoA$ + H—:B → $H_3C-C(=O)(-O-SER-ENZ)$ + H–S–CoA + :B → (– CoA) → $H_3C-C(=O)(-O-SER-ENZ)$ + H–S–ACP + :B → $H_3C-C(O^-)(-O-SER-ENZ)-S-ACP$ + H—:B → $H_3C-C(=O)-S-ACP$ + O—H (SER–ENZ) + :B

Malonyl Transferase

$^-O-C(=O)-CH_2-C(=O)-S-CoA$ + O—H (SER–ENZ) + :B → $^-O-C(=O)-CH_2-C(O^-)(-O-SER-ENZ)-S-CoA$ + H—B → $^-O-C(=O)-CH_2-C(=O)(-O-SER-ENZ)$ + H–S–CoA + :B → (– CoA) → $^-O-C(=O)-CH_2-C(=O)(-O-SER-ENZ)$ + H–S–ACP + :B → $^-O-C(=O)-CH_2-C(O^-)(-O-SER-ENZ)-S-ACP$ + H—B → $^-O-C(=O)-CH_2-C(=O)-S-ACP$ + O—H (SER–ENZ) + :B

right. The next step is the transfer of the acetyl group to the sulfhydryl of the acyl carrier protein (ACP). This acyl group is then transferred one more time to a cysteine residue of β-ketoacyl-ACP synthase as shown below. This frees the acyl carrier protein to acquire the malonyl group from the malonyl transferase. The next step is the condensation reaction in which decarboxylation of the malonyl-ACP generates a highly reactive nucleophile that attacks the carbonyl of acetyl-S-KSase.

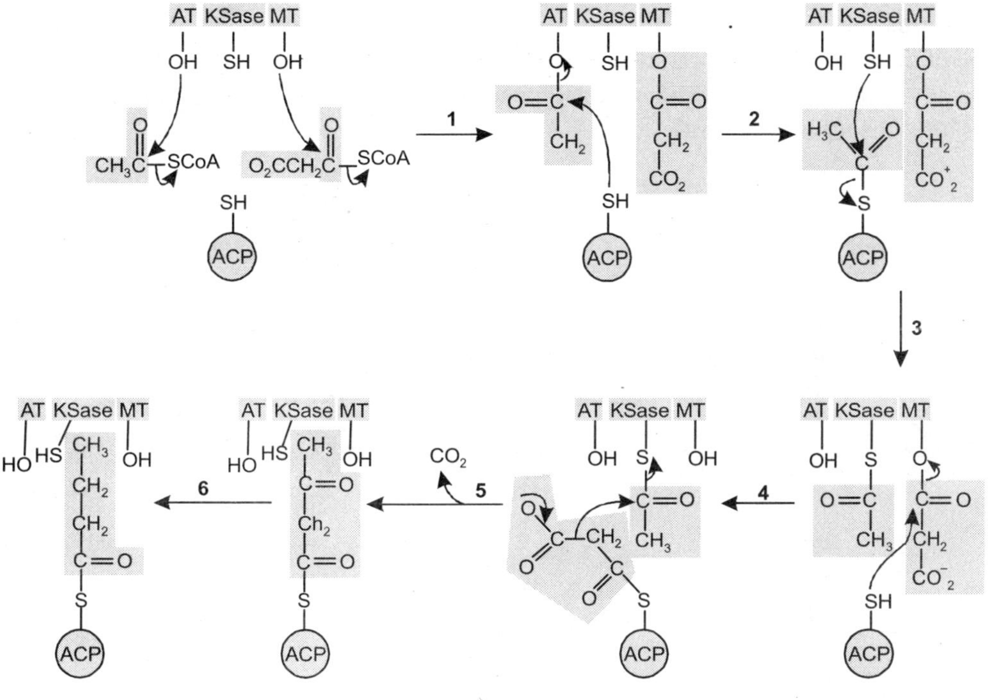

AT KSase MT
OH SH OH
CH_3C–SCoA O_2CCH_2C–SCoA
SH
ACP
1
2
3
4
5
6
CO_2

Fig. 5.42 : The Reaction of Multiple Enzyme Complex Concept

The next three steps are the reduction of the carbonyl to the alcohol, the dehydration and the reduction of the alkene to form saturated butyryl-ACP. A second malonyl group is transferred from malonyl CoA to the active site serine of malonyl transferase. The butyryl-ACP is then transferred to the cysteine residue of KSase. This frees the acyl carrier protein to acquire the malonyl group from the malonyl transferase. The next step is the condensation reaction in which decarboxylation of the malonyl-ACP generates a highly reactive nucleophile that attacks the carbonyl of butyryl-S-KSase. This cycle continues until palmitoyl ACP is formed. The thioester bond is hydrolyzed by thioesterase to form palmitate.

(*i*) Glucorticoids depress update of glucose and there is less α-glycerol-P available. Thus decreases rate of esterification.

(*ii*) Stimulates synthesis of adenylate cyclase thus increasing C-AMP level in the cells.

(*iii*) Facilitates adipokinetic property of growth hormones and

(*iv*) Increases synthesis of new lipase protein by a c-AMP independent pathway.

METABOLISM OF CHOLESTEROL

Cholesterol is oxidized by the liver into a variety of bile acids. These in turn are conjugated with glycine, taurine, glucuronic acid, or sulfate. A mixture of conjugated and non-conjugated bile acids along with cholesterol itself is excreted from the liver into the bile. Approximately 95% of the bile acids are reabsorbed from the intestines and the remainder lost in the feces. The excretion and reabsorption of bile acids forms the basis of the enterohepatic circulation which is essential for the digestion and absorption of dietary fats. Under certain circumstances, when more concentrated, as in the gallbladder, cholesterol crystallizes and is the major constituent of most gallstones, although lecithin and bilirubin gallstones also occur less frequently.

BIOSYNTHESIS OF CHOLESTEROL

Essentially all tissues form cholesterol, liver is the major site of cholesterol biosynthesis also other tissues are active in this regard e.g. adrenal cortex gonads within intestine are also most active.

Low order of synthesis-adipose tissue, muscle aorta and neutral tissues. Brain of new born synthesize cholesterol while the adult brain cannot synthesis cholesterol.

Efficiency of Formation of Cholesterol from Labeled C^{14} Acetate

Tissues	Efficiency of Cholesterol formation liver (100)
Liver	100
Adult Skin	90
Small Intestine	60
Kidney	4
Adult brain	0
New born Brain	185

Enzymes involved in the cholesterol biosynthesis are:

1. Cytoplasmic particles "microsomes"
2. Soluble fraction - crystal

'Active' acetate (acetyl CoA) is the starting material and principal precursor. The entire carbon – skeleton, all 27C of cholesterol in humans can be synthesis from the active acetate.

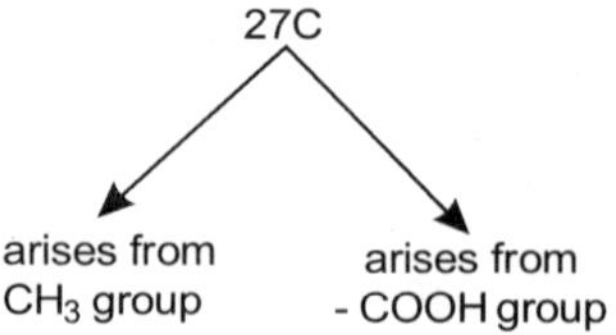

Cholesterol biosynthesis takes place in five groups of reactions. They are:

1. Synthesis of Mevalnoate a-6-C compound from Acetyl CoA.
2. Formation of 'Iso-Pernoid units (C-5) from Melvonate by successive phosphorylation and followed by loss of CO_2

3. Formation of Squalene A 30 carbon aliphatic chain formed by condensation of six isoprenoid units*
4. Cyclization of Squalene to form Lanosterol
5. Conversion of Lanosterol to Cholesterol

Synthesis of Mevalnoate From Acetyl CoA

(*a*) Formation of HMG – CoA (β-OH methyl glutaryl CoA) HMG-CoA can be formed in the crystal from acetyl CoA in two steps catalyzed by the enzyme "Thiolase" and "HMG-CoA synthase"**

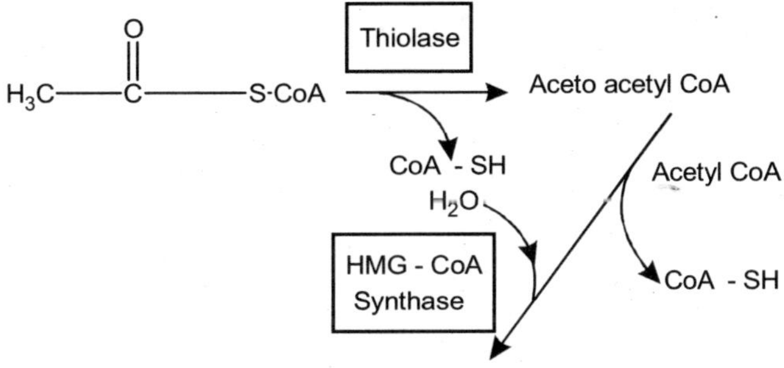

(*b*) In the next step, which is the "rate of limiting" step HMG CoA is converted to Mevalnoic acid (Mevalnoate) catalyzed by the enzyme HMG – CoA reductase.

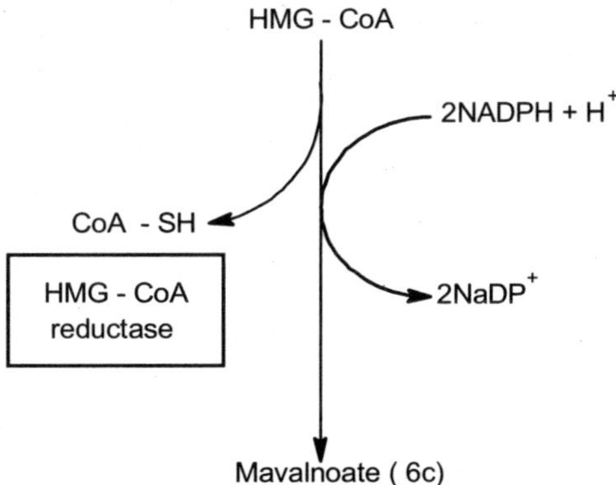

*The isoprenoid units are regarded as the building blocks of the steroids nucleus

**1. MG-CoA may also be produced as an intermediate in the metabolic degradation of ammonical L-Leucine.

2. There are two pools of HMG-CoA

(*i*) Mitochondial-concerned with Ketogenesis

(*ii*) Extramitochondiral (systolic) concerned with synthesis of Mevalnoate and isopernoid.

Characteristic of this Reaction

1. Most important and "rate limiting" step
2. Irreversible reaction
3. Enzyme contains - SH group
4. NADPH required as co-factor supplied by HMP-pathway
5. Enzyme activity not affected in Diabetic patients.
6. Diatory cholesterol and endogenously synthesized cholesterol inhibits this rate limiting step.
7. Hormones-Insulin and Thyroid increases the reductase activity Glucagon and glucocorticoids reduces the activity.

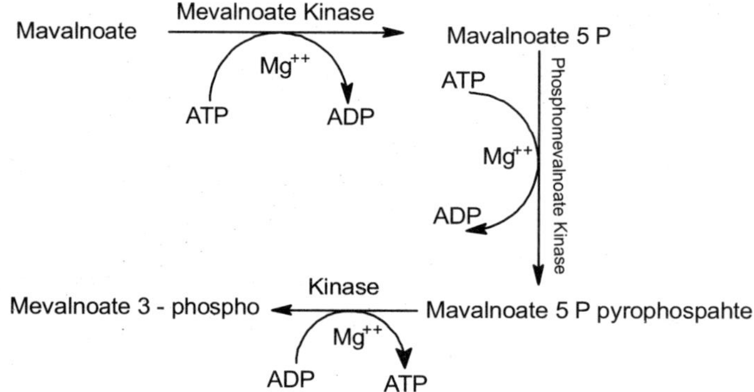

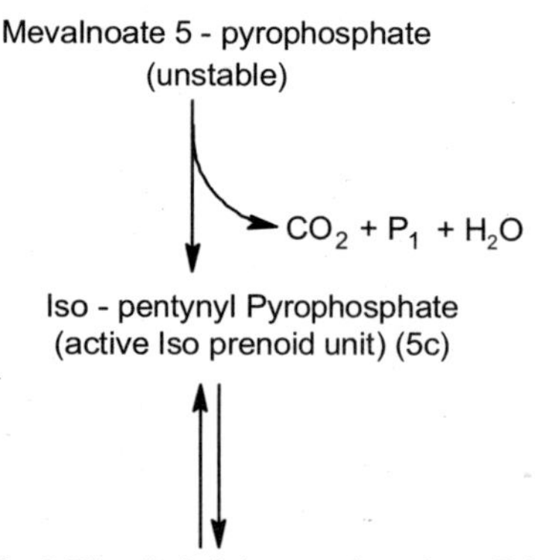

Formation of Isoprenoid Units

1. Mevalnoate is phosphorylated by ATP to form several 'active' phosphorylated intermediates.

2. Three such phosphorylated compounds are formed and it is followed by decarboxylation to form first "active" iso-prenoid unit Iso-pentenyl Pyrophosphate (5c). One of the intermediate pyrophosphorylated compound is "Mevalnoate-3-phosphor-5-Pyrophosphate" which is unstable
3. Iso-pentyl pyrophosphate under goes isomerization to form another 5 C iso-prenoid unit called "3-3 Dimethyl allyl Pyrophosphate"

Formation of Squalene

(*a*) The Pyrophosphorylated isoprenoid units condense to form ultimately a 30-carbon aliphatic chain called Squalene.

(*b*) The condensation occurs in three steps:

(*i*) One molecule of iso-pentenyl pyrophosphate first condenses with one molecule of 3, 3 dimethyl allyl prophopshate to form a 10°C compound called "generyl pyrophosphate" the reaction is catalyzed by the enzyme pyrophosphate synthase.

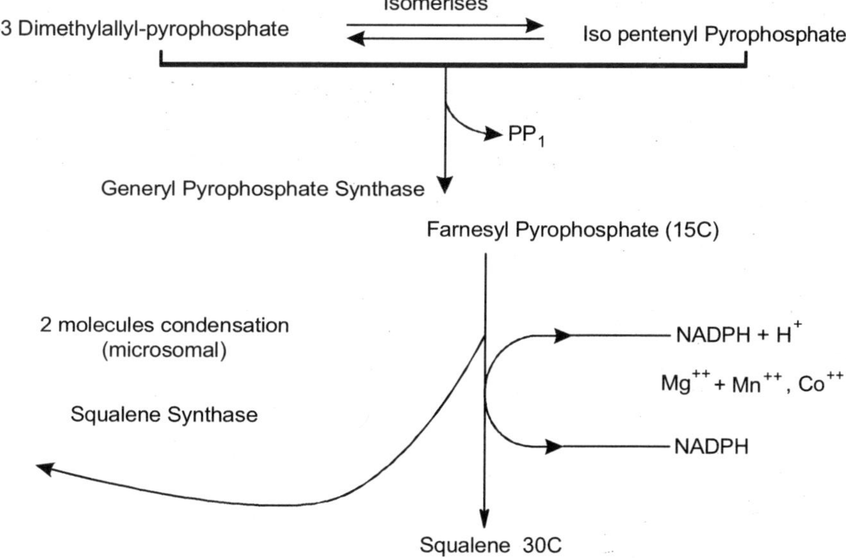

Cyclization of Squalene to Lanosterol Formation

Generally this formation requires two steps to complete

(1) In the first step Squalene-2-3-expoxide is formed catalyzed by the enzyme Squalene mono-oxygenated which requires NADPH and molecular O_2

(2) In the next step an enzyme cyclase brings about the Cyclization of Squalene to form Lanosterol.

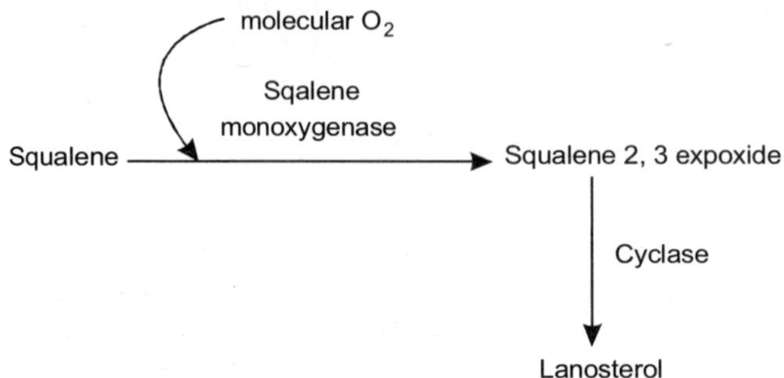

Conversion of Lanosterol to Cholesterol

Main changes are as follows:

(1) Removal of three angular CH_3 group. This involves a series of reaction mechanism-demethylation is not properly known. CH_3 group at C14 is first eliminated

(2) Shift of double bond between C_8 and C_9 to C_5 and C_6

(3) Saturation of double bond in side chain

Therefore two possible pathways are been illustrated

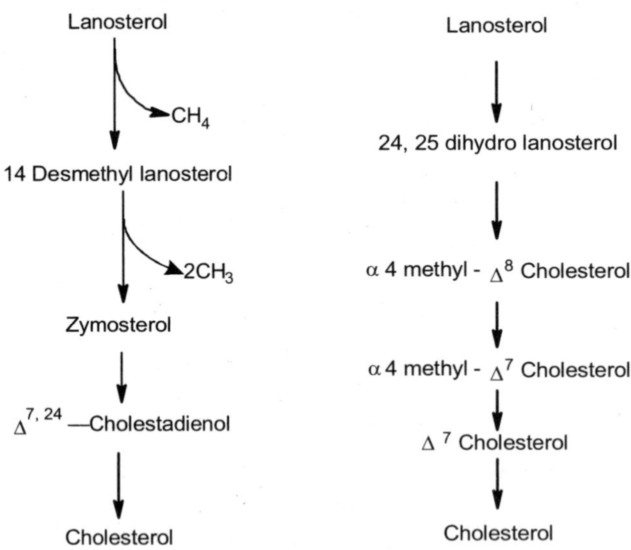

CONTROL OF CHOLESTEROL BIOSYNTHESIS

1. The step in the biosynthesis of cholesterol to HMG CoA are reversible. The formation of Mevalnoate however in the next step is "irreversible" and is the "commited step".

2. The "rate limiting step in the biosyntheiss of cholesterol is the conversion of cholesterol is the conversion of HMG CoA to Mevalnoate by enzyme HMG CoA reductase*
3. Fast starving also inhibits the enzyme and activate HMG – CoAlyase to form ketone bodies.
4. A second control point appears to be at the Cyclization of sqwalene and conversion to Lanosterol, but details of the regulation at this step is not clear.
5. The feeding of cholesterol reduces the hepatic biosynthesis of cholesterol by reducing the activity of HMG-CoA reductase.

   In contrast feeding of diets high in fat or carbohydrates tend to increase hepatic cholesterol biosynthesis.
6. Role of cyclic AMP

   HMG-CoA reductase may exist in active and 'inactive' forms which is 'reversely' modified by phosphorylation dephosphorylation mechanisms which mechanisms which may be C-AMP dependant protein kinases Cyclic AMP inhibits cholesterol biosynthesis by converting HMF-CoA reductase to inactivate form.

CHOLESTEROL LOWERING DRUGS THEIR MECHANISM OF ACTION

Several drugs are known to block the formation of cholesterol are various stages in the biosynthetic pathway. Some may increase the catabolism/excretion of cholesterol also; many of the drugs have harmful side effects. The table below shows the drugs used cholesterol limiting.

Drugs	Mechanism of action
1	2
1. Aromatically substituted carboxylic acids—e.g. p-phenyl butyrate, p-bi-phenyl butyrate, p-bi-phenyl butyrate	• Inhibits acetate incorporation. Experiments in human volunteers-disappointing (Not used).
2. Triparanol compounds related to non-steroidal estrogens and estrogen antagonists	• Inhibits reduction of desmosterol. After widespread use-the drug was withdrawn due to side effects like—cataract, alopecia, change in hair colour, Leucopenia. (Not used)
3. Pro-adifen HCl	• Blocks pathway between mevalonate to squalene (Not used-due to side efects)

*Cholesterol itself inhibits the enzymes producing an effective product feedback inhibition controlling the synthesis.

1	2
4. *Nicotinic acid*— In large doses has hypocholesterolaemic effect.	• Reduces the flux of FFA by inhibiting adipose issue lipolysis, thereby inhibiting VLDL production in Liver. In larges does—may produce fatty Liver.
5. Oestrogen	• Lower cholesterol level and increases HDl.
6. Sitosterol	• Acts by blocking esterification of cholesterolin gut thus reducing cholesterol absorption. (Synthesis may be increased later on).
7. Dexitrothyroxine (cholaxin), and Neomycin	• Increases faecal excretion of cholesterol and bile acids.
8. Clofibrate (Atromid S) Gemfibrozil CPI B (Ethyl-p-cholorophenoxy isobutyrate)	• Acts by various ways— (*i*) inhibiting secretion of VLDL by liver, (*ii*) inhibiting hepatic cholesterol synthesis, (*iii*) probbly also increases faecal excretion, (*iv*) they facilitate hydrolysis of VLDL triacylglycerol by lipoprotein lipase (commonly used)
9. Certain Resins, e.g. • colestipol • cholestyramine (Questran)	• Prevent the reabsorption of bile salts by combining with them, increasing their faecal loss.
10. Probucol	• Increases catabolism of LDL by receptor—independant pathway.
11. Mevastatin • Lovostatin (Recent drugs—obtained from fungi)	• Reducess LDL cholesterol level—Few adverse effects (most commonly used)

TRIGLYCERIDE AND PHOSPHOLIPID BIOSYNTHESIS

Fatty acids are stored for future use as triacylglycerols (TAGs) in all cells, but primarily in adipocytes of adipose tissue. TAGs constitute molecules of glycerol to which three fatty acids have been esterified. The fatty acids present in TAGs are predominantly saturated. The major building block for the synthesis of TAGs, in tissues other than adipose tissue, is glycerol. Adipocytes lack glycerol kinase, therefore, dihydroxyacetone phosphate (DHAP), produced during glycolysis, is the precursor for TAG synthesis in adipose tissue. This means that adipoctes must have glucose to oxidize in order to store fatty acids in the form of TAGs. DHAP can also serve as a backbone precursor for TAG synthesis in tissues other than adipose, but does so to a much lesser extent than glycerol.

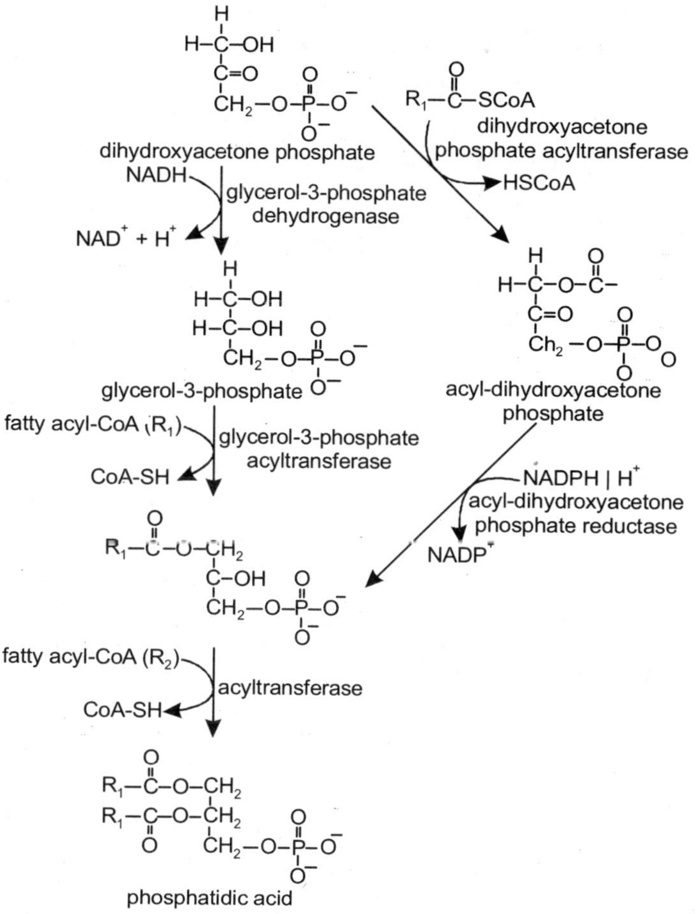

Phosphatidic acid Synthesis 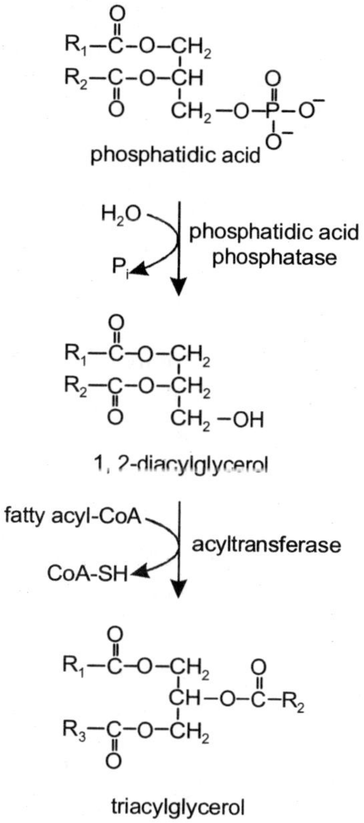 **Triglyceride Synthesis**

The glycerol backbone of TAGs is activated by phosphorylation at the C-3 position by glycerol kinase. The utilization of DHAP for the backbone is carried out through either of two pathways depending upon whether the synthesis of triglycerides is carried out in the mitochondria and ER or the ER and the peroxisomes. In the former case the action of glycerol-3-phosphate dehydrogenase, a reaction that requires NADH (the same reaction as that used in the glycerol-phosphate shuttle), converts DHAP to glycerol-3-phosphate. Glycerol-3-phosphate acyltransferase then esterifies a fatty acid to glycerol-3-phosphate generating the monoacylglycerol phosphate structure called lysophosphatidic acid. The second reaction pathway utilizes the peroxisomal enzyme DHAP acyltransferase to fatty acylate DHAP to acyl-DHAP which is then reduced by the NADPH-requiring enzyme acyl-DHAP reductase. An interesting feature of the latter pathway is that DHAP acyltransferase is one of only a few enzymes that are targeted to the peroxisomes through the recognition of a peroxisome targeting sequence 2 (PTS2) motif in the enzyme. Most peroxisomal enzymes contain a PTS1 motif. For more information on peroxisome enzymes

The fatty acids incorporated into TAGs are activated to acyl-CoAs through the action of acyl-CoA synthetases. Two molecules of acyl-CoA are esterified to glycerol-3-phosphate to yield 1,2-diacylglycerol phosphate (commonly identified as phosphatidic acid). The phosphate is then removed, by phosphatidic acid phosphatase (PAP1), to yield 1,2-diacylglycerol, the substrate for addition of the third fatty acid. Intestinal monoacylglycerols, derived from the hydrolysis of dietary fats, can also serve as substrates for the synthesis of 1,2-diacylglycerols.

Recent studies have identified a critical role for the enzyme PAP1 in overall TAG and phospholipid homeostasis. In the yeast Saccharomyces cerevisiae, the PAP1 gene was identified as Smp2p and the encoded protein was shown to be the yeast ortholog of the mammalian protein called lipin-1. The fission yeast lipin-1 ortholog is identified as Ned1p. Lipin-1 is only one of four lipin proteins identified in mammals. The lipin-1 gene (symbol = LPN1) was originally identified in a mutant mouse called the fatty liver dystrophy (fld) mouse. The mutation causing this disorder was found to reside in the LPN1 gene. There are three lipin genes with the LPN1 gene encoding two isoforms derived through alternative splicing. These two lipin-1 isoforms are identified as lipin-1A and lipin-1B. Mutations in the LPN2 gene have recently been associated with Majeed syndrome which is characterized by chronic recurrent osteomyelitis, cutaneous inflammation, recurrent fever, and congenital dyserythropoietic anemia. In addition to the obvious role of lipin-1 in TAG synthesis, evidence indicates that the protein is also required for the development of mature adipocytes, coordination of peripheral tissue glucose and fatty acid storage and utilization, and serves as a transcriptional co-activator. The latter function has significance to diabetes as it has been shown that some of the effects of the thiazolidinedione (TZD) class of drugs used to treat the hyperglycemia associated with type 2 diabetes are exerted via the effects of lipin-1.

Phospholipid Structures

Phospholipids are synthesized by esterification of an alcohol to the phosphate of phosphatidic acid (1,2-diacylglycerol 3-phosphate). Most phospholipids have a saturated fatty acid on C-1 and an unsaturated fatty acid on C-2 of the glycerol backbone. The most commonly added alcohols (serine, ethanolamine and choline) also contain nitrogen that may be positively charged, whereas, glycerol and inositol do not. The major classifications of phospholipids are:

Structure	Name
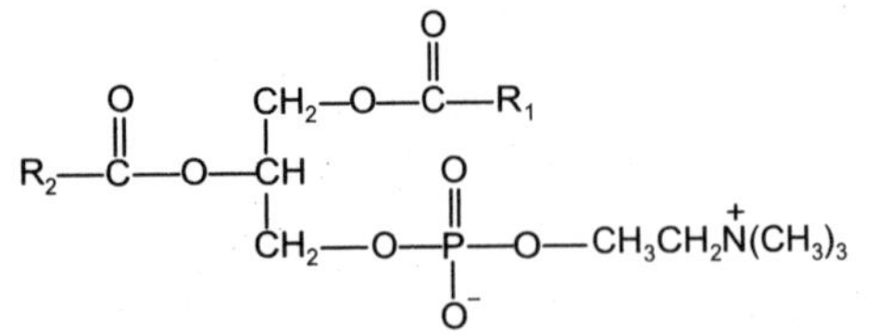 $CH_2{-}O{-}C({=}O){-}R_1$ $R_2{-}C({=}O){-}O{-}CH$ $CH_2{-}O{-}P({=}O)(O^-){-}O{-}CH_3CH_2\overset{+}{N}(CH_3)_3$	Phosphatidylcholine (PC)
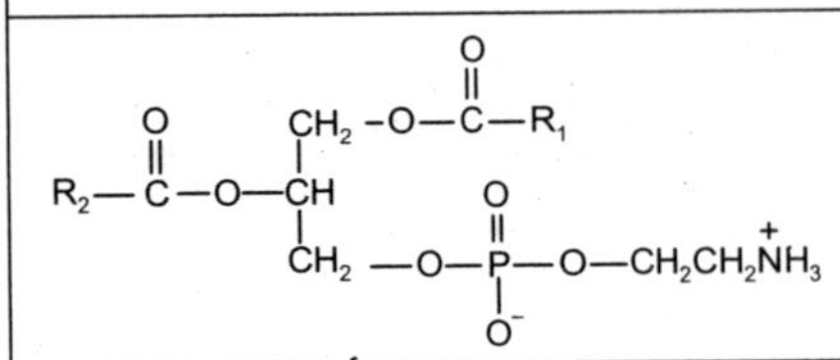 $CH_2{-}O{-}C({=}O){-}R_1$ $R_2{-}C({=}O){-}O{-}CH$ $CH_2{-}O{-}P({=}O)(O^-){-}O{-}CH_2CH_2\overset{+}{N}H_3$	Phosphatidylethanolamine (PE)

Structure	Name
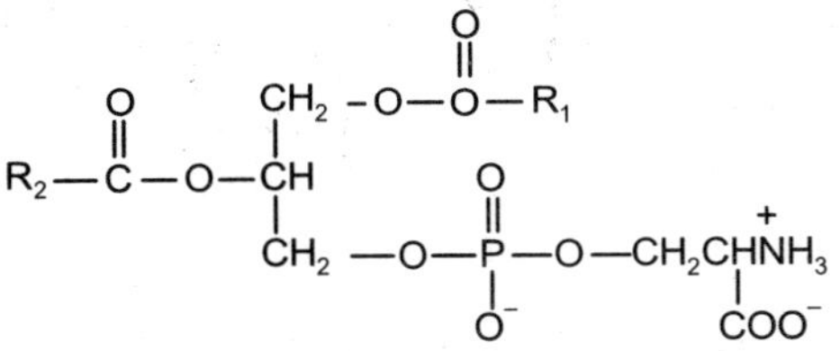 $R_2—\overset{O}{\overset{\|\|}{C}}—O—CH$; $CH_2—O—\overset{O}{\overset{\|\|}{C}}—R_1$; $CH_2—O—\overset{O}{\overset{\|\|}{P}}(O^-)—O—CH_2CH(COO^-)\overset{+}{N}H_3$	Phosphatidylserine (PS)
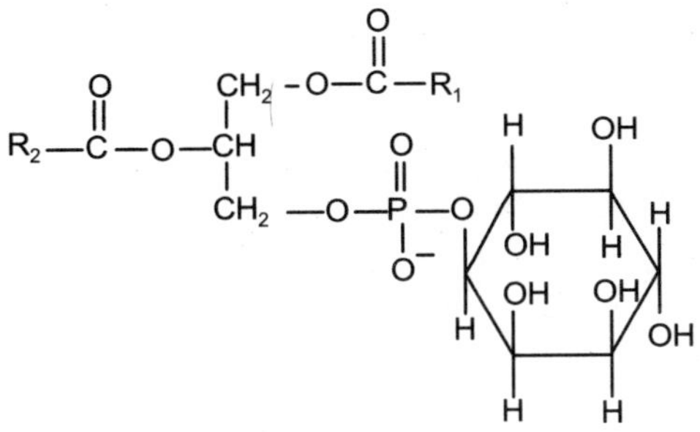 $R_2—\overset{O}{\overset{\|\|}{C}}—O—CH$; $CH_2—O—\overset{O}{\overset{\|\|}{C}}—R_1$; $CH_2—O—\overset{O}{\overset{\|\|}{P}}(O^-)—O$–inositol ring (H, OH, H, OH, OH, H, OH, OH, H, H, H)	Phosphatidylinositol (PI)
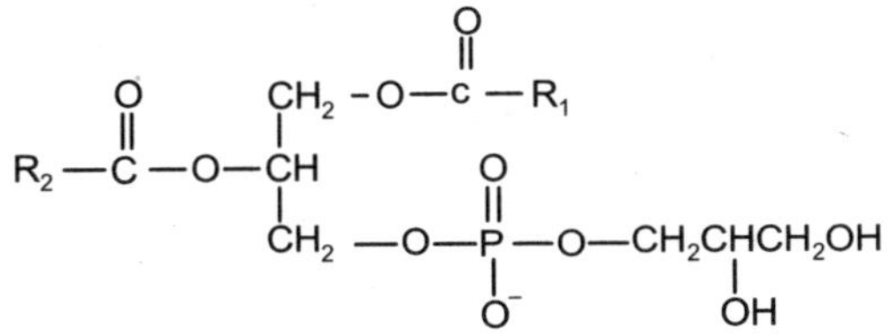 $R_2—\overset{O}{\overset{\|\|}{C}}—O—CH$; $CH_2—O—\overset{O}{\overset{\|\|}{C}}—R_1$; $CH_2—O—\overset{O}{\overset{\|\|}{P}}(O^-)—O—CH_2CH(OH)CH_2OH$	Phosphatidylglycerol (PG)
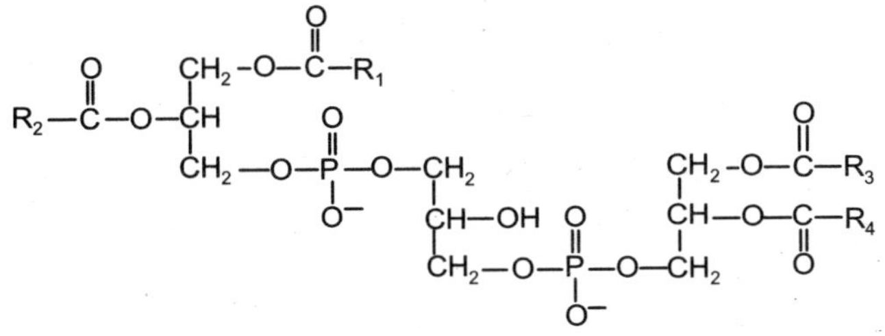 $R_2—\overset{O}{\overset{\|\|}{C}}—O—CH$; $CH_2—O—\overset{O}{\overset{\|\|}{C}}—R_1$; $CH_2—O—\overset{O}{\overset{\|\|}{P}}(O^-)—O—CH_2—CH(OH)—CH_2—O—\overset{O}{\overset{\|\|}{P}}(O^-)—O—CH_2—CH(—O—\overset{O}{\overset{\|\|}{C}}—R_4)—CH_2—O—\overset{O}{\overset{\|\|}{C}}—R_3$	Diphosphatidylglycerol (DPG)

Phospholipid Synthesis

Phospholipids can be synthesized by two mechanisms. One utilizes a CDP-activated polar head group for attachment to the phosphate of phosphatidic acid. The other utilizes CDP-activated 1,2-diacylglycerol and an inactivated polar head group.

- **PC:** This class of phospholipids is also called the lecithins. At physiological pH, phosphatidylcholines are neutral zwitterions. They contain primarily palmitic or stearic acid at carbon 1 and primarily oleic, linoleic or linolenic acid at carbon 2. The lecithin dipalmitoyllecithin is a component of lung or pulmonary surfactant. It contains palmitate at both carbon 1 and 2 of glycerol and is the major (80%) phospholipid found in the extracellular lipid layer lining the pulmonary alveoli. Choline is activated first by phosphorylation and then by coupling to CDP prior to attachment to phosphatidic acid. PC is also synthesized by the addition of choline to CDP-activated 1,2-diacylglycerol. A third pathway to PC synthesis, involves the conversion of either PS or PE to PC. The conversion of PS to PC first requires decarboxylation of PS to yield PE; this then undergoes a series of three methylation reactions utilizing S-adenosylmethionine (SAM) as methyl group donor.

- **PE:** These molecules are neutral zwitterions at physiological pH. They contain primarily palmitic or stearic acid on carbon 1 and a long chain unsaturated fatty acid (e.g. 18:2, 20:4 and 22:6) on carbon 2. Synthesis of PE can occur by two pathways. The first requires that ethanolamine be activated by phosphorylation and then by coupling to CDP. The ethanolamine is then transferred from CDP-ethanolamine to phosphatidic acid to yield PE. The second involves the decarboxylation of PS.
- **PS:** Phosphatidylserines will carry a net charge of –1 at physiological pH and are composed of fatty acids similar to the phosphatidylethanolamines. The pathway for PS synthesis involves an exchange reaction of serine for ethanolamine in PE. This exchange occurs when PE is in the lipid bilayer of the a membrane. As indicated above, PS can serve as a source of PE through a decarboxylation reaction.
- **PI:** These molecules contain almost exclusively stearic acid at carbon 1 and arachidonic acid at carbon 2. Phosphatidylinositols composed exclusively of non-phosphorylated inositol exhibit a net charge of –1 at physiological pH. These molecules exist in membranes with various levels of phosphate esterified to the hydroxyls of the inositol. Molecules with phosphorylated inositol are termed polyphosphoinositides. The polyphosphoinositides are important intracellular transducers of signals emanating from the plasma membrane. The synthesis of PI involves CDP-activated 1,2-diacylglycerol condensation with *myo*-inositol. PI subsequently undergoes a series of phosphorylations of the hydroxyls of inositol leading to the production of polyphosphoinositides. One polyphosphoinositide (phosphatidylinositol 4,5-bisphosphate, PIP_2) is a critically important membrane phospholipid involved in the transmission of signals for cell growth and differentiation from outside the cell to inside.
- **PG:** Phosphatidylglycerols exhibit a net charge of –1 at physiological pH. These molecules are found in high concentration in mitochondrial membranes and as components of pulmonary surfactant. Phosphatidylglycerol also is a precursor for the synthesis of cardiolipin. PG is synthesized from CDP-diacylglycerol and glycerol-3-phosphate. The vital role of PG is to serve as the precursor for the synthesis of diphosphatidylglycerols (DPGs).
- **DPG:** These molecules are very acidic, exhibiting a net charge of –2 at physiological pH. They are found primarily in the inner mitochondrial membrane and also as components of pulmonary

surfactant. One important class of diphosphatidylglycerols is the cardiolipins. These molecules are synthesized by the condensation of CDP-diacylglycerol with PG.

The fatty acid distribution at the C-1 and C-2 positions of glycerol within phospholipids is continually in flux, owing to phospholipid degradation and the continuous phospholipid remodeling that occurs while these molecules are in membranes. Phospholipid degradation results from the action of phospholipases. There are various phospholipases that exhibit substrate specificities for different positions in phospholipids.

In many cases the acyl group which was initially transferred to glycerol, by the action of the acyl transferases, is not the same acyl group present in the phospholipid when it resides within a membrane. The remodeling of acyl groups in phospholipids is the result of the action of phospholipase A1 (PLA_1) and phospholipase A2 (PLA_2).

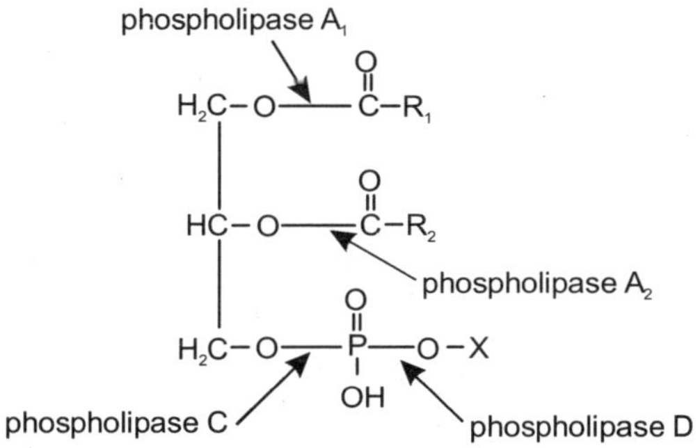

phospholipase A_1
$H_2C-O-\overset{O}{\overset{\|}{C}}-R_1$
$HC-O-\overset{O}{\overset{\|}{C}}-R_2$
phospholipase A_2
$H_2C-O-\overset{O}{\overset{\|}{P}}(OH)-O-X$
phospholipase C
phospholipase D

Sites of Action of the Phospholipases A1, A2, C and D

The products of these phospholipases are called lysophospholipids and can be substrates for acyl transferases utilizing different acyl-CoA groups. Lysophospholipids can also accept acyl groups from other phospholipids in an exchange reaction catalyzed by lysolecithin:lecithin acyltransferase (LLAT).

PLA2 is also an important enzyme, whose activity is responsible for the release of arachidonic acid from the C-2 position of membrane phospholipids. The released arachidonate is then a substrate for the synthesis of the eicosanoids. In fact there is not just a single PLA2 enzyme. At least 19 enzymes have been identified with PLA2activity. There are 10 isozymes that are in the secretory pathway and these PLA2 isozymes are abbreviated sPLA2. These secretory enzymes are low molecular weight proteins that are $Ca2^+$-requiring and are involved in numerous processes including modification of eicosanoid generation, host defense, and inflammation. Like the sPLA2 enzymes, the cPLA2 enzymes are tightly regulated by $Ca2^+$. In addition, this class of PLA2 enzyme is regulated by phosphorylation. There is an additional family of two PLA2 isozymes that are not dependent on $Ca2^+$ for activity and they identified as iPLA2. This latter class of enzyme is involved primarily with the remodeling of phospholipids.

METABOLISMS OF PROTEINS

Diatary Requirements

Proteins are broken down in the stomach during digestion by enzymes known as proteases into smallerpolypeptides to provide amino acids for the organism, including the essential amino acids that the organism cannot biosynthesize itself. Aside from their role in protein synthesis, amino acids are also important nutritional sources of nitrogen.

Proteins, like carbohydrates, contain 4 kilocalories per gram as opposed to lipids which contain 9 kilocalories and alcohols which contain 7 kilocalories. The liver, and to a much lesser extent the kidneys, can convert amino acids used by cells in protein biosynthesis into glucose by a process known asgluconeogenesis. The amino acids leucine and lysine are exceptions.

Dietary sources of protein include meats, eggs, nuts, grains, legumes, and dairy products such as milk and cheese. Of the 20 amino acids used by humans in protein synthesis, 11 "non-essential" amino acids can be synthesized in sufficient quantities by the adult body, and are not required in the diet (though there are exceptions for some in special cases). The nine essential amino acids, plus arginine for the young, cannot be created by the body and must come from dietary sources.

Most animal sources and certain vegetable sources have the complete complement of all the essential amino acids in adequate proportions. However, it is not necessary to consume a single food source that contains all the essential amino acids, as long as all the essential amino acids are eventually present in the diet: see complete protein and protein combining.

Different proteins have different levels of biological availability (BA) to the human body. Many methods have been introduced to measure protein utilization and retention rates in humans. They include biological value, net protein utilization, and PDCAAS (Protein Digestibility Corrected Amino Acids Score) which was developed by the FDA as an improvement over the Protein Efficiency Ratio (PER) method. These methods examine which proteins are most efficiently used by the body. In general they conclude that animal complete proteins that contain all the essential amino acids such as milk, eggs, and meat, and the complete vegetable protein soy are of most value to the body.

Egg whites have been determined to have the standard biological value of 100 (though some sources may have biological values higher), which means that most of the absorbed nitrogen from egg white protein can be retained and used by the body. The biological value of plant protein sources is usually considerably lower than animal sources. For example, corn has a BA of 70 while peanuts have a relatively low BA of 40.

According to the recently updated US/Canadian Dietary Reference Intake guidelines, women aged 19-70 need to consume 46 grams of protein per day, while men aged 19-70 need to consume 56 grams of protein per day to avoid a deficiency. The difference is because men's bodies generally have more muscle mass than those of women, or this may be attributed to weight difference by taking 0.8 g (of protein)/kg of lean body weight.

Because the body is continually breaking down protein from tissues, even adults who do not fall into the above categories need to include adequate protein in their diet every day. If enough energy is not taken in through diet, as in the process of starvation, the body will use protein from the muscle mass to meet its energy needs, leading to muscle wasting over time. If the body does not consume adequate protein in nutrition, then muscle will also waste as more vital cellular processes (e.g. respiration enzymes, blood cells) recycle muscle protein for their own requirements. Other recommendations suggest 0.8 gram of protein per kilogram of lean bodyweight per day while other sources suggest that higher intakes of 1-1.4 grams of protein per kilogram of bodyweight for enhanced athletes or those with a large muscle mass. How much protein needed in a person's daily diet is determined in large part by overall energy intake, as well as by the body's need for nitrogen and essential amino acids. Physical activity and exertion as well as enhanced muscular mass increase the need for protein.

Requirements are also greater during childhood for growth and development, during pregnancy or when breast-feeding in order to nourish a baby, or when the body needs to recover from malnutrition or trauma or after an operation. Protein deficiency is a serious cause of ill health and death in developing countries. Protein deficiency plays a part in the disease kwashiorkor. War, famine, overpopulation and other factors can increase rates of malnutrition and protein deficiency. Protein deficiency can lead to reduced intelligence or mental retardation, see deficiency in proteins, fats, carbohydrates. In countries that suffer from widespread protein deficiency, food is generally full of plant fibers, which makes adequate energy and protein consumption very difficult.

Symptoms of kwashiorkor include apathy, diarrhea, inactivity, failure to grow, flaky skin, fatty liver, and edema of the belly and legs. This edema is explained by the normal functioning of proteins in fluid balance and lipoprotein transport. Researcher claims that malnutrition is a frequent cause of death and disease in third world countries. Protein-energy malnutrition (PEM) affects 500 million people and kills 10 million annually. In severe cases white blood cell numbers decline and the ability of leukocytes to fight infection decreases. Protein deficiency is relatively rare in developed countries but some people have difficulty getting sufficient protein due to poverty.

Protein deficiency can also occur in developed countries in people who are dieting or crash dieting to lose weight, or in older adults, who may have a poor diet. Convalescent people recovering from surgery, trauma, or illness may become protein deficient if they do not increase their intake to support their increased needs. The body is unable to store excess protein. Protein is digested into amino acids which enter the bloodstream. Excess amino acids are converted to other usable molecules by the liver in a process called deamination. Deamination converts nitrogen from the amino acid into ammonia which is converted by the liver into urea in the urea cycle. Excretion of urea is performed by the kidneys. These organs can normally cope with any extra workload but if kidney disease occurs, a decrease in protein will often be prescribed.

Many researchers think excessive intake of protein forces increased calcium excretion. If there is to be excessive intake of protein, it is thought that a regular intake of calcium would be able to stabilize, or even increase the uptake of calcium by the small intestine, which would be more beneficial in older women. Specific proteins are often the cause of allergies and allergic reactions to certain foods. This is because the structure of each form of protein is slightly different; some may trigger a response from the immune system while others remain perfectly safe.

Many people are allergic to casein, the protein in milk; gluten, the protein in wheat and other grains; the particular proteins found in peanuts; or those in shellfish or other sea foods. The classic assay for protein concentration in food is the Kjeldahl method. This test determines the total nitrogen in a sample. The only major component of most food which contains nitrogen is protein (fat, carbohydrate and dietary fibre do not contain nitrogen).

If the amount of nitrogen is multiplied by a factor depending on the kinds of protein expected in the food the total protein can be determined. On food labels the protein is given by the nitrogen multiplied by 6.25, because the average nitrogen content of proteins is about 16%. The Kjeldahl test is used because it is the method the AOAC International has adopted and is therefore used by many food standards agencies around the world. The limitations of the Kjeldahl method were at the heart of the 2008 Chinese Milk Scandal in which the toxic chemical melamine was added to the milk to increase the measured "protein".

BIOLOGICAL VALUE

Biological value (BV) is a measure of the proportion of absorbed protein from a food which becomes incorporated into the proteins of the organism's body. It summarizes how readily the broken down protein can be used in protein synthesis in the cells of the organism. Proteins are the major source of nitrogen food, unlike carbohydrates and fats. This method assumes protein is the only source of nitrogen and measures the proportion of this nitrogen absorbed by the body which is then excreted. The remainder must have been incorporated into the proteins of the organisms body.

A ratio of nitrogen incorporated into the body over nitrogen absorbed gives a measure of protein 'usability' - the BV. Unlike some measures of protein usability, biological value does not take into account how readily the protein can be digested and absorbed (largely by the small intestine). This is reflected in the experimental methods used to determine BV.

BV, confusingly, uses two similar scales:

1. The true percentage utilization (usually shown with a percent symbol).
2. The percentage utilization relative to a readily utilizable protein source, often egg (usually shown as unit less).

These two values will be similar but not identical.

The BV of a food varies greatly, and depends on a wide variety of factors. In particular the BV value of a food varies depending on its preparation and the recent diet of the organism. This makes reliable determination of BV difficult and of limited use - fasting prior to testing is universally required in order to make the values reliable. BV is commonly used in nutrition science in many mammalian organisms, and is a relevant measure in humans. It is a popular guideline in body building in protein choice.

For accurate determination of BV

1. the test organism must only consume the protein or mixture of proteins of interest (the test diet).
2. the test diet must contain no non-protein sources of nitrogen.
3. the test diet must be of suitable content and quantity to avoid use of the protein primarily as an energy source.

These conditions mean the tests are typically carried out over the course of over one week with strict diet control. Fasting prior to testing helps produce consistency between subjects (it removes recent diet as a variable). There are two scales on which BV is measured; percentage utilization and relative utilization. By convention percentage BV has a percent sign (%) suffix and relative BV has no unit.

Percentage Utilization

Biological value is determined based on this formula.

$$BV = (N_r / N_a) * 100$$

Where:

N_a = nitrogen absorbed in proteins on the test diet

N_r = nitrogen incorporated into the body on the test diet

However direct measurement of Nr is essentially impossible. It will typically be measured indirectly from nitrogen excretion in urine. Faecalexcretion of nitrogen must also be taken into account - this protein is not absorbed by the body and so not included in the calculation of BV.

$$BV = ((N_i - N_{e(f)} - N_{e(u)} - N_b) / N_i - N_{e(f)}) * 100$$

Where:

N_i = nitrogen intake in proteins on the test diet

$N_{e(f)}$ = nitrogen excreted in faeces whilst on the test diet

$N_{e(u)}$ = nitrogen excreted in urine whilst on the test diet

N_b = nitrogen excreted on a protein free diet

Note:

$$N_r = N_i - N_{e(f)} - N_{e(u)} - N_b$$

$$N_a = N_i - N_{e(f)}$$

This can take any value of 100 or less, including negative. A BV of 100% indicates complete utilization of a dietary protein, i.e. 100% of the protein ingested and absorbed is incorporated into proteins into the body. Negative values are possible if excretion of nitrogen exceeds intake in proteins. All non-nitrogen containing diets have negative BV. The value of 100% is an absolute maximum, no more than 100% of the protein ingested can be utilized (in the equation above Ne(u), Ne(f) and Nb cannot go negative, setting 100% as the maximum BV).

Due to experimental limitations BV is often measured relative to an easily utilizable protein. Normally egg protein is assumed to be the most readily utilizable protein and given a BV of 100. For example:

Two tests of BV are carried out on the same person; one with the test protein source and one with a reference protein (egg protein).

$$\text{relative BV} = (BV_{(test)} / BV_{(egg)}) * 100$$

Where:

$BV_{(test)}$ = percentage BV of the test diet for that individual

$BV_{(egg)}$ = percentage BV of the reference (egg) diet for that individual

This is not restricted to values of less than 100. The percentage BV of egg protein is only 93.7% which allows other proteins with true percentage BV between 93.7% and 100% to take a relative BV of over 100. For example, whey protein* takes a relative BV of 104, while its percentage BV is under 100%.

The principal advantage of measuring BV relative to another protein diet is accuracy; it helps account for some of the metabolic variability between individuals. In a simplistic sense the egg diet is testing the maximum efficiency the individual can take up protein, the BV is then provided as a percentage taking this as the maximum.

Providing it is known which protein measurements were made relative to it is simple to convert from relative BV to percentage BV:

$$BV_{(percentage)} = (BV_{(relative)} / BV_{(reference)}) * 100$$

$$BV_{(relative)} = (BV_{(percentage)} / 100) * BV_{(reference)}$$

Where:

$BV_{(relative)}$ = relative BV of the test protein

$BV_{(reference)}$ = percentage BV of reference protein (typically egg: 93.7%).

$BV_{(percentage)}$ = percentage BV of the test protein

• Whey protein is a mixture of globular proteins isolated from whey, the liquid material created as a by-product of cheese production. Some preclinical studies in rodents have suggested that whey protein may influence glutathione production and possess anti-inflammatory or anti-cancer properties; however, human data are lacking. Whey protein is commonly marketed and ingested as a dietary supplement, and various health claims have been attributed to it in the alternative medicine community. Whey is a common allergen, and is responsible for a significant proportion of cow milk allergies. In people without milk allergies, whey protein is generally considered safe.

While this conversion is simple it is not strictly valid due to the differences between the experimental methods. It is, however, suitable for use as a guideline.

The determination of BV is carefully designed to accurately measure some aspects of protein usage whilst eliminating variation from other aspects. When using the test (or considering BV values) care must be taken to ensure the variable of interest is quantified by BV. Factors which affect BV can be grouped into properties of the protein source and properties of the species or individual consuming the protein.

Three major properties of a protein source affect its BV:

- Amino acid composition, and the limiting amino acid, which is usually lysine;
- Preparation (cooking); and
- Vitamin and mineral content.

Amino acid composition is the principal effect. All proteins are made up of combinations of the 21 biological amino acids. Some of these can be synthesised or converted in the body, whereas others cannot and must be ingested in the diet. These are known as essential amino acids (EAAs), of which there are 9 in humans. The number of EAAs varies according to species. EAAs missing from the diet prevent the synthesis of proteins that require them. If a protein source is missing critical EAAs, then its biological value will be low as the missing EAAs form a bottleneck in protein synthesis. For example, if a hypothetical muscle protein requiresphenylalanine (an essential amino acid), then this must be provided in the diet for the muscle protein to be produced.

If the current protein source in the diet has no phenylalanine in it the muscle protein cannot be produced, giving a low usability and BV of the protein source. In a related way if amino acids are missing from the protein source which are particularly slow or energy consuming to synthesise this can result in a low BV. Methods of food preparation also have an impact on availability of amino acids in a food source. Some of food preparation may damage or destroy some EAAs, reducing the BV of the protein source. Many vitamins and minerals are vital for the correct function of cells in the test organism.

If critical minerals or vitamins are missing from the protein source this can result in a massively lowered BV. Many BV tests artificially add vitamins and minerals (for example in yeast extract) to prevent this. Variations in BV under test conditions are dominated by the metabolism of the individuals or species being tested. In particular differences in the essential amino acids (EAAs) species to species has a significant impact, although even minor variations in amino acid metabolism individual to individual have a large effect.

The fine dependence on the individual's metabolism makes measurement of BV a vital tool in diagnosing some metabolic diseases. The principal effect on BV in everyday life is the organism's current diet, although many other factors such as age, health, weight, sex, etc. all have an effect. In short any condition which can affect the organism's metabolism will vary the BV of a protein source. In particular, whilst on a high protein diet the BV of all foods consumed is reduced - the limiting rate at which the amino acids may be incorporated into the body is not the availability of amino acids but the rate of protein synthesis possible in cells. This is a major point of criticism of BV as a test; the test diet is artificially protein rich and may have unusual effects.

BV is designed to ignore variation in digestibility of a food - which in turn largely depends on the food preparation. For example compare raw soy beans and extracted soy bean protein. The raw soy beans, with tough cell walls protecting the protein, have a far lower digestibility than the purified, unprotected, soy bean protein extract. As a foodstuff far more protein can be absorbed from the extract than the raw beans, however the BV will be the same. The exclusion of digestibility is a point of misunderstanding and leads to misrepresentation of the meaning of a high or low BV

BV provides a good measure of the usability of proteins in a diet and also plays a valuable role in detection of some metabolic diseases. BV is, however, a scientific variable determined under very strict and unnatural conditions. It is not a test designed to evaluate the usability of proteins whilst an organism is in everyday life - indeed the BV of a diet will vary greatly depending on age, weight, health, sex, recent diet, current metabolism, etc. of the organism. In addition BV of the same food varies significantly species to species. Given these limitations BV is still relevant to everyday diet to some extent. No matter the individual or their conditions a protein source with high BV, such as egg, will always be more easily used than a protein source with low BV.

There are many other major methods of determining how readily used a protein is, including:

- Net protein Utilization (NPU)
- Protein Efficiency Ratio (PER)
- Nitrogen Balance (NB)
- Protein digestibility (PD)
- Protein Digestibility Corrected Amino Acid Score (PDCAAS)

These all hold specific advantages and disadvantages over BV, although in the past BV has been held in high regard.

The Biological Value method is also used for analysis in animals such as cattle, poultry, and various laboratory animals such as rats. It was used by the poultry industry to determine which mixtures of feed were utilized most efficiently by developing chicken. Although the process remains the same, the biological values of particular proteins in humans differs from their biological values in animals due to physiological variations.

Common foodstuffs and their values: Note: this scale uses 100 as 100% of the nitrogen incorporated.

• Isolated Whey	100
• Whole bean	96
• Whole Soy Bean	96
• Human milk	95
• Chicken egg	94

Note: These values use "whole egg" as a value of 100, so foodstuffs that provide even more nitrogen than whole eggs, can have a value of more that 100, 100, does not mean that 100% of the nitrogen in the food is incorporated into the body, and not excreted, as in in other.

Food	Value
Soybean milk	91
Cow milk	90
Cheese	84
Rice	83
Defatted soy flour	81
Fish	76
Beef	74.3
Immature bean	65
Full-fat soy flour	64
Soybean curd (tofu)	64
Whole wheat	64
White flour	41

Common foodstuffs and their values:

Whey protein concentrate:	104
Whole egg	100
Cow milk	91
Beef	80
Casein	77
Soy	74
Wheat gluten	64

Since the method measures only the amount that is retained in the body critics have pointed out what they perceive as a weakness of the biological value methodology. Critics have pointed to research that indicates that because whey protein isolate is digested so quickly it may in fact enter the bloodstream and be converted into carbohydrates through a process called gluconeogenesis much more rapidly than was previously thought possible, so while amino acid concentrations increased with whey it was discovered that oxidation rates also increased and a steady-state metabolism, a process where there is no change in overall protein balance, is created. They claim that when the human body consumes whey protein it is absorbed so rapidly that most of it is sent to the liver for oxidation. Hence they believe the reason so much is retained is that it is used for energy production not protein synthesis. This would bring into question whether the method defines which proteins are more biologically utilizable.

A further critique published in the Journal of Sports Science and Medicine states that the BV of a protein does not take into consideration several key factors that influence the digestion and interaction of protein with other foods before absorption, and that it only measures a proteins maximal potential quality and not its estimate at requirement levels. Also, the study which is often cited to demonstrate the superiority of whey protein hydrolysate by marketers, measured nitrogen balance in rats after three days of starvation, which corresponds to a longer period in humans. The study found that whey protein hydrolysate led to better nitrogen retention and growth than the other proteins studied. However the study's flaw is in the BV method used, as starvation affects how well the body will store incoming protein (as does a very high caloric intake), leading to falsely elevated BV measures.

So, the BV of a protein is related to the amount of protein given. BV is measured at levels below the maintenance level. This means that as protein intake goes up, the BV of that protein goes down. For example, milk protein shows a BV near 100 at intakes of 0.2 g/kg. As protein intake increases to roughly maintenance levels, 0.5 g/kg, BV drops only around 70.[20] Pellet et al., concluded that "biological measures of protein quality conducted at suboptimal levels in either experimental animals or human subjects may overestimate protein value at maintenance levels." As a result, while BV may be important for rating proteins where intake is below requirements, it has little bearing on individuals with protein intakes far above requirements.

This flaw is supported by the FAO/WHO/UNU, who state that BV and NPU are measured when the protein content of the diet is clearly below that of requirement, deliberately done to maximize existing differences in quality as inadequate energy intake lowers the efficiency of protein utilization and in most N balance studies, calorie adequacy is ensured. And because no population derives all of its protein exclusively from a single food, the determination of BV of a single protein is of limited use for application to human protein requirements.

Another limitation of the use of Biological Value as a measure of protein quality is that proteins which are completely devoid of one essential amino acid (EAA) can still have a BV of up to 40. This is because of the ability of organisms to conserve and recycle EAAs as an adaptation of inadequate intake of the amino acid.

Lastly, the use of rats for the determination of protein quality is not ideal. Rats differ from humans in requirements of essential amino acids. This has led to a general criticism that experiments on rats lead to an over-estimation of the BV of high-quality proteins to man because human requirements of essential amino acids are much lower than those for rats (as rats grow at a much faster rate than humans). Also, because of their fur, rats are assumed to have relatively high requirements of sulphur-containing amino acids (methionine and cysteine).

As a result, the analytical method that is universally recognized by the FAO/WHO as well as the FDA, USDA, United Nations University (UNU) and the National Academy of Sciences when judging the quality of protein in the human is not PER or BV but the Protein Digestibility Corrected Amino Acid Score (PDCAAS), as it is viewed as accurately measuring the correct relative nutritional value of animal and vegetable sources of protein in the diet.

DIGESTION OF PROTEIN

Digestion in Digestive System

There are no proteolytic enzymes present in the mouth. After mastification and chewing the bolus of food reaches stomach where it meets the gastric juice, then begins the digestion of food in stomach.

Gastric juice contains a number of proteolytic enzymes. They are :

1. Pepsin,
2. Rennin,
3. Gastrisin,
4. Gelatenase.

Pepsin is a important proteolytic enzyme and is present in gastric juice of different species including the mammals. Pepsin is secreted as inactive zymogen form pepsinogen having a molecular weight 42,

500 approx. It is synthesized in 'chief cells' of stomach and 99% is poured gastric juice as pepsinogen. Remaining 1% is separated in the blood stream from where it is ultimately excreted in urine. The urinary pepsin is called Uropepsin.

Pepsinogen is hydrolyzed in stomach, with the help of HCl or pepsin itself (auto catalytically) to form the 'active pepsin having a molecular weight of (34,500). In the process of activation (*i*) an inactive peptide called 'pepsin inhibitor' and (*ii*) 5 smaller peptide are liberated.

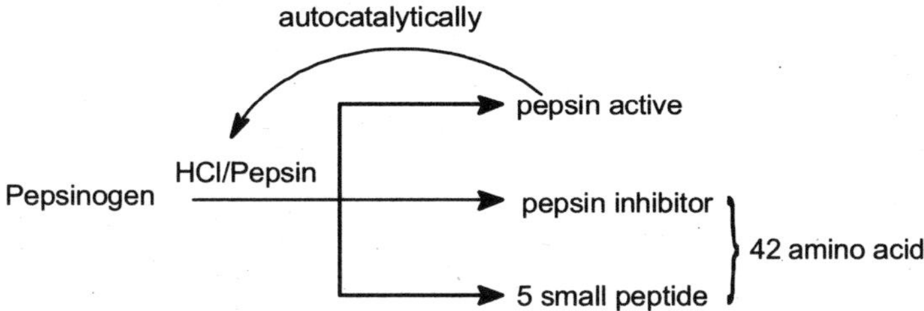

HCl maintains the gastric pH between 1 to 2 to ensure the proper pepsin activity. Optimum pH for pepsin is 1.6 to 2.5 and pepsin gets denaturized if the pH is greater than 5. Pepsin is protenase a non-specific endopeptidase and hydrolyses peptide bonds well inside the protein molecule and produced proteoses and peptones.

$$\text{Proteins} \xrightarrow{\text{Pepsin}} \text{Proteoses and Peptones}$$

It is particularly active on a peptide bond, which connects the – COOH group of an aromatic amino acid like Phe, Tyr and Tyrp with the amino group of either dicarboxylic or aromatic amino acid.

It can also hydrolyze the peptide bonds of

(*i*) COOH group of methionine and leucine

(*ii*) Leucine and glutamic acid

(*iii*) Glutamic acid and asparagines

(*iv*) Leucine valine and

(*v*) Valine and cystine

Pepsin cannot act on protein like terapenes, silfibrions mucoproteins mucoids and protamines.

$$\text{Casein (Soluble)} \xrightarrow[\text{Chymotrypsin}]{\text{Pepsin/Renin}} \text{Paracasein + Protease (whey protein)}$$

$$\text{Paracasein (Soluble)} + Ca^{++} \xrightarrow[\text{Spontaniousity}]{} \text{Calcium Paracaseinate (insolublecurd)}$$

Pepsin can also act on milk. It hydrolyzes the soluble phosphoprotein "casein" of milk to procedure "paracasein and proteose the latter is the whey protein. Paracasein is then precipitated as caparacaseinate which is further digested by pepsin to peptones.

Rennin not present in adult humans, nevertheless infants have little amount of rennin the optimum pH for its activity is 4 and the specificity for action of rennin is like pepsin, here rennin hydrolyses peptide bonds connected with L-aromatic amino acids. Like pepsin rennin also reacts with casein of milk to form paracasein which is immediately precipitated by Ca^{++}. Thus it is also coagulates milk pepsin.

Gastrisin is secreted in the gastric juice of humans as inactive zymogen form, which activated in the presence of HCl. The optimum activity of this enzyme is pH 3 to 4. It acts as a Protenase and requires an acidic medium for its activity.

After stomach the food stuff reaches Duodenum, the small intestine here it mixes with pancreatic juice. The pancreatic juice consists of several proteolytic enzymes which act on protein and partly digested food.

The main enzymes are:

1. Trypsin.
2. Chymotrypsin.
3. Carboxypeptidases.
4. Elastases.
5. Collagenases.

Trypsin (EC 3.4.21.4) is a serine protease found in the digestive system of many vertebrates, where it hydrolyses proteins. Trypsin is produced in the pancreas as the inactive proenzyme trypsinogen. Trypsin predominantly cleaves peptide chains at thecarboxyl side of the amino acids lysine and arginine, except when either is followed byproline. It is used for numerous biotechnological processes. The process is commonly referred to as trypsin proteolysis or trypsinisation and proteins that have been digested/treated with trypsin are said to have been trypsinized.

Trypsin is secreted into the duodenum, where it acts to hydrolyse peptides into their smaller building blocks, namely amino acids (these peptides are the result of the enzyme pepsin breaking down the proteins in the stomach). This is necessary for the uptake of protein in the food as though peptides are smaller than proteins, they are still too big to be absorbed through the lining of the ileum. Trypsin catalyses the hydrolysis of peptide bonds.

The enzymatic mechanism is similar to other serine proteases. These enzymes contain acatalytic triad consisting of histidine-57, aspartate-102, and serine-195. These three residues form a charge relay which serves to make the active site serine nucleophilic. This is achieved by modifying the electrostatic environment of the serine. The enzymatic reaction that trypsins catalyze is thermodynamically favourable but requires significantactivation energy (it is "kinetically unfavorable"). In addition, trypsin contains an "oxyanion hole" formed by the backbone amide hydrogen atoms of Gly-193 and Ser-195 which serves to stabilize the developing negative charge on the carbonyl oxygen atom of the cleaved amide.

The aspartate residue (Asp 189) located in the catalytic pocket (S1) of trypsins is responsible for attracting and stabilizing positively-charged lysine and/or arginine, and is thus responsible for the specificity of the enzyme. This means that trypsin predominantly cleaves proteins at the carboxyl side (or "C-terminal side") of the amino acids lysine andarginine, except when either is followed by proline. Trypsins are consideredendopeptidases, i.e., the cleavage occurs within the polypeptide chain rather than at the terminal amino acids located at the ends of polypeptides. Trypsins have an optimal operating pH of about 8 and optimal operating temperature of about 37°C.

Trypsin is produced in the pancreas in the form of inactive zymogen, trypsinogen. When the pancreas is stimulated by cholecystokinin, it is then secreted into the small intestine. Once in the small intestine, the enzyme enteropeptidase activates it into trypsin by proteolytic cleavage. The resulting trypsins themselves activate more trypsinogens (autocatalysis), so only a small amount of enteropeptidase is necessary to start the reaction. This activation mechanism is common for most serine proteases, and serves to prevent autodigestion of the pancreas.

The activity of trypsins is not affected by the inhibitor tosyl phenylalanyl chloromethyl ketone TPCK, which deactivates chymotrypsin. This is important because, in some applications, like mass spectrometry, the specificity of cleavage is important.

Trypsin is available in high quantity in pancreases, and can be purified rather easily. Hence it has been used widely in various biotechnological processes. In a tissue culture lab, trypsins are used to re-suspend cells adherent to the cell culture dish wall during the process of harvesting cells. Trypsin can also be used to dissociate dissected cells (for example, prior to cell fixing and sorting).

Trypsins can be used to break down casein in breast milk. If trypsin is added to a solution of milk powder, the breakdown of casein will cause the milk to become translucent. The rate of reaction can be measured by using the amount of time it takes for the milk to turn translucent.

Trypsin is commonly used in biological research during proteomics experiments to digest proteins into peptides for mass spectrometry analysis, e.g. in-gel digestion. Trypsin is particularly suited for this, since it has a very well defined specificity, as it hydrolyzes only the peptide bonds in which the carbonyl group is contributed either by an Arg or Lys residue. Trypsin can also be used to dissolve blood clots in its microbial form and treat inflammation in its pancreatic form. Trypsin is used in baby food to pre-digest it. It can break down the protein molecules which helps the baby to digest it as its stomach is not strong enough to digest bigger protein molecules.

Trypsin and chymotrypsin, like most proleotytic enzymes, are synthesized as inactive zymogen precursors (trypsinogen and chymotrypsinogen) to prevent unwanted destruction of cellular proteins, and to regulate when and where enzyme activity occurs. The inactive zymogens are secreted into the duodenum, where they travel the small and large intestines prior to excretion. Zymogens also enter the bloodstream, where they can be detected in serum prior to excretion in urine. Zymogens are converted to the mature, active enzyme by proteolysis to split off a pro-peptide, either in a subcellular compartment or in an extracellular space where they are required for digestion.

Trypsin and chymotrypsin are structurally very similar, although they recognise different substrates. Trypsin acts on lysine and arginine residues, while chymotrypsin acts on large hydrophobic residues such as tryptophan, tyrosine and phenylalanine, both with extraordinary catalytic efficiency. Both enzymes have a catalytic triad of serine, histidine and aspartate within the S1 binding pocket, although the hydrophobic nature of this pocket varies between the two, as do other structural interactions beyond the S1 pocket.

The human pancreas secretes three isoforms of trypsinogen: cationic (trypsinogen-1), anionic (trypsinogen-2) and mesotrypsinogen (trypsinogen-3). Cationic and anionic trypsins are the major isoforms responsible for digestive protein degradation, occurring in a ratio of 2:1, while mesotrypsinogen accounts for less than 5% of pancreatic secretions. Mesotrypsin is a specialised protease known for its resistance to trypsin inhibitors. It is thought to play a special role in the degradation of trypsin inhibitors, possibly to aid in the digestion of inhibitor-rich foods such as soybeans and lima beans. An alternatively spliced mesotrypsinogen in which the signal peptide is replaced with a different exon 1 is expressed in the human brain; the function of this brain trypsinogen is unknown.

There are two isoforms of pancreatic chymotrypsin, A and B, which are known to cleave proteins selectively at specific peptide bonds formed by the hydrophobic residues tryptophan, phenylalanine and tyrosine.

Chymotrypsin (bovine α-chymotrypsin: PDB 1AB9, EC 3.4.21.1) is a digestive enzyme that can perform proteolysis. Chymotrypsin cleaves peptides at the carboxyl side oftyrosine, tryptophan,

R NH NH O O R CH_3 OH N NH O O R R R Ser[195] His[57] Asp[102]

R NH NH O O R CH_3 HN N OH O R R R Ser[195] His[57] Asp[102] H_2O O R H_2N CH_3

R NH O O H O H N NH O O R R R Ser[195] His[57] Asp[102]

R NH O OH O HN N OH O R R R Ser[195] His[57] Asp[102] R NH O OH

OH N NH O O R R R Ser[195] His[57] Asp[102]

and phenylalanine because these three amino acids containaromatic rings, which fit into a 'hydrophobic pocket' in the enzyme. Over time, chymotrypsin also hydrolyzes other amide bonds, particularly those with leucine-donated carboxyls.

Chymotrypsin is synthesized in the pancreas by protein biosynthesis as a precursorcalled chymotrypsinogen that is enzymatically inactive. On cleavage by trypsin into two parts that are still connected via an S-S bond, cleaved chymotrypsinogen molecules can activate each other by removing two small peptides in a trans-proteolysis. The resulting molecule is active chymotrypsin, a three polypeptide molecule interconnected via disulfide bonds. *n vivo*, chymotrypsin is a proteolytic enzyme acting in the digestive systems of mammals and other organisms. It facilitates the cleavage of peptide bonds by ahydrolysis reaction, a process which albeit thermodynamically favourable, occurs extremely slowly in the absence of a catalyst. The main substrates of chymotrypsin include tryptophan, tyrosine, phenylalanine, leucine, and methionine, which are cleaved at the carboxyl terminal. Like many proteases, chymotrypsin will also hydrolyse ester bonds in vitro, a virtue that enabled the use of substrate analogs such as N-acetyl-L-phenylalanine p-nitrophenyl ester for enzyme assays.

Mechanism of Peptide Bond Cleavage in α-chymotrypsin

Chymotrypsin cleaves peptide bonds by attacking the unreactive carbonyl group with a powerful nucleophile, the serine 195 residue located in the active site of the enzyme, which briefly becomes covalently bonded to the substrate, forming an enzyme-substrate intermediate.

It was found that the reaction of chymotrypsin with its substrate takes place in two stages, an initial "burst" phase at the beginning of the reaction and a steady-state phase following Michaelis-Menten kinetics. It is also called "ping-pong" mechanism. The mode of action of chymotrypsin explains this as hydrolysis takes place in two steps. First acylation of the substrate to form an acyl-enzyme intermediate and then deacylation in order to return the enzyme to its original state.

Carboxypeptidase (EC number 3.4.16 - 3.4.18) is an enzyme that hydrolyzes the carboxy-terminal (C-terminal) end of a peptide bond. Carboxypeptidase is secreated from the pancreas and reaches the intestine Humans, animals, and plants contain several types of carboxypeptidases with diverse functions ranging from catabolism to protein maturation.

The first carboxypeptidases studied were those involved in the digestion of food (pancreatic carboxypeptidases A1, A2, and B). However, most of the known carboxypeptidases are not involved in catabolism; they help to mature proteins or regulate biological processes. For example, the biosynthesis of neuroendocrine peptides such as insulin requires a carboxypeptidase. Carboxypeptidases also function inblood clotting, growth factor production, wound healing, reproduction, and many other processes.

Carboxypeptidases are usually classified into one of several families based on their active site mechanism.

- Enzymes that use a metal in the active site are called "metallo-carboxypeptidases" (EC number 3.4.17).
- Other carboxypeptidases that use active site serine residues are called "serine carboxypeptidases" (EC number 3.4.16).

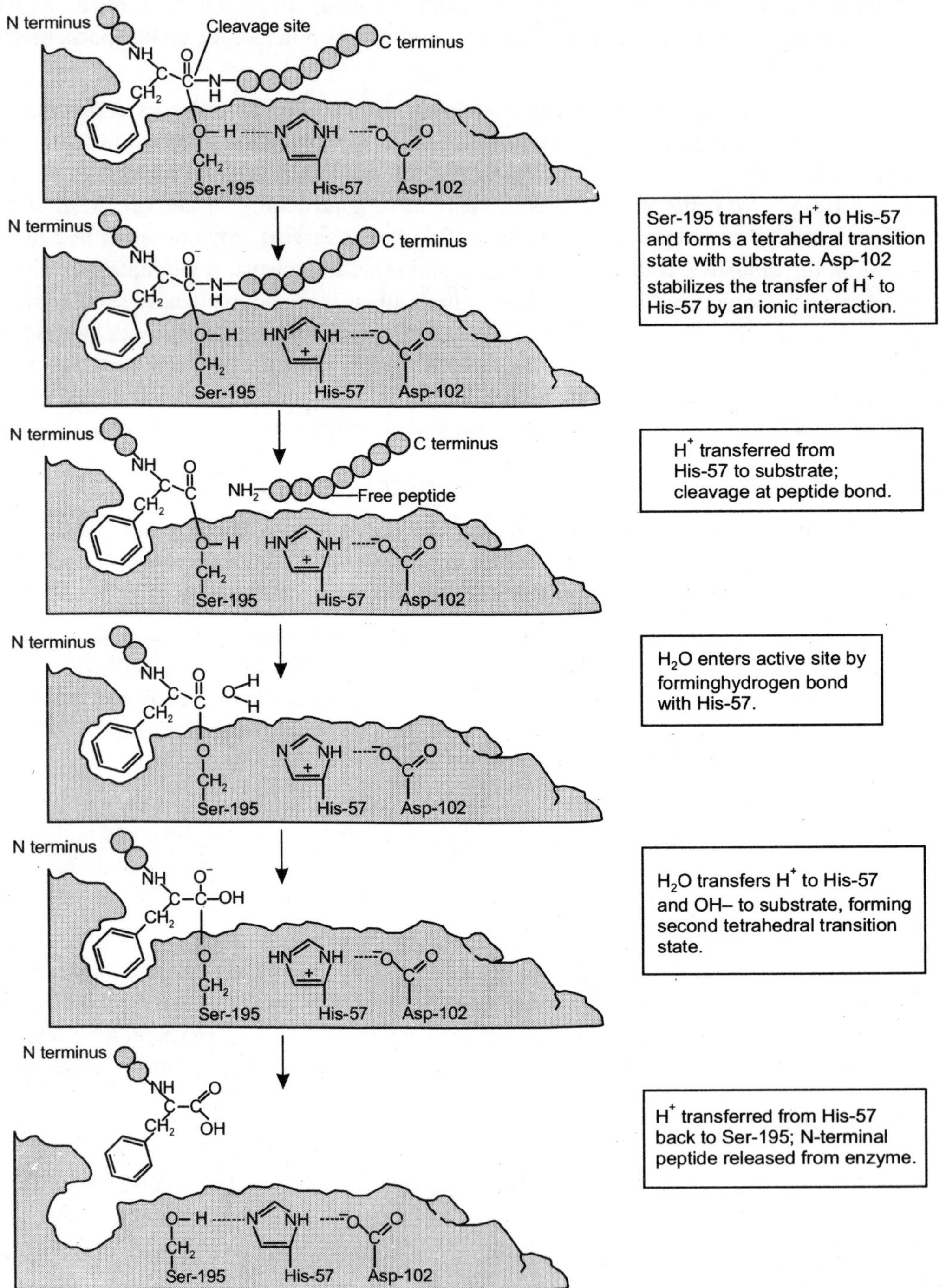

Fig. 5.43 : The Catalytic Reaction of Chymotrypsin

- Those that use an active site cysteine are called "cysteine carboxypeptidase" (or "thiol carboxypeptidases") (EC number 3.4.18).

These names do not refer to the selectivity of the amino acid that is cleaved.

Another classification system for carboxypeptidases refers to their substrate preference.

- In this classification system, carboxypeptidases that have a stronger preference for those amino acids containing aromatic or branchedhydrocarbon chains are called carboxypeptidase A (A for aromatic/aliphatic).
- Carboxypeptidases that cleave positively charged amino acids (arginine, lysine) are called carboxypeptidase B (B for basic).

A metallo-carboxypeptidase that cleaves a C-terminal glutamate from the peptide N-acetyl-L-aspartyl-L-glutamate is called "glutamate carboxypeptidase".

A serine carboxypeptidase that cleaves the C-terminal residue from peptides containing the sequence -Pro-Xaa (Pro is proline, Xaa is any amino acid on the C-terminus of a peptide) is called "prolyl carboxypeptidase".

Some, but not all, carboxypeptidases are initially produced in an inactive form; this precursor form is referred to as a procarboxypeptidase. In the case of pancreatic carboxypeptidase A, the inactive zymogen form, pro-carboxypeptidase A, is converted to its active form - carboxypeptidase A - by the enzyme enteropeptidase. This mechanism ensures that the cells wherein pro-carboxypeptidase A is produced are not themselves digested.

Trypsin inhibitors are chemicals that reduce the availability of trypsin, an enzyme essential to nutrition of many animals, includinghumans.

There are four commercial sources of trypsin inhibitors:

Source	Molecular weight	Inhibitatory power	Details
Serum (α_1-antitrypsin)	52 kDa		Also known as serum trypsin inhibitor
Lima beans	8-10 kDa	2.2 times weight	There are six different lima bean inhibitors.
Bovinepancreas	6.5 kDa	2.5 times weight	*Kunitz inhibitor* is the best known pancreatic inhibitor. Chymotrypsin is also inhibited by this chemical, but less tightly. When extracted from lung tissue, this is known as aprotinin.
Ovomucoid	8-10 kDa	1.2 times weight	Ovomucoids are the glycoprotein protease inhibitors found in raw avian egg white. There are other protease inhibitors in ovomucoids as well.
Soybeans	20.7-22.3 kDa	1.2 times weight	Soybeans contain several inhibitors; the one in the chart is considered the primary one. All of them bind chymotrypsin to a lesser degree.

ABSORPTION OF AMINO ACIDS

Under normal circumstances, the dietary proteins are almost completely digested to their constituent amino acids. But some amounts of oligopeptides like tri and dipeptides may remain as such. The above products are rapidly absorbed. The amino acids are generally absorbed in between ileum and distant jejunum. Olegopeptides like di and tri peptides are absorbed from duodenum and proximal jejunum.

Amino acid and other products of digestions like di and tri peptides are carried by portal blood* to liver.

There is a marked rise of the amino acid in the portal blood after a protein meal. The rate of absorption of amino acids is different for e.g. The L-amino acids and the L-isomers are absorbed faster than the D-amino acids by the active transport process. The L-amino acids are actively transported across the intestine from mucosa to serosal surface.

Pyridoxal (P) (B_6-PO_4) is probably involved in this process. In the case of D-amino acids, they are absorbed slowly and also by simple passive diffusion process.

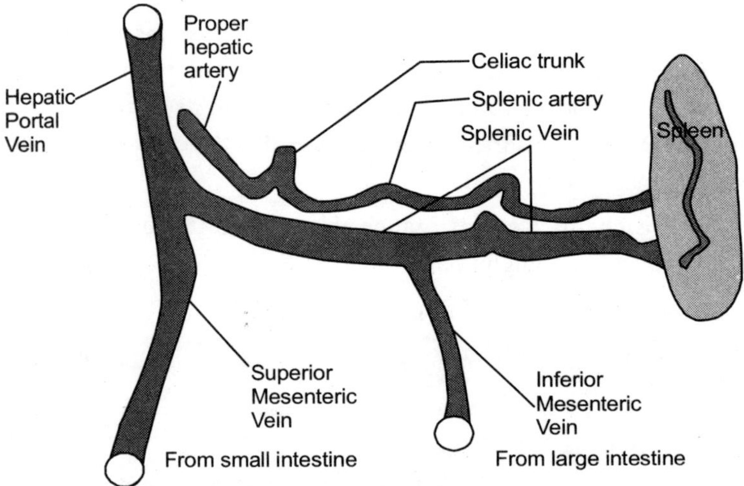

Fig. 5.44 : The Portal Vein and the Portal Blood

Mechanism of L-amino Acid Absorption

This process is also called Ion – gradiant hypothesis. Here amino acids are absorbed from small intestine by sodium (Na^+) dependant and energy is provided by ATP.

1. L-Amino acids and Na^+ combine with a common "carrier" protein molecule present on the outer or mucosal surface of the microvillus membrane to form a carrier amino acid Na^{++} complex

*the blood that is carried to the liver by portal vein from small intestine to liver.

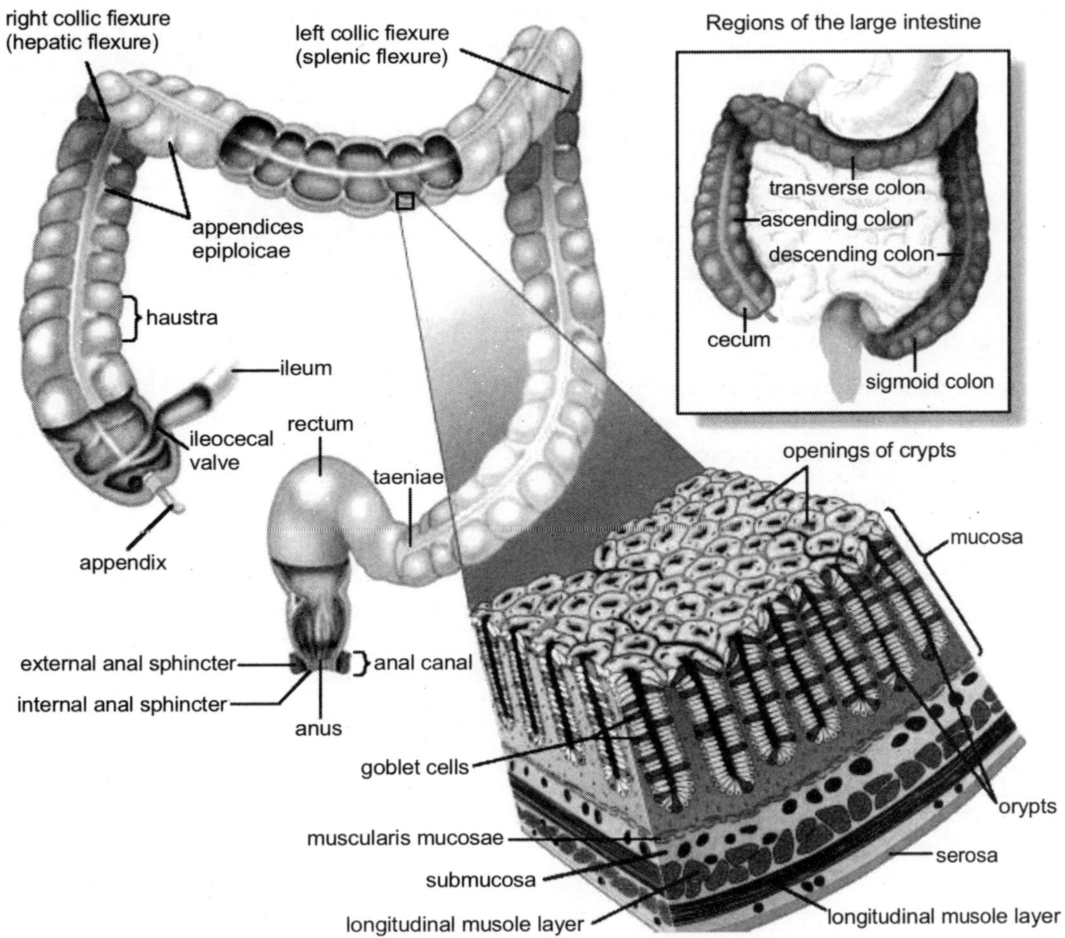

Fig. 5.45 : The Mucosa and the Serosal Surface

2. The complex passes to the inner or cytoplasmic surface of the same membrane. There it dissolved to generate the free amino acid and Na^{++}
3. Na^+ is actively carried out through the cell membrane by a sodium pump mechanism with the help of transport ATPase to that intracellular Na^+ concentration is always maintained below.
4. Carrier protein molecule comes back to the brush border again.
5. The amino acid ultimately passes out through the serosal membrane of the cell by diffusion down an outward concentration gradiant of amino acid is taken by portal blood to the liver.

Evidences of "Active" Absorption

1. Rate and extent of absorption of L-amino acids is considerably higher than D-Isomers.

2. If Na^+ is replaced by lithium and/or K^+ in bathing fluid the rate of absorption is depressed and may be practically nil.
3. Inhibitors like dinitrophenol (DNP) or cyanide depress L-amino acid absorption. DNP acts an uncoupler in oxidative phosphorylation and thus interferes with the ATP formation.
4. Rates of absorption of different L-amino acid are different from each other and are independent for their diffusion and concentration gradients.

Different classes of amino acids such as di–amino acids small neutral amino acids, imino acids and large neutral amino acids are believed to be absorbed by different 'carrier' protein molecules present in the microvillus membrane of intestinal cells.

5. High concentration of one L-amino acid sometimes reduces the rate of absorption of some other L-amino acids indicating several L-amino acids may share common " carrier molecule and may complete with each other.
6. Basic and dicarboxylic acid are generated are also slowly absorbed than neutral amino acids.

L-Oligo peptides are also actively absorbed. Intramolecular peptides hydrolyze them into amino acids. This hydrolyses within the intestinal epithelial cells, is rapid enough to keep peptide concentration low in these cells. Transport mechanisms for L-peptides appear to be independent of L-amino acids.

GAMMA—GLUTAMYL CYCLE

Meister has proposed that glutathione participates in an "active group" translocation of amino acids (except L-proline) into the cells of small intestine, kidneys, seminal vessels, epididymis and brain. He proposed a cyclic pathway in which the Glutathione is generated again, and it is called **γ-Glutamyl Cycle**. (see Fig. 5.46).

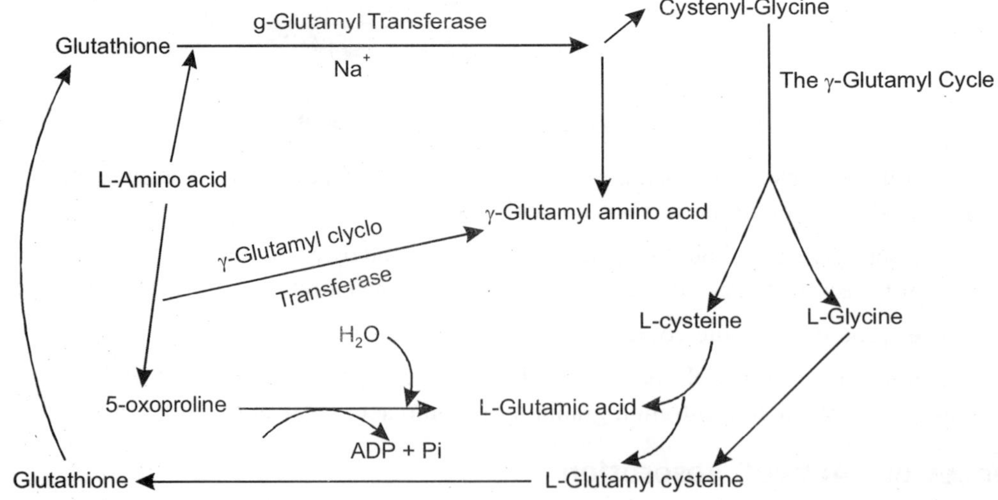

Fig. 5.46 : The Gamma-Glutamyl Cycle.

Some individuals show sensitivity to dietary proteins. Proteins to be antigenic so that can stimulate an immunological response it should be relatively high molecule. Normally proteins are digested in the G I tract to small peptides and amino acids. Digestions of protein to small peptides and amino acids stage destroy the antigenicity. How dietary proteins can be antigenic is a puzzling feature? Hence it is suggested that in some individuals, there must be absorption of some "unhydrolyzed" protein probably by pinocytosis. It has been shown that to be antigenic, the protein must be polypeptide containing 6 to 7 amino acids having mol wt of 828-928 Daltons and presence of glutinamine and proline in the protein molecule is a must.

FATE OF AMINO ACIDS AFTER

Absorption Amino Acids after absorption from intestine are carried out to liver by portal blood. They are taken up by liver cells to some extent and remaining enters the systemic circulation and diffuse throughout the body fluids and taken up by the tissue cells.

If the cells intake of amino acid is equal to Its loss It Is In a state of dynamic equilibrium and if the loss is greater the cells wastes and if the gain is greater the cells grows

At the same time the tissue proteins both "structural" and functional proteins (including plasma) proteins are commonly undergoing disintegration undergoing to release amino acids which likewise enter the circulation. There is also a continuous flow of amino acids (except the essential amino acids). Amino acids from all the different sources get mixed up to constitute what is known as general amino acids pool of the body. Though amino acid pool has now anatomical reality, but represents the availability of the amino acid building units. No functional distinction can be drawn between the fate of the amino acids derived from the dietary sources and those derived from the tissue breakdown.

All tissues including exocrine* and endocrine§ glands draw freely from the amino acids pool to synthesize the tissue proteins, enzymes and protein hormones. Amino acids is taken up by each cell according to its specific needs, to be built inside the cell structure and materials required.

If the cells intake of amino acid is equal to its loss it is in a state of dynamic equilibrium and if the loss is greater the cells wastes and if the gain is greater the cells grows. In man protein turnover involves the breakdown and resynthesis of 80 to 100 g of tissue protein per day about 1 of it occurring in liver. On average plasma proteins are completely replaced every 15 days. The pool is constantly undergoing depletion because:

1. Large deamination of presumably surplus amino acids takes place.

---

* Exocrine glands are glands that secrete their products (hormones) into ducts (duct glands). They are the counterparts to endocrine glands, which secrete their products (hormones) directly into the bloodstream (ductless glands) or release hormones (paracrines) that affect only target cells nearby the release site.

§ Endocrine glands are glands of the endocrine system that secrete their products, hormones, directly into the blood rather than through a duct. The main endocrine glands include the pituitary gland, pancreas, ovaries, testes, thyroid gland, and adrenal glands.

2. Amino acids and their derivatives, e.g. urea creatinine are lost in the urine and the other excretions
3. Amino acids are commonly and continually building up into those proteins, which are not the part of the dynamic system. On the other hand, amino acid pool is being always re-established by amino acid derived from the following.
4. Reamination of certain non-nitrogenous residues.
5. Amination of appropriate fragments which are present in the common metabolic pool (and therefore derived from fat and carbohydrate breakdown)
6. Amino acid split off from dietary protein and absorbed from the intestine into the blood.

This state is called "continuing metabolism of the amino acids"

All amino acids occur in blood in varying concentrations and make a total of 30 to 50 mg/10ml in the post absorption period.

In terms of amino acid N2 it is 4 to 5mg/100ml.

Circadian Changes in the Plasma Amino Acid Level

The plasma levels of most amino acids do not remain constant though out 24 hours period but rather change by varying in circadian rhythm about a "mean value". This was first noted in amino acid tyrosine and latter on confirmed on the many other amino acids. In general, plasma amino acid levels are lowest in the early morning (4 am) and rise about 15 to 35% by noon to early afternoon.

Amino acids present at the highest mean concentration e.g. glycine, alanine, valine, serine etc, show most striking changes in level. The exact physiological significance of the circadian changes occurring in plasma amino acid levels remains elucidated. The amino acids are transferred and transported to tissues actively. Pyridoxal P (B6 – P) is one of the requirements for this active transport which has been discussed in the subsection 5.25e.

PROCESS OF TRANSAMINATION

It was first discovered by Brounstein and Kritzmann it is a process of combined Amination and deamination. Transamination is a reversible reaction in which α-NH_2 group of one aminoacid is transferred to a keto acid resulting in formation of a new amino acid and a new keto acid.

The general process of Transamination may be represented as follows:

OH, O=, CH—NH_2, R_1 + OH, O=, C=O, R_2 ⇌ OH, O=, C=O, R_1 + OH, O=, CH—NH_2, R_2

L - amino acid (donor) | L - keto acid (recipient) | new keto acid | new amino acid

> The process represents only and intermolecular transfer of NH_2 group with the splitting out of NH_3 also NH_3 formation does not take place by transamination.

Donor amino acid (i) thus becomes a new keto acid (i) after losing the α - NH_2 group, and the recipient keto acid (II) after receiving the NH_2 group. Trans Amination reaction is reversible reaction and catalyzed by enzymes. Transamination takes place in the liver, kidney heart and brain. But the enzyme is present in almost all the mammalian tissues and transamination can be carried out in all tissues to some extent. Transamination requires enzymes called transaminases, better known as aminotransferases.

ROLE OF PYRIDOXAL PHOSPHATE

The CO-enzyme required for the transamination is the pyridoxal P (B6 –P). In the process of transamination, the amino acid reacts with the enzyme bond, pyridoxal-P to form an enzyme bond Schiff base complex, when then decomposes forming a second new amino acid and regenerates the pyridoxal-P. The reaction is shown below.

amino acid (donor)

pyridoxal - P

new amino acid

TRANSAMINASES

new keto acid

PYRODOXAMINE - P

new keto acid recipient

Limitations of Transamination

1. Group Transaminases and Specific Transaminases. (a) While most amino acids may act as Donor 1 the recipient keto acids may be either keto oxoglutartae, or oxaloacetate or pyruvate.
2. It is to be noted that all are components of "TCA cycle" and hence there are common metabolities of cell and are easily available. The amino acids formed from these recipients ketoacid are respectively glutamic acid aspartic and alanine.

3. Specific transaminases. But there are two transaminases of clinical importance in the body in that they use specific amino acid and specipic keto acid.

These specific transaminases are (or Asparate amino transferase) Previously used to be called SGOT (Serum Glutamate Oxaloacetate Transaminases).

In this Aspartic acid is the donor amino acid and α - oxoglutarate is the recipient keto-acid. New amino acid formed is again always glutamic acid.

Asparate + α – oxoglutarate $\rightleftarrows$ Oxalocetate + Glutamate

DEAMINATION

In the human body, deamination takes place primarily in the liver, however Glutamate is also deaminated in the kidneys. Deamination is the process by which amino acids are broken down when too much protein has been taken in. The amino group is removed from the amino acid and converted to ammonia. The rest of the amino acid is made up of mostly carbon and hydrogen, and is recycled or oxidized for energy. Ammonia is toxic to the human system, and enzymes convert it to urea or uric acid by addition of carbon dioxide molecules (which is not considered a deamination process) in the urea cycle, which also takes place in the liver. Urea and uric acid can safely diffuse into the blood and then be excreted in urine.

Deamination is the removal of an amine group from a molecule

Deamination is a process by which N – of amino acid is removed as NH_3 and this is of two types (i) Oxidative (ii) Non-oxidative deamination.

Oxidative Deamination

This type of deamination is mainly in the kidney and liver. The enzymes like D and L amino acid oxidases are present in the tissues which can act on the D and the L amino acid respectively and also can liberate oxidatively NH_3 form amino acids.

The essential difference between these two enzymes are shown in the table below:

D-amino acid oxidase	L-amino oxidase
1. Can action D-amino acids only	1. Can action L-amino acid
2. Can be readily extracted with water free from	2. Bound to tissue particles cannot be extracted with water
3. Contains FP(FAD)	3. Contains FP(FMN)

L - Amino acid oxidase

R—CH(NH2)—COOH ← α-amino acid; FP / FP H2; H2O2 / O2; Catalease; 1/2 O2

α-amino acid → (H_2O) → α-keto acid

Serine → (H_2O) → Imino acid ↔ Imino acid → (NH_3) → Pyruvate

Non-Oxidative Deamination

There are certain amino acids which can be non-oxidatively deaminated by specific enzymes and can form NH_3. These reaction do contribute to NH_3 formation, but does not actively take part in the NH_3 formation. Only three types of non-oxidative deamination has been discussed.

Amino acid Dehydrate: The hydroxyl amino acids, e.g. serine, threonine and homoserine are deaminated by specific enzymes called amino acid dehydrates which requires Pyridoxal-P-(B_6-P) co-enzyme.

The enzyme catalyze a primary dehydration followed by spontaneous deamination. An example will make the things clear. Histidine is non-oxidatively deaminated by the enzyme histidase to form NH_3 and urocanic acid.

Histidine —(histidase, −NH_3)→ urocanic acid

Amino acid Desulphydrases: The amino acids consisting SH group e.g. cysteine and homosysteine are deaminated by a primary desulphydration process (removal of H_2S) forming an amino acid which is then spontaneously hydrated.

L - cysteine —(−H_2S)→ intermediate (imino acid) → imino acid —(+H_2O, −NH_3)→ Pyruvic acid

Transdeamination

It is to be noted that L-Glutamic acid is not deaminated by L-amino acid oxidase but by a specific enzyme called L-glutamate dehydrogenase.

L-Glutamate ⇌(L-Glutamate dehydrogenase) 2-iminopentanedioic acid α-iminoglutamic acid ⇌(+H_2O, −NH_3) α-Keto glutaric acid

Characteristics of the Enzyme L-Glutamate Dehydrogenase

1. The enzyme has four polypeptide chains
2. A Zn^{+} containing metalloenzyme, one atom Zn^{++} present in each peptide chain.
3. It is widely distributed in tissues in humans and has high activity.
4. Specific for L-Glutamate.
5. It requires NAD^{+} or $NADP^{+}$ as co-enzymes.
6. It is a regulated enzyme whose activity is affected by allosteric modifiers at ATP GTP NADH which inhibits the enzyme.
7. Certain hormones appear to influence the enzyme activity in vitro.

Reaction the enzyme L-Glutamate dehydrogenase the deamination of L-Glutamate to from α-Iminoglutaric acid, which on addition of a molecule of water forms ammonia and α-ketoglutarate.

It is to be noted that the reaction is reversible and the equibrium constant favours glutamate formation but the quick removal of NH_3 to form urea in urea cycle and α - ketoglutarate to TCA cycle favours an onward reaction of NH_3 formation.

TRANSMETHYLATION

Transmethylation is a biologically important organic chemical reaction in which a methyl group is transferred from one compound to another.

An example of transmethylation is the recovery of methionine from homocysteine. In order to sustain sufficient reaction rates during metabolic stress, this reaction requires adequate levels of vitamin B12 and folic acid. Methyl tetrahydrofolate delivers methyl groups to form the active methyl form of vitamin B12 that is required for methylation of homocysteine. Deficiencies of vitamin B12 or folic acid cause increased levels of circulatiing homocysteine. Elevated homocysteine is a risk factor for cardiovascular disease and is linked to the metabolic syndrome (insulin insensitivity).

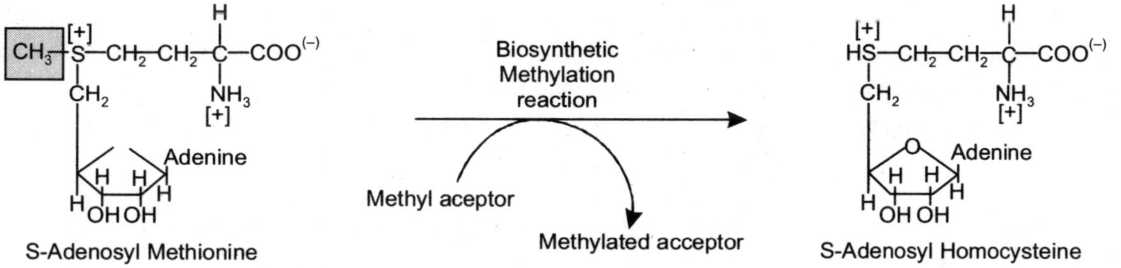

FORMATION OF CREATININE

The importance of creatine phosphate as a reservoir of high-energy phosphate readily convertible to A TP in muscles and other tissues has been pointed out. The synthesis of creatine involves the amino acids glycine, arginine, and methionine (as S-adenosylmethionine). Guanidoacetic acid is first formed by a transamidination reaction between glycine and arginine, in which the amidine group of arginine is transferred to glycine. The transamidinase enzyme is found in mammalian kidney and pancreas but appears to be absent from such other tissues as liver heart. skeletal muscles spleen blood brain

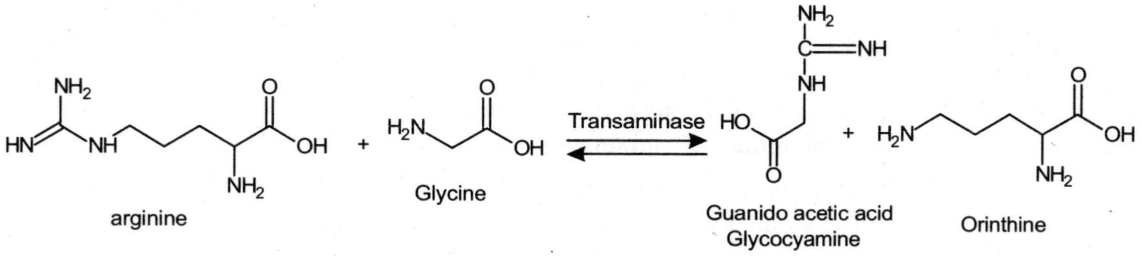

It is of interest that Walker ohtained evidence that an amidine derivative of the transamidinase enzyme is an intermediate in the reaction.

The process may be represented as follows:

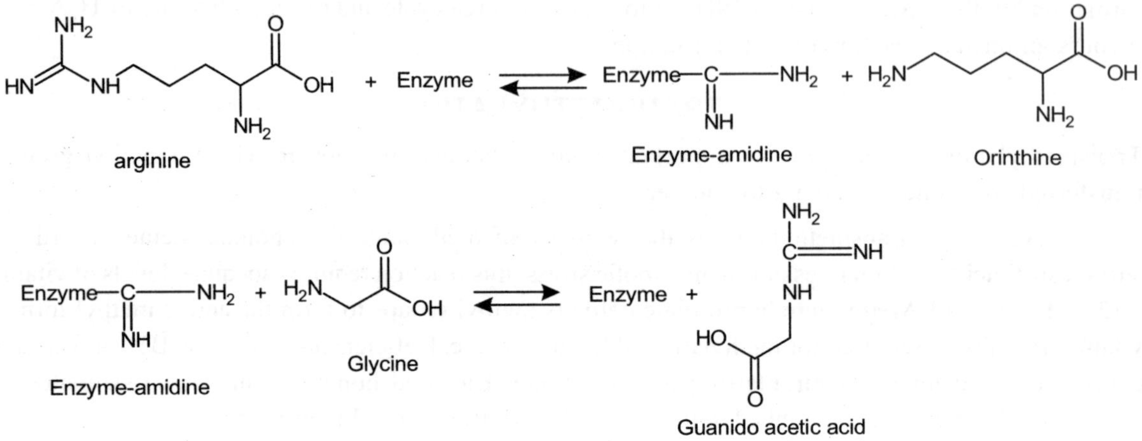

arginine + Enzyme ⇌ Enzyme-amidine + Orinthine

Enzyme-amidine + Glycine ⇌ Enzyme + Guanido acetic acid

In the second stage of creatine synthesis methionine is converted to "active" methionine, S-adenosylmethionine, by ATP and the "methionine-activating" enzyme, and then guanidoacetic acid is methylated to creatine by S-adenosylmethionine and the enzyme "guanidoacetic methylpherase". The second stage takes place in the liver.

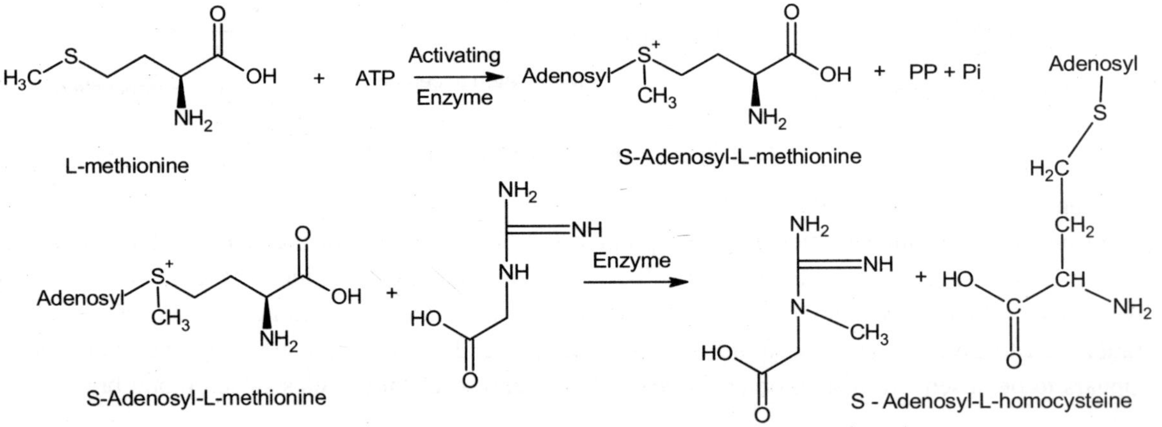

L-methionine + ATP —(Activating Enzyme)→ S-Adenosyl-L-methionine + PP + Pi

S-Adenosyl-L-methionine + guanidoacetic acid —(Enzyme)→ creatine + S-Adenosyl-L-homocysteine

The methyl group of creatine is not transferred to other compounds, as is the methyl of S-adenosylmethionine and of betaine. Also, the methyl group of creatine is not oxidized to formaldehyde or formate and, consequently, does not contribute to the one-carbon pool. However, the methyl groups of creatine may be formed from the one-carbonpool (through methionine).

Creatine occurs generally in the tissues of the body, though in uneven distribution. The highest concentrations are found in striated muscle, heart muscle, testes, liver, and kidneys, and somewhat lesser quantities in the brain. Very small amounts are present in blood, about 0.2 to 0.6 mg per cent in plasma and around 3 mg per cent in the cells. The muscles contain about 98 per cent of the total body creatine (*ca.* 0.5 per cent).

Most of the creatine in normal tissues (red cells excepted) occurs as the high-energy compound creatine phosphate.

The urine of normal adult persons contains only small amounts of creatine, but much creatinine, the anhydride of creatine. Creatinine apparently serves no useful function in the body but simply represents a waste product of creatine metabolism. Researchers showed that most of the creatinine formed in the body arises from the spontaneous decomposition of creatine phosphate according to the equation:

HN—O—P(=O)(—O^-)—O^- / HN=C / N—CH_3 / H_2C—C(=O)—OH → HN / HN=C / N—CH_3 / H_2C—C(=O)—O + inorganic phosphate

Creatinine Phosphate — creatinine

The blood plasma contains about 0.6 to 1.0 mg per cent of creatinine, while the concentration in the red cells has been found to be less, about 0.5 to 0.65 mg per cent. It appears that creatinine, unlike creatine, is distributed between the plasma and cells in proportion to their respective water con-tents, indicating free diffusibility. Researchers found 0.1 to 4.7 mg of creatinine p.er 100 g to be present in rat tissues, amounts proportional to the creatine present. Creatinine is present in sweat, in bile, and in all the gastrointestinal secretions.

Fate of Ingested and Injected Creatine

The oral and parenteral administration of creatine into animals or adult human males is followed by the excretion of only a fraction in the urine. A single dose of creatine causes no increase in urinary creatinine, though extra creatinine is found in the urine after prolonged feeding of large amounts of creatine and continues to be excreted for several weeks after creatine ingestion is discontinued.

Chanutin Researchers gave large amounts of creatine to rats and mice and found that the muscle creatine increased sharply for about a day, after which it remained practically constant. Liver and kidney creatine also increased. Upon cessation of creatine feeding, the excess creatine rapidly disappeared from the tissues except the muscles, which retained excessive amounts for some time.

Researchers showed that N15-labelled creatine fed to rats was deposited in the tissues, and the urinary creatinine excreted contained amounts of N15 indicating its origin from creatine. These workers also fed isotopic creatinine and found the prompt excretion of most of it in the urine. Also, no N15 was found in the body creatine, indicating that the conversion of creatine to creatinine is biologically irreversible.

Researchers labelled the tissue creatine of a human subject on a creatine-free diet with N15 by feeding isotopic creatine for 38 days and then isotopic guanidoacetic acid for ten days. Nonisotopic creatine then was given for five days and the urine analyzed for 28 days. The rate of turnover of endogenous creatine, as determined by the N15 concentration of excreted creatinine .(and creatine) after nonisotopic creatine was given, showed that 1.64 per cent of the tissue creatine (endogenous creatine) turned over per day. A balance between the amount of creatine retained in the body, and the amounts deposited in the tissues or excreted as extra creatinine in the urine could not be achieved. These workers suggest that the synthesis of endogenous creatine is retarded by the presence of exogenous creatine (administered). The amount of creatinine excreted per day as determined from its N15 content was directly proportional to the amount of creatine in the body.

Hoberman and associates have shown, by the use of N15-labeled creatine, that the administration of the hormone methyltestosterone increases the rate of creatine synthesis in the body.

Bloch and Schoenheimer gave isotopic creatinine (N15) to rats and found 75 per cent of the N15 in urinary creatinine. The remaining 25 per cent could not be accounted for in either tissues or urine.

Creatinuria

Creatine occurs only in small amounts in the urine of normal adults. It is present, along with creatinine, during the process of growth, and it, decreases to very low values as maturity is approached. Creatinuria is found in fevers, starvation, on a carbohydrate-free diet, and in diabetes mellitus. Its presence may be the result of excessive tissue de-struction and liberation of creatine, or failure under these conditions to keep the creatine properly phosphorylated, permitting the free creatine to diffuse from the tissues into the blood.

Excessive amounts of creatine may be excreted in the urines of patients with muscular dystrophy or hyperthyroidism.

Constancy of creatinine—creatine excretion. Shaffer showed that the daily excretion of creatinine by the adult male and of creatinine + creatine in subjects with physiologic creatinuria (such as growing children) is remarkably constant. It is little affected by diet (excluding diets high in creatine and creatinine), exercise, or urine volume. The creatinine (or creatinine + creatine) excretion apparently is characteristic of a healthy individual and is proportional to size and particularly to muscle mass. The daily excretion of creatinine is sufficiently constant for a given individual under ordinary conditions that the accuracy of 24-hour urine collections may be checked by its determination.

Creatinine Coefficient

The creatinine coefficient (or creatinine + creatine coefficient) represents the milligrams of creatinine (or creatinine + creatine) per kilogram of body weight excreted daily. The creatinine coefficient normally averages 20 to 26 for men and 14 to 22 for women.

The urine of premature infants and of normal infants during the first few days of life contains little creatine, and the creatinine coefficients are low.

Relations of Arginine and Creatine

Creatinuria and creatinine coefficients gradually increase during the first month of life:

Researchers found the daily creatinine excretion of obese persons to be low in relation to weight, but normal in relation to their ideal weight. Upon weight reduction through dietary control, the creatinine excretion remained constant. Underweight persons were found to have abnormally low creatinine coefficients. Researchers also consider the difference in creatinine coefficients of males and females to be related to the differences in muscular development rather than sex differences, since they found women with unusual muscular development to have creatinine coefficients comparable with those of males.

The relations between the metabolism of arginine and of creatine are summarized in the diagram below :

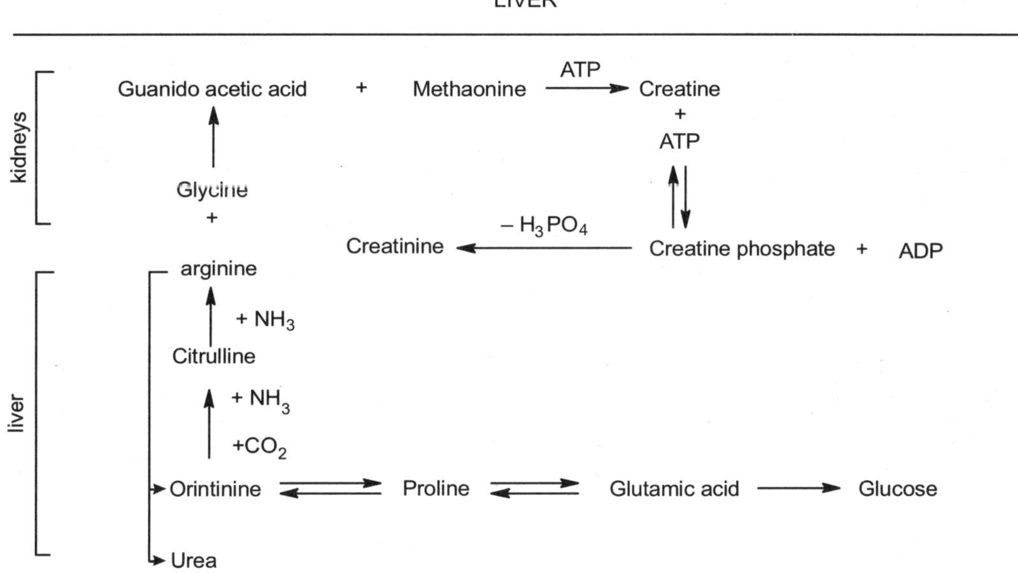

FORMATIONS AND DISPOSAL OF AMMONIA

The ammonia in the system is generated by the deamination process of amino acids as the ammonia is harmful it is disposed of from the system by the formation of uric acid and urea.

THE UREA CYCLE

The **urea cycle** (also known as the **ornithine cycle**) is a cycle of biochemical reactions occurring in many animals that produces urea$(NH_2)_2CO$ from ammonia (NH_3). This cycle was the first metabolic cycle discovered. In mammals, the urea cycle takes place only in the liver (Fig. 5.47).

Organisms that cannot easily and quickly remove ammonia usually have to convert it to some other substance, like urea or uric acid, which are much less toxic. Insufficiency of the urea cycle occurs in some genetic disorders (inborn errors of metabolism), and in liver failure. The result of liver failure is accumulation of nitrogenous waste, mainly ammonia, which leads to hepatic encephalopathy.

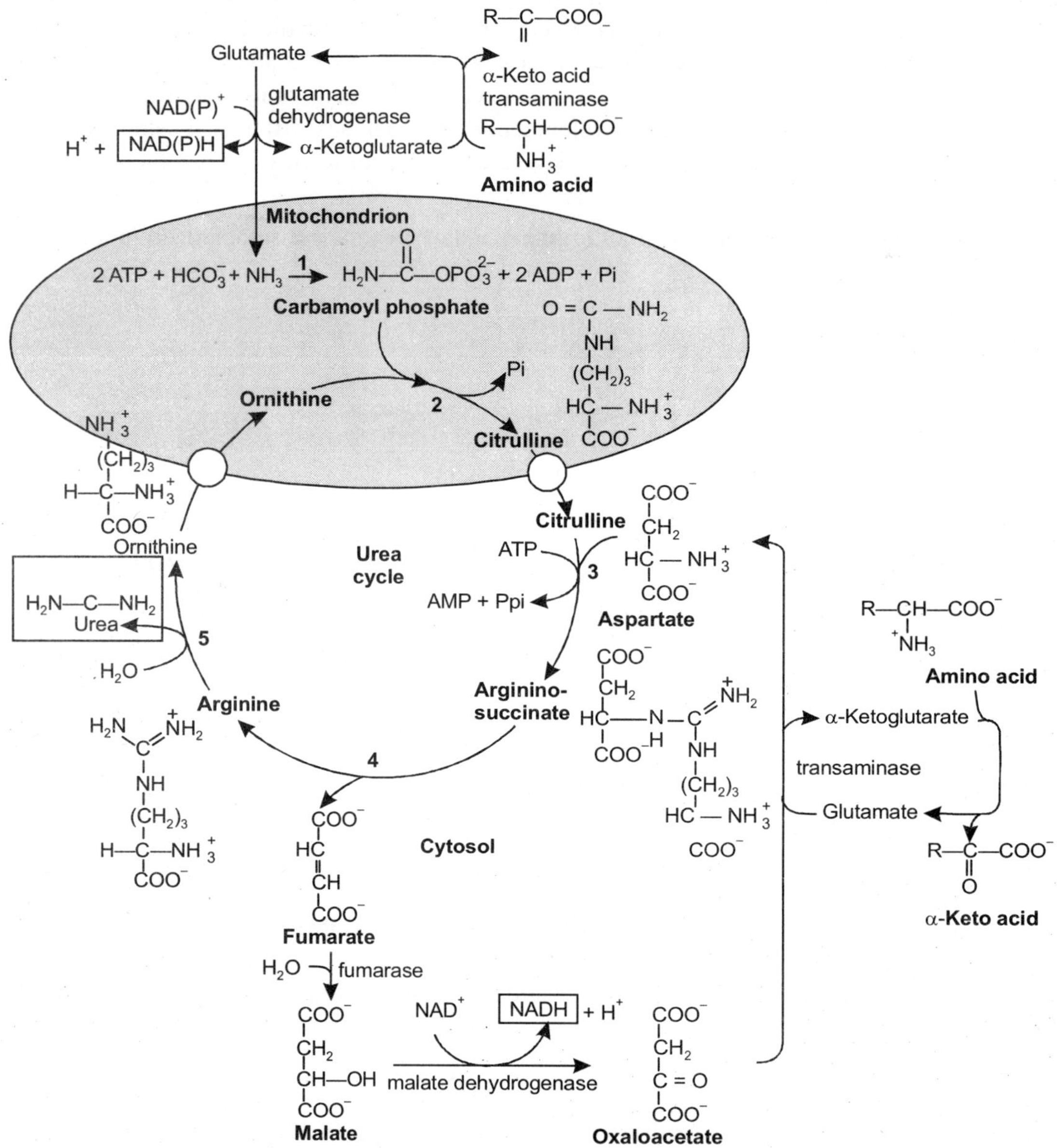

Fig. 5.47 : The Urea Cycle

The urea cycle consists of five reactions - two mitochondrial and three cytosolic. The cycle converts two amino groups, one from NH_4^+ and one from Asp, and a carbon atom from HCO_3^-, to relatively nontoxic excretion product, urea, at the cost of four "high-energy" phosphate bonds (3 ATP hydrolyzed to 2 ADP and one AMP). Orn is the carrier of these carbon and nitrogen atoms.

Reactions of urea cycle:

Step	Reactant	Product	Catalyzed by	Location
1.	$2ATP + HCO_3^- + NH_4^+$	carbamoyl phosphate + $2ADP + P_i$	CPS1	mitochondria
2.	carbamoyl phosphate + ornithine	citrulline + P_i	OTC	mitochondria
3.	citrulline + aspartate + ATP	argininosuccinate + AMP + PP_i	ASS	cytosol
4.	argininosuccinate	Arg + fumarate	ASL	cytosol
5	Arg + H_2O	ornithine + urea	ARG1	cytosol

Overall energy requirement:

- $NH_3 + CO_2$ + Aspartate + 3 ATP + 2 $H_2O \rightarrow$ urea + Fumarate + 2 ADP + 4 P_i + AMP

Overall equation of the urea cycle:

- 2 $NH_3 + CO_2$ + 3 ATP + $H_2O \rightarrow$ urea + 2 ADP + 4 P_i + AMP + 2H

Note that reactions related to the urea cycle also cause the reduction of 2 NADH, so the urea cycle releases slightly more energy than it consumes. These NADH are produced in two ways:

- One NADH molecule is reduced by the enzyme glutamate dehydrogenase in the conversion of glutamate to ammonium and α-ketoglutarate. Glutamate is the non-toxic carrier of amine groups. This provides the ammonium ion used in the initial synthesis of carbamoyl phosphate.
- The fumarate released in the cytosol is converted to malate by cytosolic fumarase. This malate is then converted to oxaloacetate by cytosolic malate dehydrogenase, generating a reduced NADH in the cytosol. Oxaloacetate is one of the keto acids preferred bytransaminases, and so will be recycled to aspartate, maintained the flow of nitrogen into the urea cycle.

The two NADH produced can provide energy for the formation of 5 ATP, a net production of one high energy phosphate bond for the urea cycle. However, if gluconeogenesis is underway in the cytosol, the latter reducing equivalent is used to drive the reversal of the GAPDH step instead of generating ATP.

The fate of oxaloacetate is either to produce aspartate via oxidative deamination or to be converted to phosphoenol pyruvate, which is a substrate to glucose.

An excellent way to memorize the Urea Cycle is to remember the phrase "Ordinarily Careless Crappers Are Also Frivolous About Urination." The first letter of each word corresponds to the order in which reactants are combined to give products or intermediates that break apart as one progresses through the cycle.

N-Acetylglutamic Acid

The synthesis of carbamoyl phosphate and the urea cycle are dependent on the presence of NAcGlu, which allosterically activates CPS1. Synthesis of NAcGlu by NAGS, is stimulated by Arg - allosteric stimulator of NAGS, and Glu - a product in the transamination reactions and one of NAGS's substrates, both of which are elevated when free amino acids are elevated. So, Arg is not only a substrate for the urea cycle reactions but also serves as an activator for the urea cycle.

Substrate Concentrations

The remaining enzymes of the cycle are controlled by the concentrations of their substrates. Thus, inherited deficiencies in the cycle enzymes other than ARG1 do not result in significant decrease in urea production (the total lack of any cycle enzyme results in death shortly after birth). Rather, the deficient enzyme's substrate builds up, increasing the rate of the deficient reaction to normal.

The anomalous substrate buildup is not without cost, however. The substrate concentrations become elevated all the way back up the cycle to NH_4^+, resulting in hyperammonemia (elevated $[NH_4^+]_P$).

Although the root cause of NH_4^+ toxicity is not completely understood, a high $[NH_4^+]$ puts an enormous strain on the NH_4^+-clearing system, especially in the brain (symptoms of urea cycle enzyme deficiencies include mental retardation and lethargy). This clearing system involves GLUD1 and GLUL, which decrease the 2OG and Glu pools. The brain is most sensitive to the depletion of these pools. Depletion of 2OG decreases the rate of TCAC, whereas Glu is both a neurotransmitter and a precursor to GABA, another neurotransmitter.

THE DISORDERS OF UREA CYCLE

An **urea cycle disorder** or **urea cycle defect** is a genetic disorder caused by a deficiency of one of the enzymes in the urea cycle which is responsible for removing ammonia from the bloodstream. The urea cycle involves a series of biochemical steps in which nitrogen, a waste product ofprotein metabolism, is removed from the blood and converted to urea. Normally, the urea is transferred into the urine and removed from the body. In urea cycle disorders, the nitrogen accumulates in the form of ammonia, a highly toxic substance, and is not removed from the body.

Urea cycle disorders are included in the category of inborn errors of metabolism. There is no cure.

Inborn errors of metabolism are generally considered to be rare but represent a substantial cause of brain damage and death among newborns and infants. Because many cases of urea cycle disorders remain undiagnosed and/or infants born with the disorders die without a definitive diagnosis, the exact incidence of these cases is unknown and underestimated. It is believed that up to 20% of Sudden Infant Death Syndromecases may be attributed to an undiagnosed inborn error of metabolism such as urea cycle disorder. In April 2000, research experts at the Urea Cycle Consensus Conference estimated the incidence of the disorders at 1 in 10000 births. This represents a significant increase in case diagnosis in the last two years.

Children with very severe urea cycle disorders typically show symptoms after the first 24 hours of life. The baby may be irritable at first, followed by vomiting and increasing lethargy. Soon after, seizures, hypotonia (poor muscle tone), respiratory distress, and coma may occur. If untreated, the child will die. These symptoms are caused by rising ammonia levels in the blood. Acute neonatal symptoms are most frequently seen in, but not limited to, boys with OTC Deficiency.

Children with mild or moderate urea cycle enzyme deficiencies may not show symptoms until early childhood, or may be diagnosed subsequent to identification of the disorder in a more severely affected relative or through newborn screening. Early symptoms may includehyperactive behaviour, sometimes accompanied by screaming and self-injurious behaviour, and refusal to eat meat or other high-protein foods. Later symptoms may include frequent episodes of vomiting, especially following high-protein meals; lethargy and delirium; and finally, if the condition is undiagnosed and untreated, coma and death. Children with this disorder may be referred to child psychologists because of their behaviour and eating problems. Childhood episodes of hyperammonemia (high ammonia levels in the blood) may be brought on by viral illnesses including chicken pox, high-protein meals, or even exhaustion. The condition is sometimes misdiagnosed as Reye's Syndrome. Childhood onset can be seen in both boys and girls.

Recently, the number of adult individuals being diagnosed with urea cycle disorders has increased at an alarming rate. Recent evidence has indicated that these individuals have survived undiagnosed to adulthood, probably due to less severe enzyme deficiencies. These individuals exhibit stroke-like symptoms, episodes of lethargy, and delirium. These adults are likely to be referred to neurologists or psychiatrists because of their psychiatric symptoms. However, without proper diagnosis and treatment, these individuals are at risk for permanent brain damage, coma, and death. Adult-onset symptoms have been observed following viral illnesses, childbirth, and use of valproic acid (an anti-epileptic drug).

There are six disorders of the urea cycle. Each is referred to by the initials of the missing enzyme.

Location	Abb.	Enyzme	Disorder	Measurements
Mitochondria	NAGS	N-Acetylglutamate synthetase	N-Acetylglutamate synthase deficiency	+Ammonia
Mitochondria	CPS1	Carbamoyl phosphate synthetase I	Carbamoyl phosphate synthetase I deficiency	+Ammonia
Mitochondria	OTC	Ornithine transcar-bamylase	Ornithine transcar-bamylase deficiency	+Ornithine, +Uracil, +Orotic acid
Cytosol	AS	Argininosuccinic acid synthetase	"AS deficiency" or citrullinemia	+Citrulline
Cytosol	AL	Argininosuccinase acid lyase	"AL deficiency" or arginino-succinic aciduria(ASA)	+Citrulline, +Argininosuccinic acid
Cytosol	AG	Arginase	"Arginase deficiency" or argininemia	+Arginine

Individuals with childhood or adult onset disease may have a partial enzyme deficiency. All of these disorders are transmitted genetically as autosomal recessive genes - each parent contributes a defective gene to the child, except for one of the defects, Ornithine Transcarbamylase Deficiency. This urea cycle disorder is acquired in one of three ways: as an X-linked trait from the mother, who may be an undiagnosed carrier; in some cases of female children, the disorder can also be inherited from the father's X-chromosome; and finally, OTC deficiency may be acquired as a "new" mutation occurring in the fetus uniquely. Recent research has shown that some female carriers of the disease may become symptomatic with the disorder later in life, suffering high ammonia levels. Several undiagnosed women have died during childbirth as a result of high ammonia levels and on autopsy were determined to have been unknown carriers of the disorder.

The treatment of urea cycle disorders consists of balancing dietary protein intake in order that the body receive the essential amino acidsresponsible for cell growth and development, but not so much protein that excessive ammonia is formed. This protein restriction is used in conjunction with medications which provide alternative pathways for the removal of ammonia from the blood. These medications are usually given by way of tube feedings, either via gastrostomy tube (a tube surgically implanted in the stomach) or nasogastric tube through the nose into the stomach. The treatment may also include supplementation with special amino acid formulas developed specifically for urea cycle disorders, multiple vitamins and calcium supplements. Frequent blood tests are required to monitor the disorders and optimize treatment, and frequently hospitalizations are necessary to control the disorder.

At the most extreme end of the spectrum, a few liver transplants have been done successfully as a cure to the disorder. This treatment alternative must be carefully evaluated with medical professionals to determine if potential of success as compared to the potential for new medical concerns.

FORMATION OF NITRIC OXIDES (NOS)

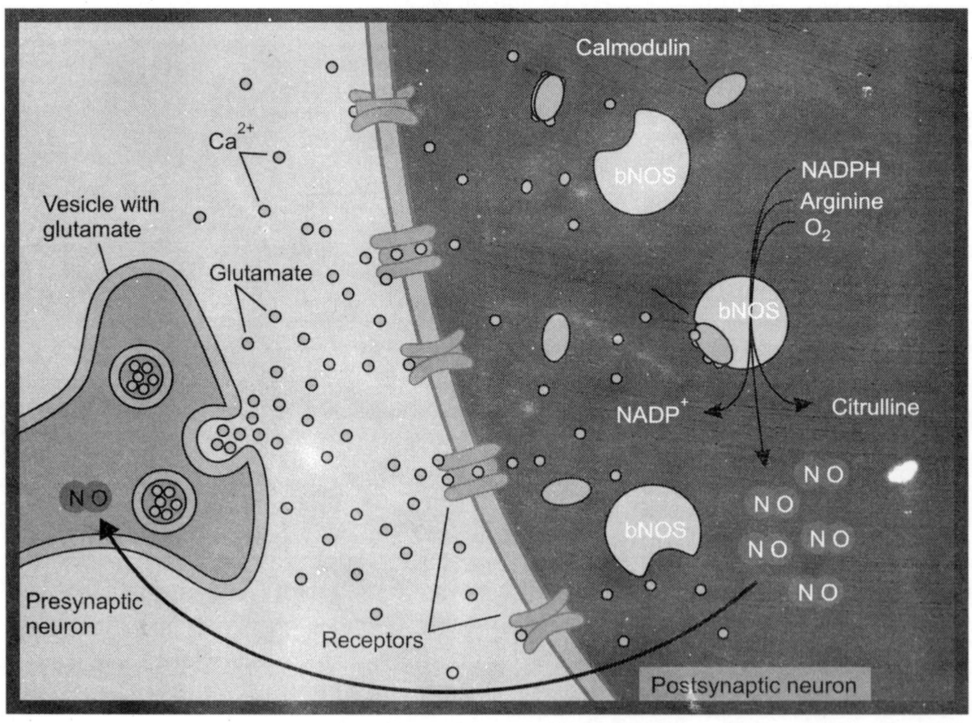

Fig. 5.48 : The formation of Nitric Oxide (NOS)

The activation of nitric oxide synthase in brain (bNOS). A neuron by a nerve centre impulse (the presynaptic neuron at left, releases glutamate into the synapse. Glutamate binds to a membrane receptor on the adjacent postsynaptic neuron. As a result of the binding, a channel opens in the receptor, allowing Ca^{2+} ions to flow into the cell and to bind to the calmodulin. The calcium-calmodulin complex then binds bNOS, thereby activating it so that it can catalyze the formation of nitric oxide. The nitric oxide thus formed can diffuse to other neurons to reinforce neural connections.

The small, free radical molecule nitric oxide (NO; N = O) has been identified as a major signal transduction molecule in vertebrates (animals). NO is derived from arginine in two steps catalyzed by nitric oxide synthase (NOS; EC 1.14.13.39). NOS catalyzes the net reaction:

$$\text{L-Arginine} + \text{n NADPH} + \text{m}O_2 = \text{Citrulline} + \text{Nitric oxide} + \text{n NADP}^+$$

with the intermediate N-(omega)-Hydroxyarginine (C05933). The catalytic activity of *nitric oxide synthase* is related to the monooxygenase activity of *cytochrome P450*. This catalytic relationship becomes apparent when comparing the gene structure of both enzymes. Nitric oxide synthase is the larger of the two enzymes. Its C-terminal domain is identical to the smaller Cyt P450 protein.

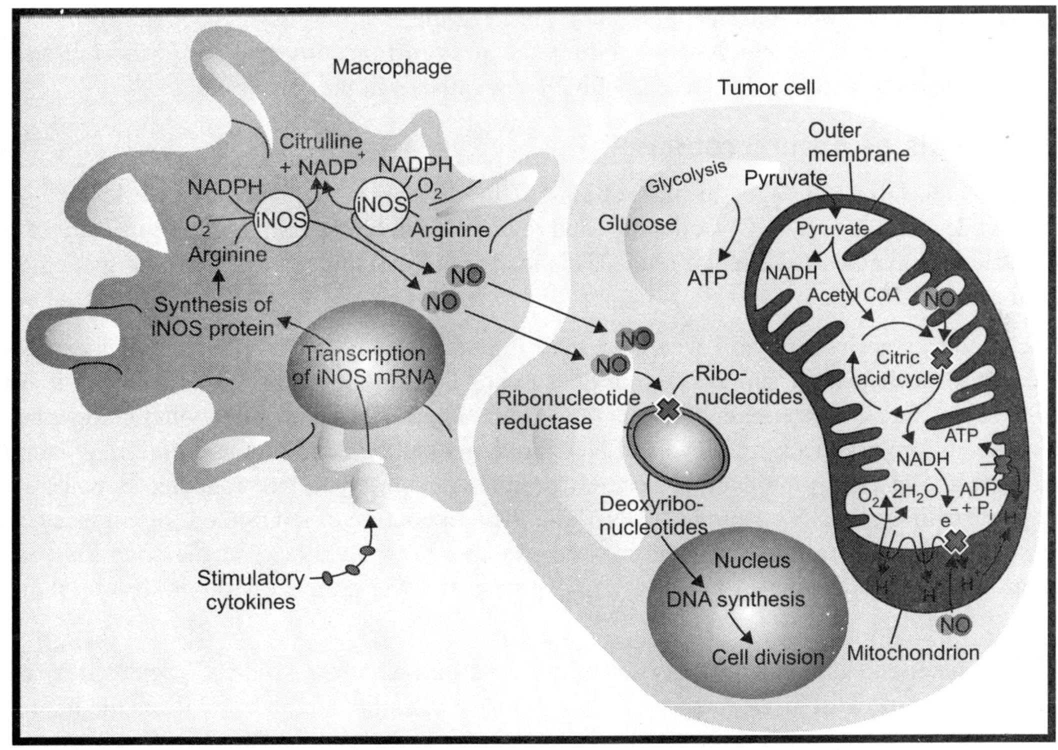

Fig. 5.49 : The Role of Nitric Acid in the Immune System

There are three forms of nitric oxide synthase - a neuronal type called nNOS, an epithelial type called eNOS, and an inducible form called iNOS. The latter is only expressed under certain conditions like immune system regulation by cytokines or pathological induction in the presence of endotoxins (bacterial lipopolysaccharide) and cytotoxins (which affect cytokine secretion). No production is a stress response and can lead to either tissue injury because of its radical chemistry, or be cytoprotective, protecting cells from damage by destroying pathogenic microorganisms first. For example, stomach ulcers have lately been associated with a bacterial infection. *Helicobacter pylori* (*H. pylori*) causes ulcers and gastric cancers. Nitric oxide and particularly its superoxide derivative peroxynitrite cause DNA damage in the bacteria.

The production of NO in the immune system takes place in macrophages (components of the immune system). When stimulatory cytokines bind to macrophages, they set off a chain of events that leads to synthesis of NOS. The DNA that encodes the gene for NOS is activated, producing the messenger RNA for the enzyme which is then synthesized by the cell. An enzyme that is produced in this way is referred to as inducible. Consequently, this form of NOS is designated as iNOS (i for inducible). When iNOS is formed it diffuses to tumor cells close to the macrophage. There it interferes with a number of cellular processes, which occurs at high rates in rapidly growing cancer cells. One site of action is inhibition of the enzyme aconitase, which catalyzes one of the early steps of the citric acid cycle. Nitric oxide also interferes with complex I of the electron transport chain. Another site of action is inhibition of ribonucleotide reductase, preventing conversion of ribonucleotide to deoxyribonucleotides and thus interfering with DNA synthesis in the cancer cells.

Free Radicals as Antimicrobials

ONOO induces DNA damage through chemical modifications (mutations) while NO inhibits ribonucleotide reductase. Both DNA damage and reductase inhibition keep the cell in a state of energy costly nucleotide synthesis and repair mode. This leads eventually to cell death by energy depletion of bacterial cells.

Such defense mechanism, however, have their draw backs. Inducible NOS, which is expressed as an emergency mechanism to suppress tumor growth in gastric epithelia, breast tissue, and the brain, is linked to septic shock. Bacterial endotoxins (e.g. from *H.pylori* or *E.coli* infections) induce the iNOS gene, which in turn produces high levels of NO damaging pathogenic DNA and inhibiting respiration (inhibits metabolic energy production needed for cell division). The free radicals, however, cannot discriminate pathogenic DNA from host DNA and overstimulation of iNOS therefore induces cell and tissue damage, sometimes leading to a fatal development (septic shock) in the course of bacterial infections. This is a well known situation in hospitals affecting patients with an alread suppressed immune system.

The neuronal and epithelial NOS isoforms are constitutively expressed and regulated by calcium concentration via calmodulin interaction. The calcium-calmodulin complex stabilizes the homodimer. Each monomer contains a reductase and oxygenase subunit containing FAD, NADPH & FMN or heme & tetrahydrobiopterine (B4H) cofactors, respectively. B4H is essential for NOS dimer formation. (H4biopterine is an important cofactor of aromatic amino acid hydroxylases.) Nitric oxide synthase is a membrane bound protein, anchored to the cytoplasmic side of endoplasmatic reticulum, Golgi, or plasma membrane by myristoylation or palmitoylation. The lipid anchored NOS are preferentially found in cholesterol and glycolipid rich membrane domains. The compartmentalization of NOS appears to be crucial for its functionality by providing local NO levels.

Nitric Oxides (NOS) Regulation

The activity of eNOS and nNOS is controlled by tetrahydrobiopterin and Ca/CaM availability because these two cofactors are needed for the proper dimer formation of an active synthetase. The dependence on calmodulin has been used as a model to explain the role of glutamate in neurotoxicity in the central nervous system. Neurotoxicity is a mechanism of glutamate induced neuronal cell death. The immediate effect of glutamate on neurons is its role in activating glutamate receptor, namely to pharmacological subtypes known as NMDA Receptors (NMDA is a methylated derivative of aspartate). Glutamate receptors are selective for calcium ions. Thus, prolonged activation of glutamate receptors stimulates eNOS via Ca/CaM complex binding to the synthetase. The formation of NO is implicated in cell death as described above: DNA damage, suppressed mitochondrial respiration, leading to energy depletion. Neurons are particularly sensitive to impaired mitochondrial ATP synthesis capacity, because neurons depend almost exclusively on the oxidative degradation of glucose and ketone bodies. The formed ATP is used by ion selective pumps to maintain the proper ion gradients for action potential generation and neurotransmitter release of presynaptic membranes.

NO can only be synthesized, however, if the amino acid arginine is available. Neuronal NOS critically depends on this substrate, which is mainly synthesized in adjacent glial cells and is transported into neurons. Arginine uptake into neurons is controlled by non-NMDA glutamate receptors. This became evident when these receptors were blocked by arginine-uptake inhibitors such as L-lysine which functions as antagonist of these glutamate receptors. The physiological role of nNOS in mechanisms such as long term potentiation has been shown to involve retrograde transport (diffusion) of NO synthesized in post synaptic neurons across the synaptic cleft into synapses, where they stimulated guanyl cyclase.

Nitric Oxide and free Radical Biochemistry

Nitric oxide is a free radical molecule and its major effect is the activation of cytoplasmic, soluble guanyl cyclase (sGC; EC 4.6.1.2). This enzyme catalyzes the cyclization of GTP to cGMP + PPi. Cyclic GMP is a signaling molecule (similar to cAMP) by virtue of activating protein kinases.

Nitric oxide binds to the heme group of cyclase. Other protein targets are metallo enzymes, where NO binds to Fe-S clusters. Aconitase is inactivated by NO, as is complex IV, the cytochrome oxidase in the inner membrane of mitochondria. Thus NO as an inhibitory effect on oxidative phosphorylation by blocking the electron transport chain and controlling the levels of citrate in the Krebs cycle essentially blocking the oxidative degradation of acetyl-CoA.

NO is a short lived chemical transmitter, which is freely diffusible across membranes. The molecule possesses a small dipole moment because of the similar electro negativity of oxygen and nitrogen, making it essentially hydrophobic. Its reactivity is due to the unpaired electron in the outer valence orbital of its oxygen constituent. NO is almost non reactive as free radical as compared to other oxygen radicals. Indeed, NO decays within seconds after its synthesis if left unbound in solution because it reacts with either molecular oxygen or superoxide.

The NO strongly interacts with molecular oxygen to form dinitro trioxide (N_2O_3), or with superoxide O_2^- to form peroxynitrite ($ONOO^-$). The NO also binds to sulphydryl groups (SH) and unsaturated fatty acids. The reaction with superoxide can be diminished by superoxide dismutase (SOD) which removes O_2^-. to form hydrogen peroxide (H_2O_2). NO can be 'stored' by covalent interaction to glutathione to form S-nitroso-glutathion. Both H_2O_2 and S-nitrosoglutathion can have a stimulatory effect on guanine cyclase. Superoxide dismutase thereby prevents the loss of nitric oxide to peroxynitrite forming hydrogen peroxide instead and increasing the cyclase stimulatory capacity of the cell.

The NO can potentially be regenerated from $ONOO^-$ in two steps; a first reduction of peroxynitrite by cytochrome C oxidase to nitrite (NO_2), followed by a reduction of nitrite to NO by the enzyme nitrate reductase. The latter enzyme exists in two isoforms, a mitochondrial type and an endoplasmatic reticulum resident protein. Both receive their electrons needed for nitrite reduction to the NO from either NADH or NADPH, and interact with flavoproteins (FAD prosthetic groups) and cytochromes (cytochrome c oxidase in mitochondrial membrane; cytochrome P450 in ER membrane).

Cytochrome C oxidase and nitrite reductase therefore reduce the concentration of highly reactive, secondary metabolites, and potentially contribute to the NO signaling. The latter has only been shown in plant cells, where nitrate reductase reaction appears to be a considerable contributor to this signaling molecule.

Peroxynitrite, hydrogen peroxide, and dinitro trioxide all have been linked to cell death (apoptosis = programmed cell death) through protein nitration and increased mutagenesis. The latter is a consequence of DNA stand breakage and guanine nitration. For example, acute neural toxicity is linked to the overproduction of peroxynitrite, which inhibits respiratory enzymes and also damages DNA by covalent bond formation to DNA and removal of bases. Inhibitors of nitric oxide synthase and antioxidants are known to have neuroprotective properties because the limit the formation of highly reactive nitrogen containing radicals.

Antioxidants

The free radical chemistry in cells can be prevented or at least diminished by adding antioxidants or free radical scavengers, molecules which have a high affinity and strongly react with these free radicals. Antioxidants are either hydrophilic or hydrophobic. Hydrophilic antioxidants include glutathione peroxidase, Fe(II) chelators like the proteins ceruloplasmin and transferrin, and hydroxylated aromatic molecules like uric acid or ascorbate (vitamin C). Hydrophobic antioxidants include flavin-nucleotide or carotene containing proteins and vitamin E.

Melatonin too is a major physiological antioxidant (and hormone) by directly reacting with hydroxyl and peroxyl radicals, or by stimulating the expression of superoxide dismutase, glutathione peroxidase, or glutathione reductase. Melatonin has also been reported to inhibit nitric oxide synthetase.

Physiological Role of NO as Neurotransmitter

In epithelial cells, the NO causes vascular dilatation by controlling smooth muscle contractility. In the central nervous system it affects synaptic transmission stimulating learning and memory capacity. Glutamate is produced and released by a synapse and activates the NMDA receptor subtype of glutamate receptors. This leads to an influx of calcium ions which in turn bind to calmodulin, activating the neuronal NOS. NOS synthesizes NO depending on the availability of L-arginine, which is mainly supplied from extra-neuronal sites (mainly glial cells). The NO not only activates the postsynaptic guanyl cyclases, but can diffuse across the synaptic cleft back into the synapse that originally released the glutamate. This retrograde transport of NO is thought to reinforce the capability of glutaminergic signaling. Such a prolonged reinforcement of synaptic stimulatory activity is known as long term potentiation and is implicated as a possible molecular mechanism promoting long term memory and learning.

In blood plasma the NO induces platelet aggregation, an important factor in wound healing and blood coagulation. It has been shown that hemoglobin is a major transport vehicle for NO in blood.

THE METABOLISM OF PHENYLALANINE AND TYROSINE

Generally mammals are unable to synthesize Phenyl alanine and tyrosine, but some microorganisms like *E coli*, neurospora etc are all capable of producing Phenylalanine and tyrosine. The amino acids being derived from common precursor, prehenic acid. Figure 5.50 will demonstrate the metabolism of Phenyl alanine and tyrosine.

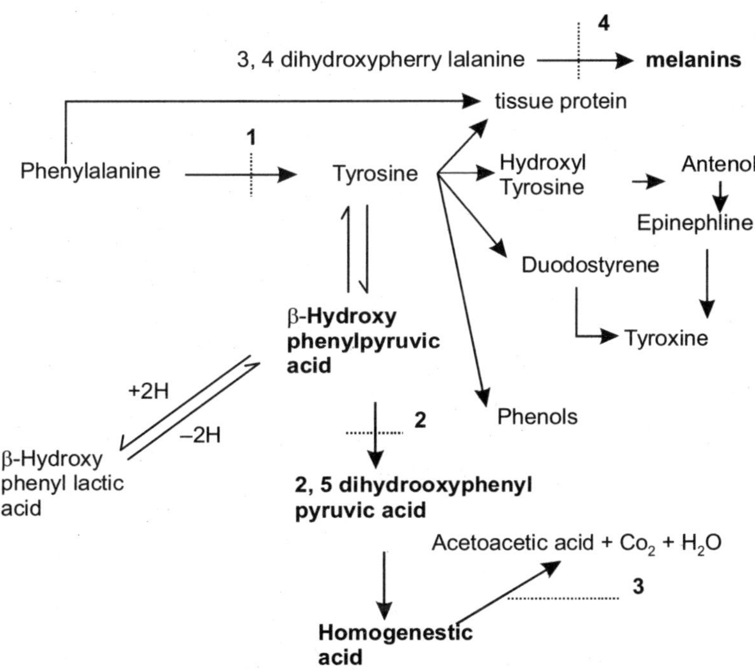

Fig. 5.50 : The Metabolism of Tyrosine and Phenylalanine

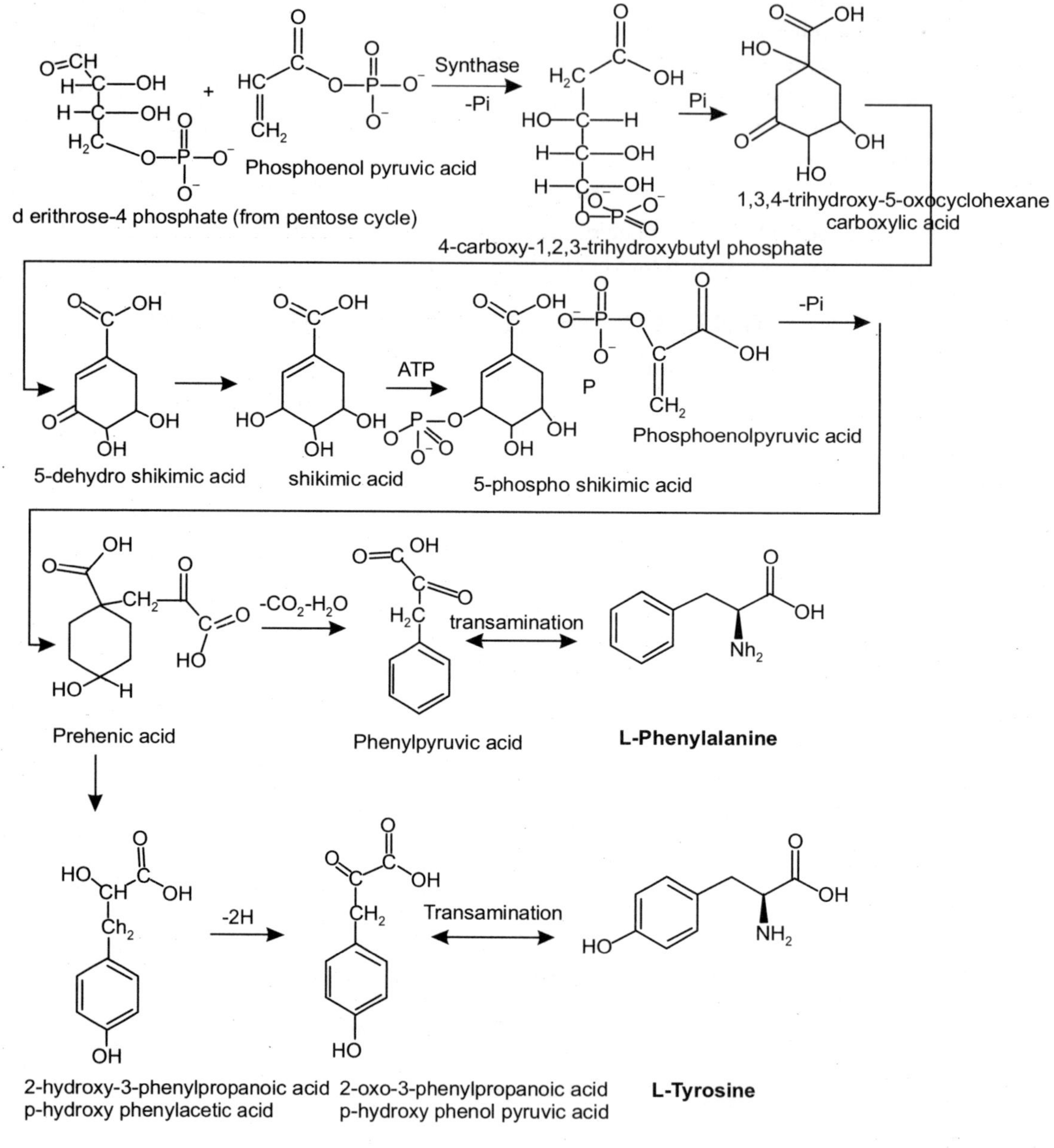

d erithrose-4 phosphate (from pentose cycle) + Phosphoenol pyruvic acid

Synthase

-Pi

4-carboxy-1,2,3-trihydroxybutyl phosphate

Pi

1,3,4-trihydroxy-5-oxocyclohexane carboxylic acid

5-dehydro shikimic acid

shikimic acid

ATP

5-phospho shikimic acid

P

Phosphoenolpyruvic acid

-Pi

Prehenic acid

$-CO_2-H_2O$

Phenylpyruvic acid

transamination

L-Phenylalanine

-2H

Transamination

2-hydroxy-3-phenylpropanoic acid
p-hydroxy phenylacetic acid

2-oxo-3-phenylpropanoic acid
p-hydroxy phenol pyruvic acid

L-Tyrosine

Fig. 5.51 : The Tyrosine and Phenylalanine Biosynthesis

FORMATION OF MELANIN

Melanin (Greek, *melani* (black); is a class of compounds found in the plant, animal, and protista kingdoms, where it serves predominantly as a pigment. The class of pigments are derivatives of the amino acid tyrosine. The most common form of biological melanin is eumelanin, a brown-black polymer of dihydroxyindole, dihydroxyindole carboxylic acid, and their reduced forms. Another common form of melanin is pheomelanin, a red-brown polymer of benzothiazine units largely responsible for red hair and freckles. The presence of melanin in the archaea and bacteria kingdoms is an issue of ongoing debate amongst researchers in the field. The increased production of melanin in human skin is called melanogenesis. It is stimulated by the DNA damages that are caused by UVB-radiation, and it leads to a delayed development of a tan. This melanogenesis-based tan takes more time to develop, but it is long lasting.

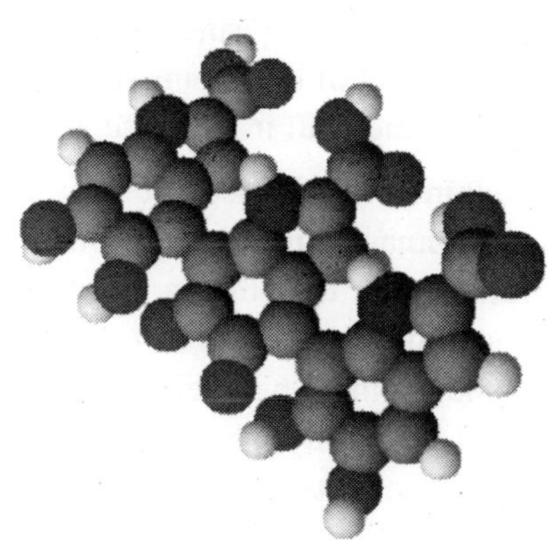

The photochemical properties of melanin make it an excellent photoprotectant. It absorbs harmful UV-radiation and transforms the energy into harmless amounts of heat through a process called "ultrafast internal conversion". This property enables melanin to dissipate more than 99.9% of the absorbed UV radiation as heat and it keeps the generation of free radicals at a minimum (see photoprotection). This prevents the indirect DNA damage which is responsible for the formation of malignant melanoma.

In humans, melanin is the primary determinant of human skin color and also found in hair, the pigmented tissue underlying the iris, the medulla and zona reticularis of the adrenal gland, the stria vascularis of the inner ear, and in pigment-bearing neurons within areas of the brain stem, such as the locus ceruleus and the substantia nigra.

Dermal melanin is produced by melanocytes, which are found in the stratum basale of the epidermis. Although human beings generally possess a similar concentration of melanocytes in their skin, the melanocytes in some individuals and ethnic groups more frequently or less frequently express the melanin-producing genes, thereby conferring a greater or lesser concentration of skin melanin. Some individual animals and humans have very little or no melanin in their bodies, a condition known as albinism.

Because melanin is an aggregate of smaller component molecules, there are a number of different types of melanin with differing proportions and bonding patterns of these component molecules. Both pheomelanin and eumelanin are found in human skin and hair, but eumelanin is the most abundant melanin in humans, as well as the form most likely to be deficient in albinism.

Eumelanin polymers have long been thought to comprise numerous cross-linked 5,6-dihydroxyindole (DHI) and 5,6-dihydroxyindole-2-carboxylic acid (DHICA) polymers; recent research into the electrical properties of eumelanin, however, has indicated that it may consist of more basic oligomers adhering to one another by some other mechanism. Thus, the precise nature of eumelanin's molecular structure is once again the object of study. Eumelanin is found in hair and skin, and colors hair grey, black, yellow, and brown. In humans, it is more abundant in peoples with dark skin. There are two different types of eumelanin, which are distinguished from each other by their pattern of polymer bonds. The two types are black eumelanin and brown eumelanin, with black melanin being darker than brown. Black eumelanin is in mostly non-Europeans and aged Europeans, while brown eumelanin is in mostly young Europeans. A small amount of black eumelanin in the absence of other pigments causes grey hair. A small amount of brown eumelanin in the absence of other pigments causes yellow (blond) colour hair.

Pheomelanin is also found in hair and skin and is both in lighter skinned humans and darker skinned humans. In general women have more pheomelanin than men, and thus women's skin is generally redder than men's. Pheomelanin imparts a pink to red hue and, thus, is found in particularly large quantities in red hair. Pheomelanin is particularly concentrated in the lips, nipples, glans of the penis, and vagina. Pheomelanin also may become carcinogenic when exposed to the ultraviolet rays of the sun. Chemically, pheomelanin differs from eumelanin in that its oligomer structure incorporates benzothiazine units which are produced instead of DHI and DHICA when the amino acid L-cysteine is present.

Neuromelanin is the dark pigment present in pigment bearing neurons of four deep brain nuclei: the substantia nigra (in Latin, literally "black substance") - Pars Compacta part, the locus ceruleus ("blue spot"), the dorsal motor nucleus of the vagus nerve (cranial nerve X), and the median raphe nucleus of the pons. Both the substantia nigra and locus ceruleus can be easily identified grossly at the time of autopsy due to their dark pigmentation. In humans, these nuclei are not pigmented at the time of birth, but develop pigmentation during maturation to adulthood. Although the functional nature of neuromelanin is unknown in the brain, it may be a byproduct of the synthesis of monoamine neurotransmitters for which the pigmented neurons are the only source. The loss of pigmented neurons from specific nuclei is seen in a variety of neurodegenerative diseases. In Parkinson's disease there is massive loss of dopamine producing pigmented neurons in the substantia nigra. A common finding in advanced Alzheimer's disease is almost complete loss of the norepinephrine producing pigmented neurons of the locus ceruleus. Neuromelanin has been detected in primates and in carnivores such as cats and dogs.

Biosynthetic Pathways

The first step of the biosynthetic pathway for both eumelanins and pheomelanins is catalysed by tyrosinase:

Tyrosine →→ DOPA → dopaquinone

Dopaquinone can combine with cysteine by two pathways to benzothiazines and pheomelanins

Dopaquinone + cysteine → 5-S-cysteinyldopa → benzothiazine intermediate → pheomelanin

Dopaquinone + cysteine → 2-S-cysteinyldopa → benzothiazine intermediate → pheomelanin

Alternatively, dopaquinone can be converted to leucodopachrome and follow two more pathways to the eumelanins

Dopaquinone → leucodopachrome → dopachrome → 5,6-dihydroxyindole-2-carboxylic acid → quinone → eumelanin

Dopaquinone → leucodopachrome → dopachrome → 5,6-dihydroxyindole → quinone → eumelanin

Tyrosine + O_2

DOPA

Dopa quinone

5-hydroxy-2,3,5,6-tetrahydro-1H-indole-2-carboxylic acid

1*H*-indole-5,6-diol

1*H*-indole-5,6-dione

melanin

Microscopic Appearance

Under the microscope melanin is brown, non-refractile and finely granular with individual granules having a diameter of less than 800 nanometers. This differentiates melanin from common blood breakdown pigments which are larger, chunky and refractile and range in color from green to yellow or red-brown. In heavily pigmented lesions, dense aggregates of melanin can obscure histologic detail. A dilute solution of potassium permanganate is an effective melanin bleach.

Melanin Deficiency in Genetic Disorders and Disease States

Melanin deficiency has been connected for some time with various genetic abnormalities and disease states.

There are approximately ten different types of oculocutaneous albinism, which is mostly an autosomal recessive disorder. Certain ethnicities have higher incidences of different forms. For example, the most common type, called oculocutaneous albinism type 2 (OCA2), is especially frequent among people of black African descent. It is an autosomal recessive disorder characterized by a congenital reduction or absence of melanin pigment in the skin, hair and eyes. The estimated frequency of OCA2 among African-Americans is 1 in 10,000, which contrasts with a frequency of 1 in 36,000 in white Americans. In some African nations, the frequency of the disorder is even higher, ranging from 1 in 2,000 to 1 in 5,000. Another form of Albinism, the “yellow oculocutaneous albinism”, appears to be more prevalent among the Amish, who are of primarily Swiss and German ancestry. People with this IB variant of the disorder commonly have white hair and skin at birth, but rapidly develop normal skin pigmentation in infancy.

Ocular albinism affects not only eye pigmentation, but visual acuity, as well. People with albinism typically test poorly, within the 20/60 to 20/400 range. Additionally, two forms of albinism, with approximately 1 in 2700 most prevalent among people of Puerto Rican origin, are associated with mortality beyond melanoma-related deaths.

Mortality also is increased in patients with Hermansky-Pudlak syndrome and Chediak-Higashi syndrome. Patients with Hermansky-Pudlak syndrome have a bleeding diathesis secondary to platelet dysfunction and also experience restrictive lung disease (pulmonary fibrosis), inflammatory bowel disease, cardiomyopathy, and renal disease. Patients with Chediak-Higashi syndrome are susceptible to infection and also can develop lymphofollicular malignancy.

The role that melanin deficiency plays in such disorders remains under study.

The connection between albinism and deafness has been well known, though poorly understood, for more than a century-and-a-half. In his 1859 treatise *On the Origin of Species*, Charles Darwin observed that "cats which are entirely white and have blue eyes are generally deaf". In humans, hypopigmentation and deafness occur together in the rare Waardenburg's syndrome, predominantly observed among the Hopi in North America. The incidence of albinism in Hopi Indians has been estimated as approximately 1 in 200 individuals. Interestingly, similar patterns of albinism and deafness have been found in other mammals, including dogs and rodents. However, a lack of melanin per se does not appear to be directly responsible for deafness associated with hypopigmentation, as most individuals lacking the enzymes required to synthesize melanin have normal auditory function. Instead the absence of melanocytes in the stria vascularis of the inner ear results in cochlear impairment, though why this is is not fully understood. It may be that melanin, the best sound absorbing material known, plays some protective function. Alternately, melanin may affect development, as Darwin suggests.

In Parkinson's disease, a disorder that affects neuromotor functioning, there is decreased neuromelanin in the substantia nigra as consequence of specific dropping out of dopaminergic pigmented neurons. This results in diminished dopamine synthesis. While no correlation between race and the level of neuromelanin in the substantia nigra has been reported, the significantly lower incidence of Parkinson's in blacks than in whites has "prompt[ed] some to suggest that cutaneous melanin might somehow serve to protect the neuromelanin in substantia nigra from external toxins."

In addition to melanin deficiency, the molecular weight of the melanin polymer may be decreased due to various factors such as oxidative stress, exposure to light, perturbation in its association with melanosomal matrix proteins, changes in pH or in local concentrations of metal ions. A decreased molecular weight or a decrease in the degree of polymerization of **ocular melanin** has been proposed to turn the normally anti-oxidant polymer into a pro-oxidant. In its pro-oxidant state, melanin has been suggested to be involved in the causation and progression of macular degeneration and melanoma.

Higher eumelanin levels also can be a disadvantage, however, beyond a higher disposition toward vitamin D deficiency. Dark skin is a complicating factor in the laser removal of port-wine stains. Effective in treating white skin, lasers generally are less successful in removing port-wine stains in people of Asian or African descent. Higher concentrations of melanin in darker-skinned individuals simply diffuse and absorb the laser radiation, inhibiting light absorption by the targeted tissue. Melanin similarly can complicate laser treatment of other dermatological conditions in people with darker skin.

Freckles and moles are formed where there is a localized concentration of melanin in the skin. They are highly associated with pale skin.

Melanin and Human Adaptation

Melanocytes insert granules of melanin into specialized cellular vesicles called melanosomes. These are then transferred into the other skin cells of the human epidermis. The melanosomes in each recipient cell accumulate atop the cell nucleus, where they protect the nuclear DNA from mutations caused by the ionizing radiation of the sun's ultraviolet rays. People whose ancestors lived for long periods in the regions of the globe near the equator generally have larger quantities of eumelanin in their skins. This makes their skins brown or black and protects them against high levels of exposure to the sun, which more frequently results in melanomas in lighter skinned people.

With humans, exposure to sunlight stimulates the skin to produce vitamin D. Because high levels of cutaneous melanin act as a natural sun screen, dark skin can be a risk factor for vitamin D deficiency

In the United Kingdom, which lies at a northern latitude, descendants of the Britons have white skin. When their skin is exposed to the meager sunlight, the scant amount of melanin their skin produces is unable to block the sunlight. Therefore, their bodies are able to make Vitamin D with the help of sunlight. Vitamin D, a vitamin found in fish oil, is necessary to prevent rickets, a bone disease caused by too little calcium.

In contrast, in Sub-Saharan Africa, which is near the equator, humans with a higher concentration of melanin absorb more intense sunlight to make Vitamin D. Africans visiting the United Kingdom during the Industrial Revolution developed symptoms of rickets, such as retarded growth, bowed legs, and fractures because sunlight at that latitude was insufficient for their melanin levels.

Fortunately, in 1930, Vitamin D was discovered and dispensed as a supplement to add to the diet. Now many common foods like milk and bread are Vitamin D fortified.

The most recent scientific evidence indicates that all humans evolved in Africa, then populated the rest of the world through successive radiations. It is most likely that the first people had relatively large numbers of eumelanin producing melanocytes and, accordingly, darker skin (as displayed by the indigenous people of Africa, today). As some of these original peoples migrated and settled in areas of Asia and Europe, the selective pressure for eumelanin production decreased in climates where radiation from the sun was less intense. Thus variations in genes involved in melanin production began to appear in the population, resulting in lighter hair and skin in humans residing at northern latitudes. Studies have been carried out to determine whether these changes were due to genetic drift or positive selection, perhaps driven by requirement for vitamin D. Of the two common gene variants known to be associated with pale human skin, does not appear to have undergone positive selection, while SLC24A5 has.

As with peoples who migrated northward, those with light skin who migrate southward acclimatize to the much stronger solar radiation. Most people's skin darkens when exposed to UV light, giving them more protection when it is needed. This is the physiological purpose of sun tanning. Dark-skinned people, who produce more skin-protecting eumelanin, have a greater protection against sunburn and the development of melanoma, a potentially deadly form of skin cancer, as well as other health problems related to exposure to strong solar radiation, including the photodegradation of certain vitamins such as riboflavins, carotenoids, tocopherol, and folate.

Melanin in the eyes, in the iris and choroid, helps protect them from ultraviolet and high-frequency visible light; people with blue and gray eyes are more at risk for sun-related eye problems and for red-eye effect in photographs. Further, the ocular lens yellows with age, providing added protection. However, the lens also becomes more rigid with age, losing most of its accommodation—the ability to change shape to focus from far to near—a detriment due probably to protein crosslinking caused by UV exposure.

Recent research by J.D. Simon suggests that melanin may serve a protective role other than photoprotection. Melanin is able to effectively ligate metal ions through its carboxylate and phenolic hydroxyl groups, in many cases much more efficiently than the powerful chelating ligand ethylenediaminetetraacetate (EDTA). It may thus serve to sequester potentially toxic metal ions, protecting the rest of the cell. This hypothesis is supported by the fact that the loss of neuromelanin observed in Parkinson's disease is accompanied by an increase in iron levels in the brain.

Physical Properties and Technological Applications

"Melanin" is the name not only of the biological substance described in the article thus far, but of a quite different class of synthetic substances. This section uses the term "melanin" chiefly to refer to the synthetic substances.

Melanins, in the synthetic sense, are "rigid-backbone" conductive polymers composed of polyacetylene, polypyrrole, and polyaniline "Blacks" and their mixed copolymers. The simplest melanin is polyacetylene, and some fungal melanins are pure polyacetylene.

In 1963, DE Weiss and coworkers reported high electrical conductivity in a melanin, iodine-doped and oxidized polypyrrole "Black". They achieved the quite high conductivity of 1 Ohm/cm. A decade later, John McGinness, and coworkers reported a high conductivity "ON" state in a voltage-controlled solid-state threshold switch made with DOPA melanin. Further, this material emitted a flash of light—electroluminescence—when it switched. Melanin also shows negative resistance, a classic property of electronically-active conductive polymers. Likewise, melanin is the best sound-absorbing material known due to strong electron-phonon coupling. This may be related to melanin's presence in the inner ear.

Melanin voltage-controlled switch, an "active" organic polymer electronic device from 1974. Now in the Smithsonian.

These early discoveries were "lost" until the recent emergence of such melanins in device applications, particularly electroluminescent displays. In 2000, the Nobel Prize in Chemistry was awarded to three scientists for their subsequent 1977 (re)discovery and development of such conductive organic polymers. In an essential reprise of the work by Weiss et al, these polymers were oxidized, iodine-doped "polyacetylene black" melanins. There is no evidence the Nobel committee was aware of the almost identical prior report by Weiss of passive high conductivity in iodinated polypyrrole black or of switching and high electrical conductivity in DOPA melanin and related organic semiconductors. The melanin organic electronic device is now in the Smithsonian Institution's National Museum of American History's "Smithsonian Chips" collection of historic solid-state electronic devices.

Melanin influences neural activity and mediates the conduction of radiation, light, heat and kinetic energy. As such, it is the subject of intense interest in biotech research and development, most notably in organic electronics (sometimes called "plastic electronics") and nanotechnology, where dopants are used to dramatically boost melanin conductivity. Pyrrole black and acetylene black are the most commonly studied organic semiconductors.

Although synthetic melanin (commonly referred to as BSM, or "black synthetic matter") is made up of 3-6 oligomeric units linked together—the so-called "protomolecule"—there is no evidence that naturally occurring biopolymer (BCM, for "black cell matter") mimics this structure. However, since there is no reason to believe that natural melanin does not belong to the category of the polyarenes and polycationic polyenes, like pyrrol black and acetylene black, it is necessary to review all the chemical and biological analytic data gathered to date in the study of natural melanins (eumelanins, pheomelanins, allomelanins)."

Evidence exists in support of a highly cross-linked heteropolymer bound covalently to matrix scaffolding melanoproteins. It has been proposed that the ability of melanin to act as an antioxidant is directly proportional to its degree of polymerization or molecular weight. Suboptimal conditions for the effective polymerization of melanin monomers may lead to formation of lower-molecular-weight, pro-oxidant melanin that is has been implicated in the causation and progression of macular degeneration and melanoma. Signaling pathways that upregulate melanization in the retinal pigment epithelium (RPE) also may be implicated in the downregulation of rod outer segment phagocytosis by the RPE. This phenomenon has been attributed in part to foveal sparing in macular degeneration.

Melanin-based Bias in Human Societies

Consequently, sampling in much of Asia and some remote localities is insufficient (see e.g. Japanese, Inuit and Tibetans which ought to be darker-skinned than represented here because mountain area).

When skin pigmentation as a characteristic of race is linked to social status or other human attributes, this phenomenon is known as racialism. Many people and societies overlay racialism with racist perceptions and systems which arbitrarily assign to groups of people a status of inherent superiority or inferiority, privilege or disadvantage based on skin color or racial classification. Apartheid-era South Africa is an example of a white supremacist society based on a system of stratification of power and privilege by skin color, as well as racial admixture. Similar examples can be found in Brazil's highly socially color-stratified society; and, in the U.S., segregation and institutional racism on the part of white-controlled institutions, and internal "color consciousness" on the part of members of some ethnic minorities.

THE FORMATION OF CATICOLAMINES

Catecholamines are chemical compounds derived from the amino acid tyrosine. Their name is derived from the fact that they contain catechol and amine moieties. Some of them are biogenic amines.

Catecholamines are water-soluble and are 50% bound to plasma proteins, so they circulate in the bloodstream. The most abundant catecholamines are epinephrine (adrenaline), norepinephrine (noradrenaline) and dopamine, all of which are produced from phenylalanine and tyrosine. Tyrosine is created from phenylalanine by hydroxylation by the enzyme phenylalanine hydroxylase. (Tyrosine is also ingested directly from dietary protein). It is then sent to catecholamine-secreting neurons. Here, many kinds of reactions convert it to dopamine, to norepinephrine, and eventually to epinephrine. Catecholamines are hormones that are released by the adrenal glands in situations of stress such as psychological stress or low blood sugar levels.

L-Tyrosine

O_2, Tetrahydro-biopterin

Tyrosine hydroxylase

H_2O, Dihydro-biopterin

L-Dihydroxyphenylalanine (L-DOPA)

DOPA decarboxylase
Aromatic L-amino acid decarboxylase

CO_2

Dopamine

O_2, Ascorbic acid

Dopamine β-hydroxylase

H_2O, Dehydro-ascorbic acid

Norepinephrine

S-adenosyl-methionine

Phenylethanolamine N-methyltransferase

Homocysteine

Epinephrine

Location

Catecholamines are produced mainly by the chromaffin cells of the adrenal medulla and the postganglionic fibers of the sympathetic nervous system. Dopamine, which acts as a neurotransmitter in the central nervous system, is largely produced in neuronal cell bodies in two areas of the brainstem: the substantia nigra and the ventral tegmental area.

Synthesis

Dopamine is the first catecholamine to be synthesised from steps shown. Norepinephrine and epinephrine, in turn, are derived from further modifications of dopamine. It is important to note that the enzyme dopamine hydroxylase requires copper as a cofactor (not shown) and DOPA decarboxylase requires PLP (not shown).

Function Modality

Two catecholamines, norepinephrine and dopamine, act as neuromodulators in the central nervous system and as hormones in the blood circulation. The catecholamine norepinephrine is a neuromodulator of the peripheral sympathetic nervous system but is also present in the blood (mostly through "spillover" from the synapses of the sympathetic system).

High catecholamine levels in blood are associated with stress, which can be induced from psychological reactions or environmental stressors such as elevated sound levels, intense light, or low blood sugar levels.

Extremely high levels of catecholamines (also known as catecholamine toxicity) can occur in central nervous system trauma due to stimulation and/or damage of nuclei in the brainstem, in particular those nuclei affecting the sympathetic nervous system. In emergency medicine, this occurrence is widely known as catecholamine dump.

Extremely high levels of catecholamine can also be caused by neuroendocrine tumors in the adrenal medulla, a treatable condition known as pheochromocytoma.

High levels of catecholamines can also be caused by monoamine oxidase A deficiency. This is the gene responsible for degradation of these neuortransmitters and thus increases the circulation of them considerably. It occurs in the absence of pheochromocytoma, neuroendocrine tumors, and carcinoid syndrome, but it looks similar to carinoid syndrome such as facial flushing, aggression, and ADHD. Effects

Catecholamines cause general physiological changes that prepare the body for physical activity (fight-or-flight response). Some typical effects are increases in heart rate, blood pressure, blood glucose levels, and a general reaction of the sympathetic nervous system. Some drugs, like tolcapone (a central COMT-inhibitor), raise the levels of all the catecholamines.

Function in Plants

"They have been found in 44 plant families, but no essential metabolic function has been established for them. They are precursors of benzo[c]phenanthridine alkaloids, which are the active principal ingredients of many medicinal plant extracts. CAs has been implicated to have a possible protective role against insect predators, injuries, and nitrogen detoxification. They have been shown to promote plant tissue growth, somatic embryogenesis from in vitro cultures, and flowering. CAs inhibit indole-3-acetic acid oxidation and enhance ethylene biosynthesis. They have also been shown to enhance synergistically various effects of gibberellins."

Structure

Catecholamines have the distinct structure of a benzene ring with two hydroxyl groups, an intermediate ethyl chain, and a terminal amine group.

Degradation

They have a half-life of approximately a few minutes when circulating in the blood. Monoamine oxidase (MAO) is the main enzyme responsible for degradation of catecholamines. Amphetamines and MAOIs bind to MAO in order to inhibit its action of breaking down catecholamines. This is primarily the reason why the effects of amphetamines have a longer lifespan than those of cocaine and other substances. Amphetamines not only cause a release of dopamine, epinephrine, and norepinephrine into the blood stream but also suppress re-absorption.

THE SULPHUR CONTAINING AMINO ACIDS

Most of the sulphur of proteins is represented by the methionine and cysteine, present though in small amounts of cysteine the reduced product of cysteine may be present also. This is indicated by the presence of free–SH groups in a large number of enzyme and other proteins.

Demethylation of methyl methionine produces homosysteine, which may be remethylated to methionine. Cysteine is reversely convertible to cystine and homosysteine by redox reaction. The chemical reactions and relations between these amino acids can be shown as follows:

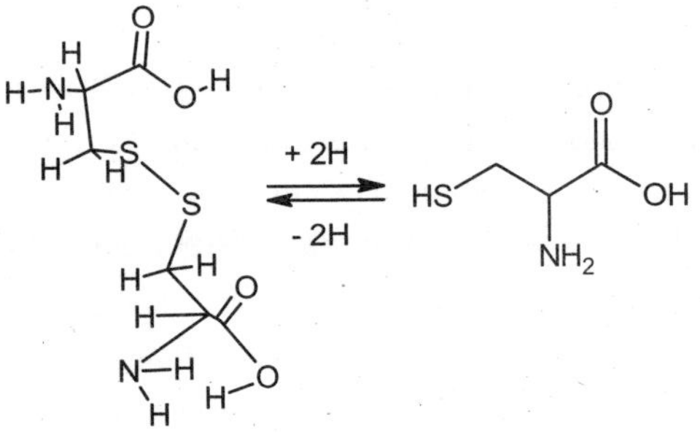

cystine $\underset{-2H}{\overset{+2H}{\rightleftharpoons}}$ cysteine

Methionine $\underset{+CH_2}{\overset{-CH_2}{\rightleftharpoons}}$ (homocysteine) $\rightleftharpoons$ Homocystine

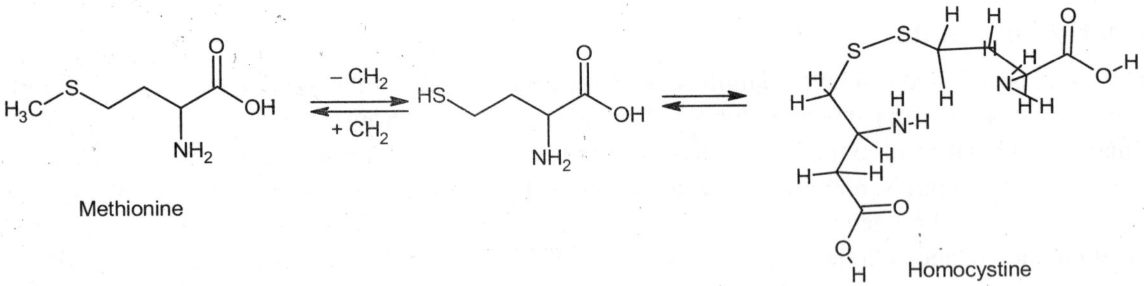

Cysteine is non enzymatically and reversibly oxidized to cystine by glutathionine present in cells.

$$\text{cysteine} + G-S-S-G \rightleftharpoons 2GSH + \text{cystine}$$

Cysteine cannot form Keto Acid

The methyl group of methionine is removed through S – adinosyl methionine in the formation of creatine and choline (A methyl group is added to homosysteine to form methionine) as pointed out in the discussion of one carbon methabolism.

Casteine cannot form keto aicd

Since cysteine, cystine and homocysteines are interconvertable readily, processes which lead to the formation of one of either pair of these compounds may be considered as to form the other pair.

Methionine but not cysteine is an essential amino acid in the diet of all met by feeding the corresponding keto acid, α-keto-γ-methylatiobutyric acid which is aminated to L-methionine through keto acid (and hydroxyl acid through the keto acid) to methionine therefore it cannot synthesize keto acid.

Since cysteine is not an essential amino acid, this indicates the formation of cysteine from methionine. The presence of cysteine in the diet reduces the methionine requirement.

Pathways of Methionine Metabolism

Methionine undergoes transamination in the liver with a - ketoglutoric acid to yield the keto acid α-keto-γ-methylthiobutaric acid

Methionine + α-Keto glutaric acid ⇌ (Transamination, Pyridoxal Sulphate) α-Keto-γ-methylthiobutyric acid + glutamic acid

The main pathway of glutamic acid-methionine metabolism leads to cysteine and homoserine.

The processes involved in the pathway of methionine may be represented in the equation below (Fig. 5.52).

It is to be noted that the reactions of methionine breakdown that only the sulphur of methionine is used in the synthesis of cysteine molecule being derived from serine. The reminder of the methionine molecule is converted to homoserine, which is breakdown through α-ketobutyric acid to propionic acid. The S of methionine goes the pathways of cysteine metabolism.

As indicated in the reaction pyridoxal phosphate is required in the formation of cystathionine and the liver preparations from pyridoxine (vitamin B_6) deficient rats do not form cysteine from homocysteine and serine unless pyridoxal phosphate is added.

In patients with severe liver disease exhibiting foul smell in breadth, methyl mercaptan is found in the urine. It appears to be formed by liver enzyme acting upon the keto acid derived from methionine (Fig. 5.53).

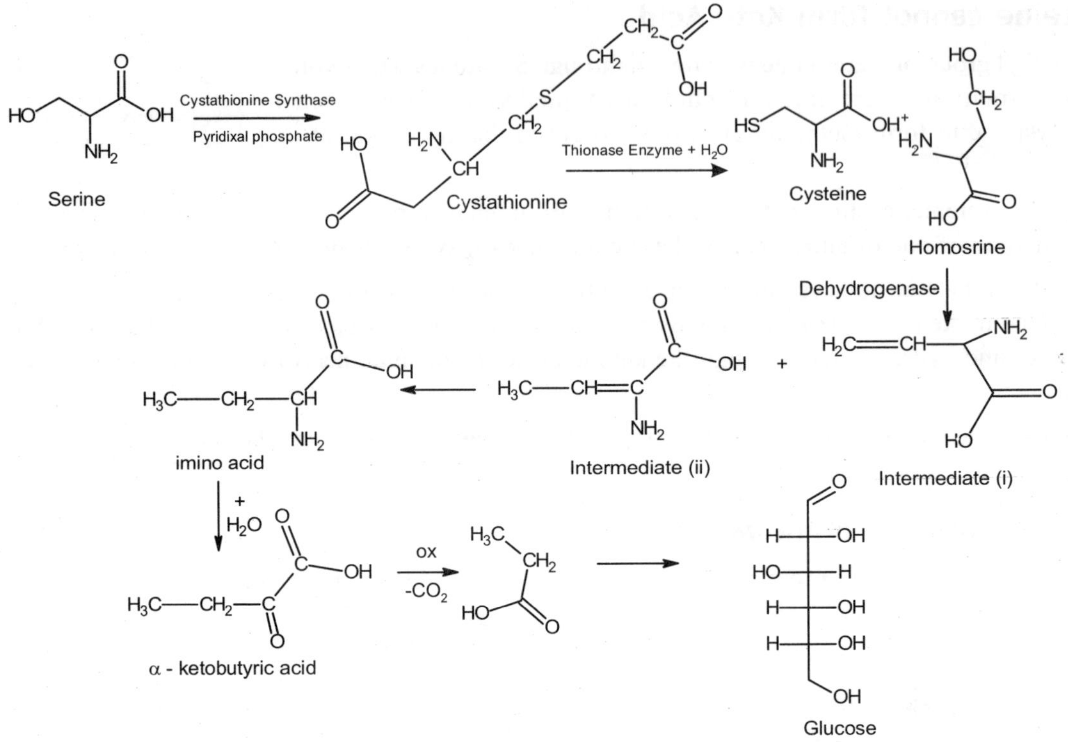

Cystathionine Synthase
Pyridixal phosphate
Serine
Cystathionine
Thionase Enzyme + H_2O
Cysteine
Homosrine
Dehydrogenase
Intermediate (i)
Intermediate (ii)
imino acid
+ H_2O
α - ketobutyric acid
ox
$-CO_2$
Glucose

Fig. 5.52 : Pathway of Methionine

a - Keto - g -hydroxybutyric acid
+ H_2O
Enzyme
methyl mercaptan
α - Keto - γ -methylthiobutyric acid
homoserine
Transamination
α - ketobutyric acid
α - Aminobutyric acid

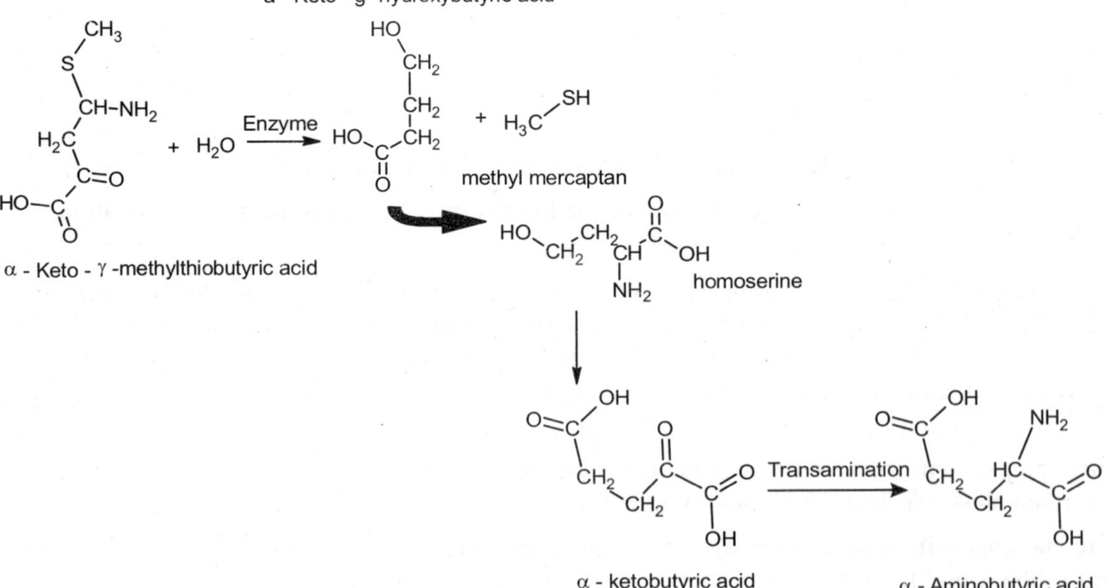

Fig. 5.53

In the fanchoni syndrome which is characterized by a low renal threshold for amino acids, amino butyric acid formed from α-ketobutyric acid and as indicated above is expected above, is excreted in the urine.

Homocysteine is convertible by a liver desuplydrase enzyme, thionase into -α-ketobutyric acid NH_3 and H_2S, most of the hydrogen sulphide formed into sulphate and a smaller portion to thiosulphate.

Metabolism of Cysteine and Cystine

Cysteine metabolism appears to proceed entirely through cysteine into which it is readily converted, when cysteine or methionine is given to the animals a large proportion of sulphur of amino acids is rapidly oxidized to sulphate and excreted in urine.

Urinary sulphur is derived almost entirely from the metabolism of the sulphur amino acids. It consists of (a) sulphur of inorganic sulphate sulphur of organic sulphide which is 15 to 20%. Oxidation of the sulphur, of the sulphur containing sulphuric acid to sulphate represents the final stage of sulphur oxidation in the body. Ethereal sulphates generally represent compounds formed in the detoxication of phenols, such as phenyl sulphuric acid, indicant, and skatoxysulphuric acid. A small portion of the organic sulphide is made up of unchanged sulphur containing sulphuric acid, and some may mercaptan sulphur, especially in certain liver diseases, but nature of most of this fraction of the fraction is still unknown. It is undoubtedly a highly complex mixture of substances.

The ethereal sulphate compounds are formed by reaction of "active sulphate" 3-phosphoadenosine -5'-phosphosulphate in with compounds.

The formation of cysteine from methionine has been discussed. Since in this process only the S of methionine is converted to the S of cysteine the metabolism of cysteine involves only the S but not the carbon chain of the methionine.

Cysteine is metabolized in animals by several pathways. Cysteine is converted to the pyruvic acid, H2S and NH_3 by a desulphydrase enzyme found in liver, kidney and pancreases.

Cysteine —(Desulphydrase, pyridoxal sulphate)→ H_2S + 2-aminoprop-2-enoic acid ⟷ imino acid + H_2O → pyruvic acid + NH_3

Cysteine transaminates with α-ketoglutoric acid to α-mercaptopyruvic acid (β-thiopyruvic acid)

cysteine + α-Keto glutaric acid —(transaminase)→ β-mercaptopyruvic acid + glutamic acid

The S is converted to H_2S by reducing agents such as cysteine and glutathione

2 Cysteine + S ⇌ cystine + H_2S

β-Mercaptopyruvic acid transfers sulphur to the cyanide ion to form sulphocyanate. The reaction is catalyzed by a liver transsulphurase enzyme.

β-mercaptopyruvic acid + CN —(Enzyme)→ SCN + Pyruvic acid

Thiosulphate is formed by reaction of β-mercaptopyruvic acid with sulphite through the action of transsulphurase enzyme found in kidney and liver.

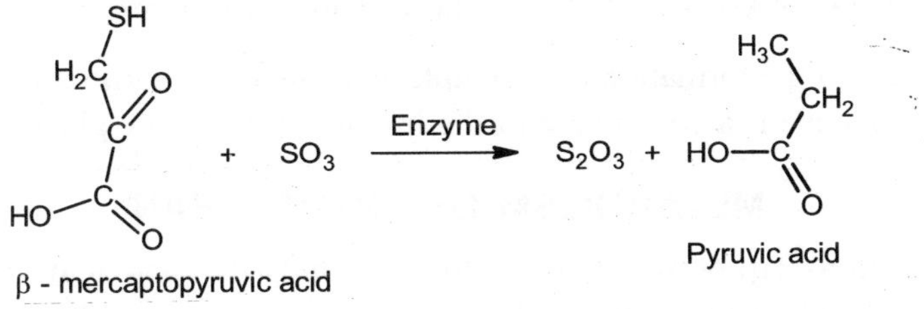

SH
H$_2$C
C=O
HO—C=O
+ SO_3 —Enzyme→ S_2O_3 + HO—C(=O)—CH_2—H_3C

β - mercaptopyruvic acid

Pyruvic acid

Sulphocyanate may also be formed through the action of a liver enzyme rhodanese from CN^- and S and form CN^- and $S_2O_3^-$

$$CN^- + S \longrightarrow SCN^-$$

$$CN^- + S_2O_3 \longrightarrow SCN^- + SO_3^-$$

It appears that a major pathway of cysteine metabolism is through cysteine sulphinic acid. Cysteine is oxidized to cysteine sulphinic acid by a liver enzyme requiring ATP, TPN and Mg^{++}

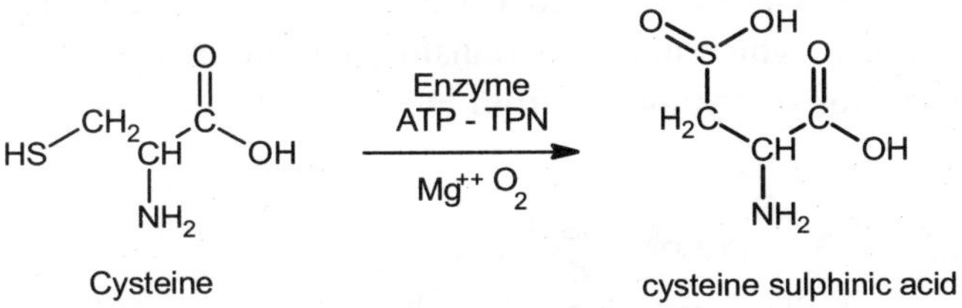

HS—CH_2—CH(NH_2)—C(=O)—OH —Enzyme, ATP - TPN, Mg^{++} O_2→ O=S(OH)—H_2C—CH(NH_2)—C(=O)—OH

Cysteine

cysteine sulphinic acid

Cysteine sulphinic acid rapidly transaminates to form β-sulphinylpyruvic acid, which to form pyruvic acid is quite analogous to the decarboxylation of oxaloacetic acid to form pyruvic acid.

The SO_2 (sulphite) is oxidized to sulphate by a liver enzyme, sulphate oxidase which appears to require hypoxanthine and lipoic acid.

Taurine appears to be formed from cysteine sulphinic acid in the liver by two major pathways. One of these is by oxidation to systemic acid then decarboxylation of the cysteic acid.

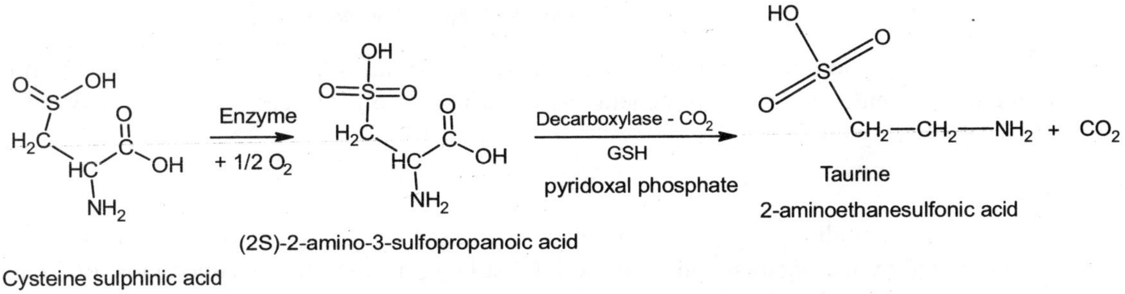

O=S(OH)—H_2C—HC(NH_2)—C(=O)—OH —Enzyme, + 1/2 O_2→ HO—S(=O)(=O)—H_2C—HC(NH_2)—C(=O)—OH —Decarboxylase - CO_2, GSH, pyridoxal phosphate→ HO—S(=O)(=O)—CH_2—CH_2—NH_2 + CO_2

(2S)-2-amino-3-sulfopropanoic acid

Taurine
2-aminoethanesulfonic acid

Cysteine sulphinic acid

Taurine is conjugated with alcohol CoA in the liver to form taurocholic acid

$$\text{Cholic acid} + \text{HS} - \text{CoA} + \text{ATP} \xrightarrow[\text{Mg}^{++}]{\text{Enzyme}} \text{Cholyl} - \text{S} - \text{CoA} + \text{AMP} + \text{PP}$$

$$\text{Taurine} + \text{Cholyl} - \text{S} - \text{CoA} \xrightarrow{\text{Enzyme}} \text{Taurocholic acid} + \text{HS} - \text{CoA}$$

The role of cysteine in the formation of mercapturic acids for detoxification has been considered. It is also known that cysteine is a constituent of the important tripeptide glutathione.

METABOLISM OF TRYPTOPHAN

Tryptophan (abbreviated as Trp or W) is one of the 20 standard amino acids, as well as an essential amino acid in the human diet. It is encoded in the standard genetic code as the codon UGG. Only the L-stereoisomer of tryptophan is used in structural or enzyme proteins, but the D-stereoisomer is occasionally found in naturally produced peptides (for example, the marine venom peptide contryphan). The distinguishing structural characteristic of tryptophan is that it contains an indole functional group.

Biosynthesis and Industrial Production

Plants and microorganisms commonly synthesize tryptophan from shikimic acid or anthranilate. The latter condenses with phosphoribosylpyrophosphate (PRPP), generating pyrophosphate as a by-product. After ring opening of the ribose moiety and following reductive decarboxylation, indole-3-glycerinephosphate is produced, which in turn is transformed into indole. In the last step, tryptophan synthase catalyzes the formation of tryptophan from indole and the amino acid, serine.

Chorismate → (Glutamine; Glutamate, Pyruvate) → Anthranilate → (PRPP; PP_i) → N-(4'-Phosphoribosyl)-anthranilate → 1-(o-Carboxyphenylamino)-1-desoxyribuose-5-phosphate → ($-CO_2$, $-HO^{\ominus}$) → Indole-3-glycerinphosphate → (-Glyceraldehyde-3-phosphate) → Indole → (Serine; H_2O) → Tryptophan

The industrial production of tryptophan is also biosynthetic and is based on the fermentation of serine and indole using either wild-type or genetically modified E. coli. The conversion is catalyzed by the enzyme tryptophan synthase.

Function

For many organisms (including humans), tryptophan is an essential amino acid. This means that it cannot be synthesized by the organism and therefore must be part of its diet. Amino acids, including tryptophan, act as building blocks in protein biosynthesis. In addition, tryptophan functions as a biochemical precursor for the following compounds (see also figure to the right at next page):

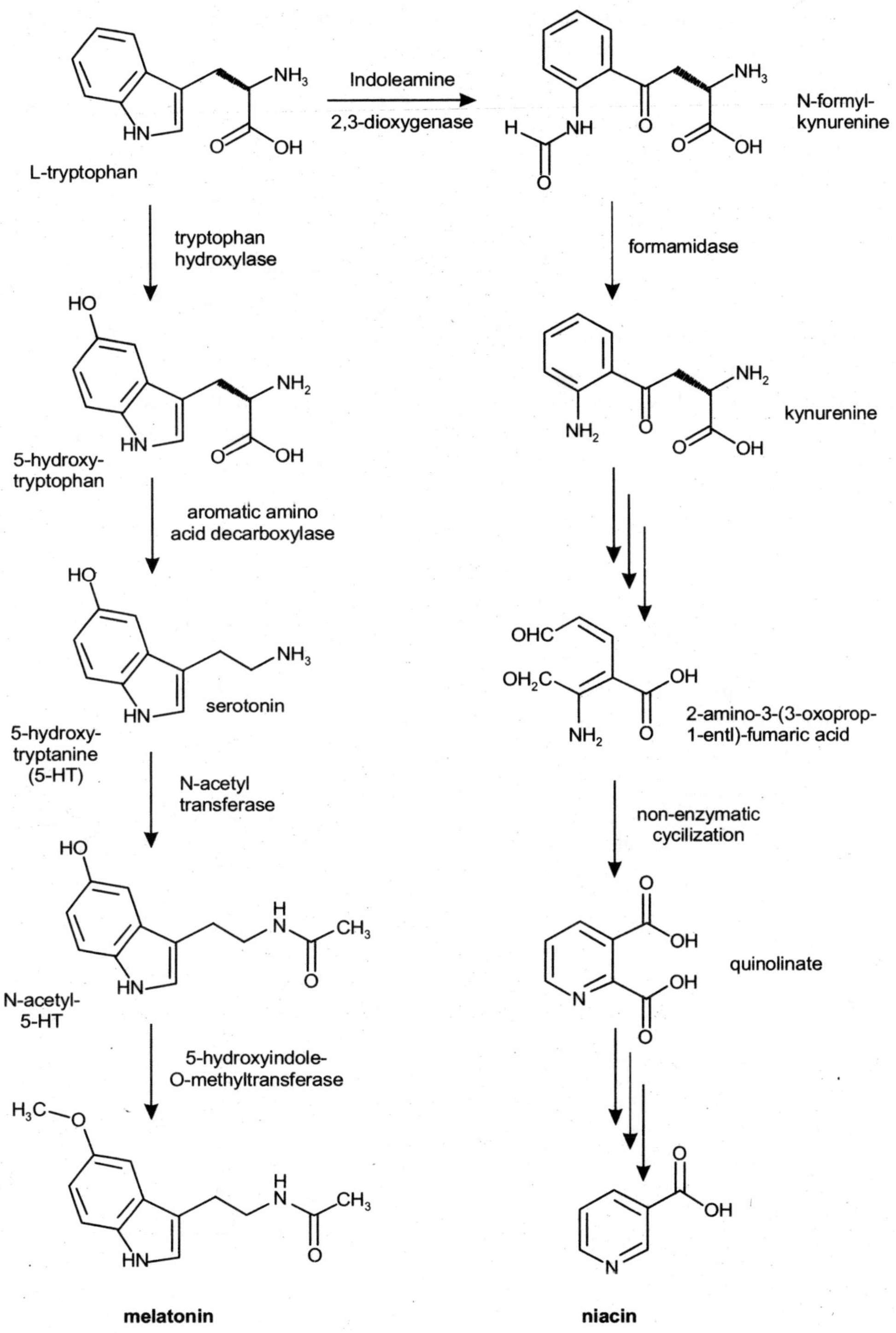

NH_3
Indoleamine
2,3-dioxygenase
HN
O OH
L-tryptophan
H NH O
O
NH_3
O OH
N-formyl-
kynurenine

tryptophan
hydroxylase

formamidase

HO
NH_2
HN
O OH
5-hydroxy-
tryptophan

NH_2
NH_2 O
O OH
kynurenine

aromatic amino
acid decarboxylase

HO
NH_3
HN
serotonin
5-hydroxy-
tryptanine
(5-HT)

OHC
OH_2C
OH
NH_2 O
2-amino-3-(3-oxoprop-
1-entl)-fumaric acid

N-acetyl
transferase

non-enzymatic
cycilization

HO
H
N
CH_3
O
HN
N-acetyl-
5-HT

O
OH
N
OH
O
quinolinate

5-hydroxyindole-
O-methyltransferase

H_3C—O
H
N
CH_3
O
HN

melatonin

O
OH
N

niacin

- Serotonin (a neurotransmitter), synthesized via tryptophan hydroxylase. Serotonin, in turn, can be converted to melatonin (a neurohormone), via N-acetyltransferase and 5-hydroxyindole-O-methyltransferase activities.
- Niacin is synthesized from tryptophan via kynurenine and quinolinic acids as key biosynthetic intermediates.
- Auxin (a phytohormone) when sieve tube elements undergo apoptosis tryptophan is converted to auxins.

The disorders Fructose Malabsorption and Lactose intolerance causes improper absorption of tryptophan in the intestine, reduced levels of tryptophan in the blood and depression.

In bacteria that synthesize tryptophan, high cellular levels of this amino acid activate a repressor **protein,** which binds to the trp operon. Binding of this repressor to the tryptophan operon prevents transcription of downstream DNA that codes for the enzymes involved in the biosynthesis of tryptophan. So high levels of tryptophan prevent tryptophan synthesis through a negative feedback loop and, when the cell's tryptophan levels are reduced, transcription from the trp operon resumes. The genetic organisation of the trp operon thus permits tightly regulated and rapid responses to changes in the cell's internal and external tryptophan levels.

Dietary Sources

Tryptophan is a routine constituent of most protein-based foods or dietary proteins. It is particularly plentiful in chocolate, oats, durians, mangoes, dried dates, milk, yogurt, cottage cheese, red meat, eggs, fish, poultry, sesame, chickpeas, sunflower seeds, pumpkin seeds, spirulina, and peanuts. It is found in turkey at a level typical of poultry in general.

Tryptophan (Trp) Content of Various Foods

Food	Protein[g/ 100 g of food]	Tryptophan[g/ 100g of food]	Tryptophan/ Protein [%]
1	2	3	4
egg, white, dried	81.10	1.00	1.23
spirulina, dried	57.47	0.93	1.62
cod, atlantic, dried	62.82	0.70	1.11
soybeans, raw	36.49	0.59	1.62
cheese, Parmesan	37.90	0.56	1.47
caribou	29.77	0.46	1.55
sesame seed	17.00	0.37	2.17
cheese, cheddar	24.90	0.32	1.29
sunflower seed	17.20	0.30	1.74

...(Contd.)

1	2	3	4
pork, chop	19.27	0.25	1.27
turkey	21.89	0.24	1.11
chicken	20.85	0.24	1.14
beef	20.13	0.23	1.12
salmon	19.84	0.22	1.12
lamb, chop	18.33	0.21	1.17
perch, Atlantic	18.62	0.21	1.12
egg	12.58	0.17	1.33
wheat flour, white	10.33	0.13	1.23
milk	3.22	0.08	2.34
rice, white	7.13	0.08	1.16
potatoes, russet	2.14	0.02	0.84
banana	1.03	0.01	0.87

Use as a Dietary Supplement

For some time, tryptophan has been available in health food stores as a dietary supplement, although it is common in dietary protein. Many people found tryptophan to be a safe and reasonably effective sleep aid, probably due to its ability to increase brain levels of serotonin (a calming neurotransmitter when present in moderate levels) and/or melatonin (a sleep-inducing hormone secreted by the pineal gland in response to darkness or low light levels). Some users of ecstasy will eat tryptophan-containing foods to shorten the 'come down' effect of having lower levels of serotonin than usual (due to an extra large release caused by the drug).Clinical research has shown mixed results with respect to tryptophan's effectiveness as a sleep aid, especially in normal patients and for a growing variety of other conditions typically associated with low serotonin levels or activity in the brain such as premenstrual dysphoric disorder and seasonal affective disorder. In particular, tryptophan has been showing considerable promise as an antidepressant alone, and as an "augmenter" of antidepressant drugs. However, the reliability of these clinical trials has been questioned. The keto acid and the hydroxy acid corresponding to tryptophan are utilized in place of tryptophan in the diet.

The following reaction reactions have been demonstrated between tryptophan, indole pyruvic acid indole acetic acid and tryptamine.

Tryptophan ?-ketoglutarate pyridoxal phopshate Indole pyruvic acid +2H -2H 1H-indol-3-ylacetic acid $-CO_2 + O_2$

O_2 Decarboxylase

tryptamine

The daily urine excretion of indole acetic acid in man generally amounts to 5 to 18 mg but may reach 200mg per day in certain pathological conditions and after tryptophan administration.

Tryptophan tyrptophan peroxidase + O_2 Formyl kynurenine kynurenine

O_2 enzyme 3 - hydroxy kynurenine Kynurenase + H_2O Alanine + 3 - hydroxyanthranilic acid Enzyme (2Z)-2-amino-3-[(1Z)-3-oxoprop-1-en-1-yl]but-2-enedioic acid Enzyme C_2O Nicotinic acid

The open chain intermediate from 3-hydroxyanthranilic acid apparently is also converted to quinolinic acid and picholinic acid.

3 - hydroxyanthranilic acid Enzyme (2 Z)-2-amino-3-[(1Z)-3-oxoprop-1-en-1-yl]but-2-enedioic acid Quinolinic acid picholinic acid

The vaso-constrictor substance* 5-hydroxytriptamine or serotonin is present in the blood is present in the blood, particularly in the gastric mucosa, intestine brain, mast cells and blood platelets. Patients with malignant carcinoid excrete large amount of the serotonin metabolite 5-hydroxy indole acetic acid in the urine. Such patients have been estimated to utilize as much as 60% of the tryptophan metabolize on the formation of serotonin as compared with 1% for the individual. Serotonin is considered to function as a neurohumoral$ agent. The formation of serotonin from tryptophan has been established by the use of L – tryptophan.

L-tryptophan —(enzyme, hydroxylation)→ 5-hydroxy L tryptophan —(decarboxylase, $-CO_2$)→ 5-hydroxytryptamine serotonin

monoamine oxidase $+ O_2 + H_2O$

→ 5-hydroxy indole acetic acid $+ NH_3 + H_2O_2$

From the above reactions it will be seen that three carbon atoms of tryptophan form alanine which may form pyruvic acid and glucose and that carbon 2 of the pyrol ring yields formic acid; thus

kynurenine —(kynureinase, pyridoxal phosphate)→ Anthranilic acid + Alanine

kynurenine —(transamination)→ Anthraanylpyruvic acid → kynurenic acid —(dehyderoxylation)→ quinoline-2-carboxylic acid

* The substances which constrict the cavity of the bold vessels

$ A substance liberated at the nerve ending that participates in the transmission of nerve impulse

tryptophan contributes to the one carbon pool. Also, the metabolism of tryptophan yields nicotinic acid which is converted to the vitamin niacin (nicotinamide) in the animal body.

Kynurenine transaminates to form the corresponding keto acid, which spontaneously forms kynurenic acid. Kynurenic acid is converted to anthranilic acid and to quinaldic acid by dehydroxylation.

3-Hydroxykynurenine undergoes reactions analogous to those of Kynurenine. For example 3-Hydroxykynurenine is converted to 3-hydroxyxanthranilic acid by kynureninase, as shown above in the main line of tryptophan metabolism.

3-Hydroxykynurenine, through the action of Kynurenine transaminases is converted to xanthuric acid, which then forms 8-hdyroxyquinaldic acid. This process is quite analogous to the Kynurenine reaction.

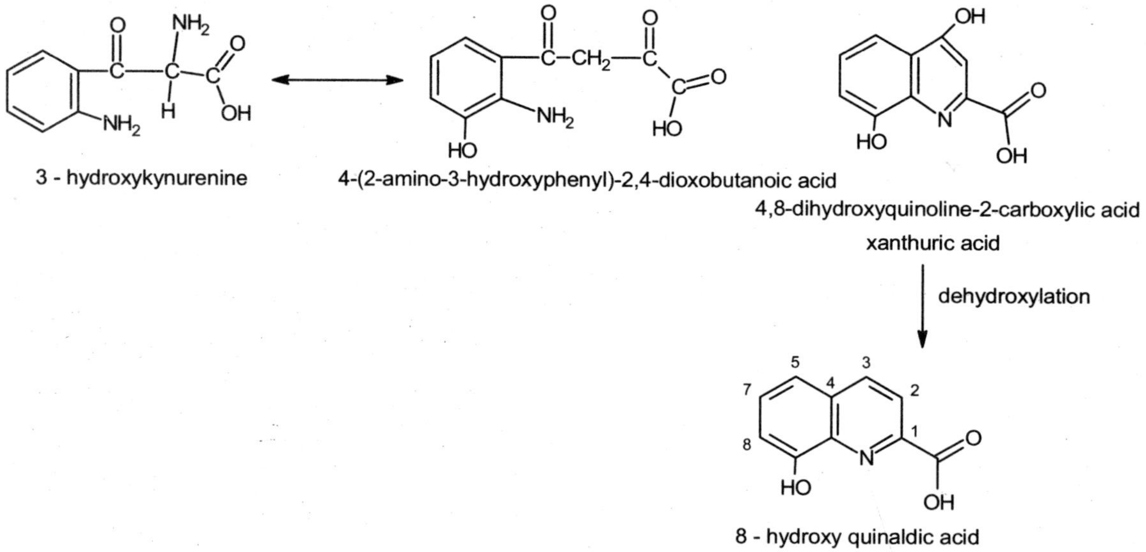

3 - hydroxykynurenine

4-(2-amino-3-hydroxyphenyl)-2,4-dioxobutanoic acid

4,8-dihydroxyquinoline-2-carboxylic acid

xanthuric acid

dehydroxylation

8 - hydroxy quinaldic acid

Bacterial Putrefaction

The action of bacteria upon tryptophan in the gut produces a large number of substances which are extracted as such or modified from (detoxified) in the urine or feces.

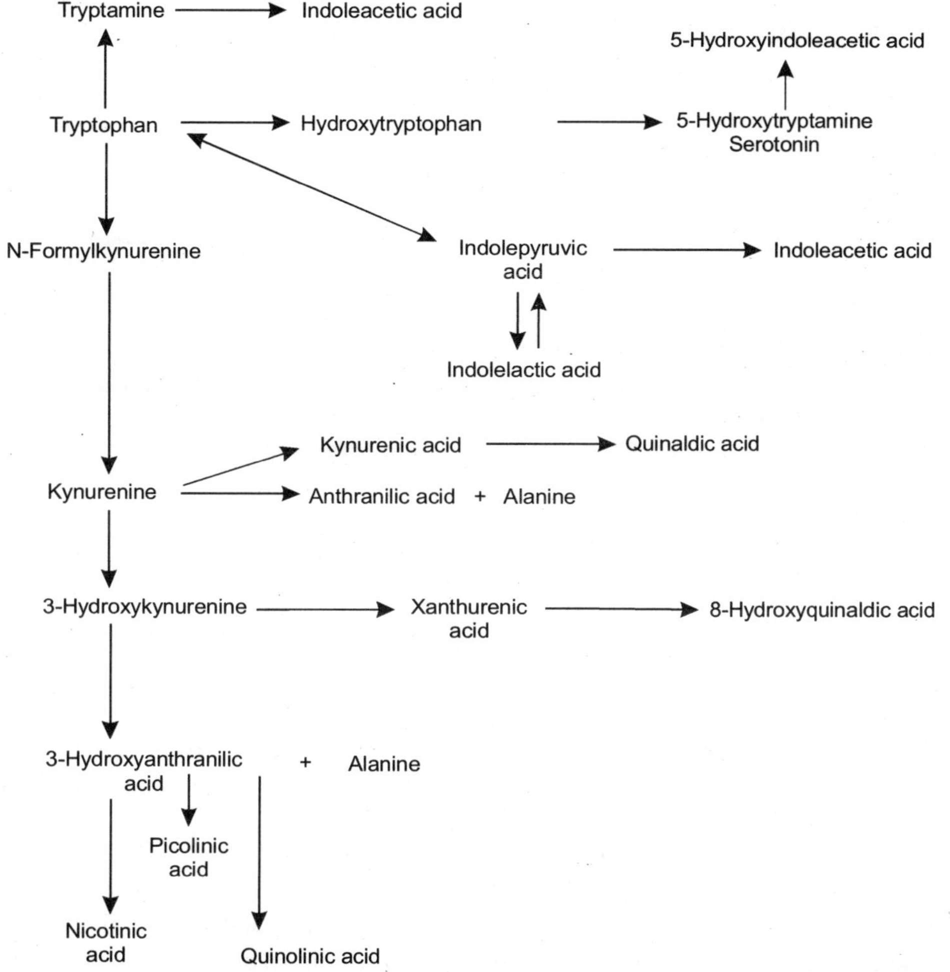

Fig. 5.54 : The Tryptophan Metabolism

THE METABOLISM OF HISTIDINE

Histidine (abbreviated as His or H) is one of the 20 standard amino acids present in proteins. Nutritionally, Histidine is considered an essential amino acid in human infants. After reaching several years of age, humans begin to synthesize it and it thus becomes a non-essential amino acid. Histidine was first isolated by German physician Albrecht Kossel in 1896..

Histidine synthesis includes the activity of imidiazoleglycerol phosphate, which is an intermediate is formed from ATP ribose 5 phosphate ATP and glutamine.

Adenosine Monophosphate + Ribose 5 P $\xrightarrow[\text{ATP, Glutarine}]{\text{Enzyme}}$ imidiazole glycerol phosphate + 5 aminodiazole 4 caroximide ribotide

imidiazole glycerol phosphate → Imidiazole acetal phopshate $\xrightarrow{\text{transamination}}$ L-histidinal Phosphate $\xrightarrow{+ H_2}$ L-histidinol $\xrightarrow{+ DPN^+}$ L-Histidinal → L-histidine

It appears that nitrogen 1 and carbon 2 of the purine ring of AMP become nitrogen 3 and carbon 2 of the imidoazole ring of histidine. It appears that nitrogen 1 and carbon 2 of the purine ring of AMP become nitrogen 3 and carbon 2 of the imidoazole ring of histidine and that the five carbon chain of histidine is derived from the ribose carbon. Nitrogen 1 of the imidoazole group of histidine apparently is derived from the amide group of glutamine.

As indicated before the adult human balances nitrogen without histidine in the diet. It has been suggested that microbial synthesis in the intestine might provide histidine for the human. However, it seems that nothing is known regarding such synthesis except that human liver apparently can incorporate formate carbon into position 2 of the imidoazole group.

Small amounts of methyl histidine, 1-methyl histidine 3-methyl histidine have been found in the urines of some animals.

THE METABOLSIM OF INORGANIC ELEMENTS

It is observed that there are least 29 different types of element in our body. Organic components such as carbohydrates, proteins and lipids form about 90% of the solid matter and mainly consist of C, H O, N.

The elements of body are divided into three major groups, (among them first two groups are being discussed) they are in the table below.

Group	Type	Name of the elements
Group 1	Nutritionally important minerals daily requirement > 100mg/day	Na, K, Cl, Ca, P, Mg and S
Group 2	Trace elements which are essential. The requirement is < 100mg/day	Cr, Co, Cu I, Fe, Mn, Mo, Se, Zn
Group 3	These are additional trace elements But their exact role is not known	Cd, Ni, Si, Sn, Vn

The total percentage requirement is given in the table below:

Element	Percentage
Oxygen	65
Carbon	18
Hydrogen	10
Nitrogen	3
Calcium	2[a]
Phosphorus	1.1[b]
Potassium	0.35
Sulfur	0.25
Sodium	0.15
Chlorine	0.15
Magnesium	0.05
Iron	0.004
Manganese	0.00013
Copper	0.00015
Iodine	0.00004
Cobalt	c
Zinc	c

Others of more doubtful status

a Estimates varywidely.

b Percentage varies with that of calcium.

c Believed to be essential, but quantitative data are not yet at hand.

IRON

Most Essential Trace Element	Daily Requirement Male: 3.8 g Female: 2.3 g

Dietary Source of Iron

Animal sources include, meat liver red marrow consist of 2.0 to 6.0 mg iron per 100 gm. Vegetable sources include 2.0 to 8.0g /100g The table below shows the iron content in different dietary sources:

Food	Iron (in mg.) per 100 gm. fresh substance
Beans, dried	10.5
Egg yolk	8.6
Peas, dried	5.7
Wheat, entire grain	4.8
Oatmeal	3.1
Eggs	4.8
Beef	3.1
Prunes	3.0
Spinach	2.8
Beefsteak, medium fat	2.5
Cheese	2.0
Beans, string, fresh	1.3
Potatoes	1.1
White flour	1.0
Rice, polished	0.9
Beets	0.8
Carrots	0.6
Bananas	0.6
Turnips	0.5
Oranges	0.5
Tomatoes	0.4
Apples	0.3
Milk	0.2

Digestion of Iron

Iron absorption is affected by the form in which iron is presented to the digestive tract, and inorganic iron ions change oxidation state during the absorption process.

There are two major forms of dietary iron.

- Heme iron, found primarily in red meats, is the most easily absorbed form.
- Other forms of iron are bound to some other organic constituent of the food. Cooking tends to break these interactions and increase iron availability.

Some iron-rich foods are poor sources of the mineral because other compounds render it non-absorbable.

- The classic example is spinach. It contains iron, but it also contains considerable oxalate, which chelates it and renders it non-absorbable.
- Phytates, present in whole grains that have not been subjected to fermentation by yeast (for example, during bread making), have a similar effect.

Iron ions undergo two important changes of oxidation state during digestion and absorption.

The first change occurs in the stomach.

- Here iron (III) is reduced to iron (II).

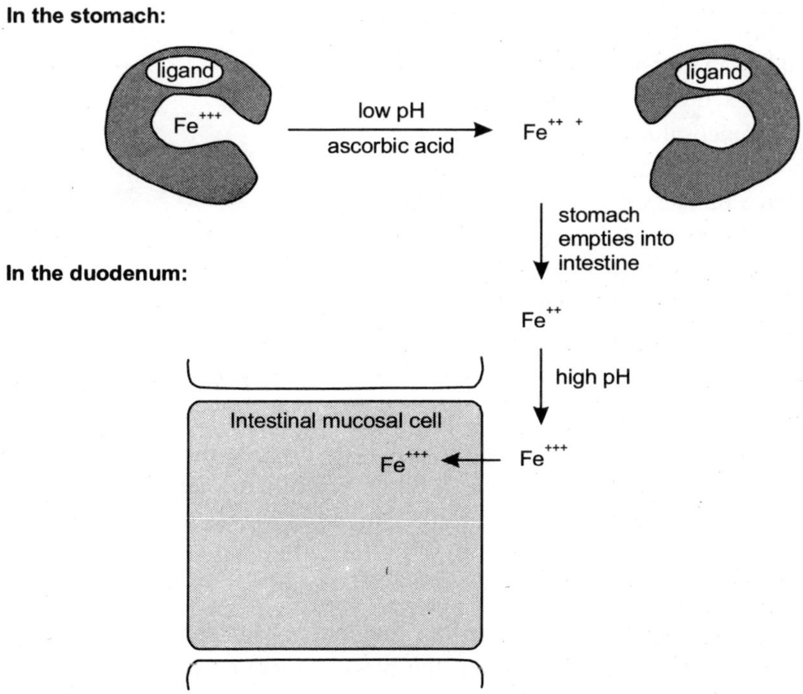

In the stomach:

In the duodenum:

Fig. 5.55 : The Fate of Iron After Digestion

- This reduction is favored by the low pH. Reducing agents, such as ascorbic acid, assist this process.
- Reduction is important because iron (II) dissociates from ligands more easily than iron (III).

The second change occurs in the duodenum.

- The duodenum is bicarbonate-rich, and alkaline.
- In the alkaline environment
  - Heme is absorbed directly by the mucosal cells. Within the cells, the iron dissociates from it.
  - Free iron (II) ions are oxidized to iron (III), which is taken up by the mucosal cells in substantial amounts under all circumstances of nutritional iron status.

Iron Absorption

Despite the fact that iron is the second most abundant metal in the earth's crust, iron deficiency is the world's most common cause of anemia. When it comes to life, iron is more precious than gold. The body hoards the element so effectively that over millions of years of evolution, humans have developed no physiological means of iron excretion. Iron absorption is the sole mechanism by which iron stores are physiologically manipulated.

The average adult stores about 1 to 3 grams of iron in his or her body. An exquisite balance between dietary uptake and loss maintains this balance. About 1 mg of iron is lost each day through sloughing of cells from skin and mucosal surfaces, including the lining of the gastrointestinal tract. Menstration increases the average daily iron loss to about 2 mg per day in premenopausal female adults. No physiologic mechanism of iron excretion exists. Consequently, absorption alone regulates body iron stores. The augmentation of body mass during neonatal and childhood growth spurts transiently boosts iron requirements.

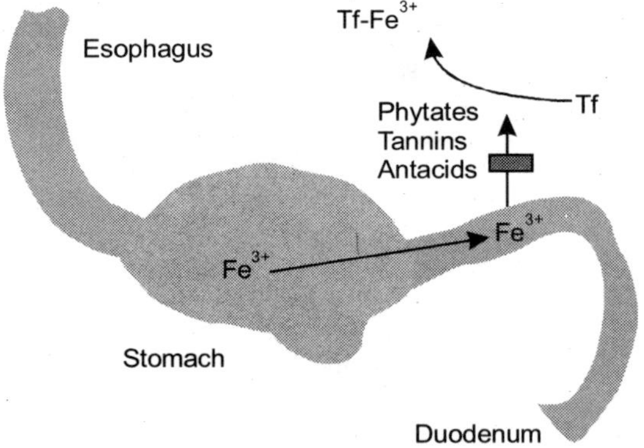

Fig. 5.56 : Iron absorption.

Iron enters the stomach from the esophagus. Iron is oxidized to the Fe^{3+} state no matter its original form when taken in orally. Gastric acidity as well as solubilizing agents such as ascorbate prevent precipitation of the normally insoluble Fe^{3+}. Intestinal mucosal cells in the duodenum and upper jejunum absorb the iron. The iron is coupled to transferrin (Tf) in the circulation which delivers it to the cells of the body. Phytates, tannins and antacids block iron absorption.

Iron absorption occurs predominantly in the duodenum and upper jejunum (Figure 5.56). The mechanism of iron transport from the gut into the blood stream remains a mystery despite intensive investigation and a few tantalizing hits. A feedback mechanism exists that enhances iron absorption in people who are iron deficient. In contrast, people with iron overload dampen iron absorption. The physical state of iron entering the duodenum greatly influences its absorption however. At physiological pH, ferrous iron (Fe^{2+}) is rapidly oxidized to the insoluble ferric (Fe^{3+}) form. Gastric acid lowers the pH in the proximal duodenum, enhancing the solubility and uptake of ferric iron (Table below).

Factors That Influence Iron Absorption

Physical State (bioavailability)	heme > Fe^{2+} > Fe^{3+}
Inhibitors	phytates, tannins, soil clay, laundry starch, iron overload, antacids
Competitors	lead, cobalt, strontium, manganese, zinc
Facilitators	ascorbate, citrate, amino acids, iron deficiency

When gastric acid production is impaired (for instance by acid pump inhibitors such as the drug, prilosec), iron absorption is reduced substantially. Heme is absorbed by machinery completely different to that of inorganic iron. The process is more efficient and is independent of duodenal pH . Consequently meats are excellent nutrient sources of iron. In fact, blockade of heme catabolism in the intestine by a heme oxygenase inhibitor can produce iron deficiency. The paucity of meats in the diets of many of the people in the world adds to the burden of iron deficiency. A number of dietary factors influence iron absorption. Ascorbate and citrate increase iron uptake in part by acting as weak chelators to help to solubilize the metal in the duodenum (Table). Iron is readily transferred from these compounds into the mucosal lining cells. Conversely, iron absorption is inhibited by plant phytates and tannins. These compounds also chelate iron, but prevent its uptake by the absorption machinery. Phytates are prominent in wheat and some other cereals, while tannins are prevalent in (non-herbal) teas. Lead is a particularly pernicious element to iron metabolism. Lead is taken up by the iron absorption machinery, and secondarily blocks iron through competitive inhibition. Further, lead interferes with a number of important iron-dependent metabolic steps such as heme biosynthesis. This multifacted attack has particularly dire consequences in children, were lead not only produces anemia, but can impair cognitive development.

Lead exists naturally at high levels in ground water and soil in some regions, and can clandestinely attack children's health. For this reason, most pediatricians in the U.S. routinely test for lead at an early age through a simple blood test. Immaturity of the gastrointestinal tract can exacerbate iron deficiency in newborns. The gastrointestinal tract does not achieve competency for iron absorption for several weeks after birth. The problem is even more severe for premature infants, who tend to be anemic for a variety of reasons. A substantial portion of iron stores in newborns are transferred from the mother late in pregnancy. Prematurity short circuits this process. Parenteral iron replacement is possible, but not often used because of the often delicate health of premature infants.

Transfusion becomes the default option in this circumstance. The mechanism by which iron enters the mucosal cells lining the upper gastrointestinal tract is unknown. Most cells in the rest of the body are believed to acquire iron from plasma transferrin (an iron-protein chelate), via specific transferrin receptors and receptor-mediated endocytosis. The hypothesis that apotransferrin (or an equivalent molecule) secreted by intestinal cells or present in bile chelates intestinal iron and facilitates its absorption is unsubstantiated. The transferrin gene is not expressed in intestinal cells. Later work indicated that transferrin found in the intestinal lumen is derived from plasma. Plasma transferrin entering bile is fully saturated with iron, obviating any intraluminal chelating function. Furthermore, hypoxia, which greatly increases iron absorption, has no effect on intestinal transferrin levels. Exogenous transferrin cannot donate iron to intestinal mucosal cells and the brush boarder membrance lacks transferrin receptors. Lastly and perhaps most compellingly, humans and mice with hypotransferrinemia paradoxically absorb more dietary iron than normal. Although the erythron is iron deficient, these individuals develop hepatic iron overload.

Mechanism of Iron Absorption

In searching for molecules involved in intestinal iron transport, Conrad and co-workers took the approach of characterizing proteins that bind iron. Their hypothesis of iron transport is based on identification of iron binding proteins at several key sites. They propose that mucins bind iron in the acid environment of the stomach, thereby maintaining it in solution for later uptake in the alkaline duodenum. According to their model, mucin-bound iron subsequently crosses the mucosal cell membrane in association with integrins. Once inside the cell, a cytoplasmic iron-binding protein, dubbed "mobilferrin", accepts the element, and shuttles it to the basolateral surface of the cell, where it is delivered to plasma. In this model mobilferrin could serve as a rheostat sensitive to plasma iron concentrations. Fully occupied mobilferrin would dampen mucosal iron uptake, and while the process would be enhanced by unsaturated mobilferrin. This model has not gained universal acceptance however. A very different scheme of iron uptake has been proposed by investigators studying iron transport in yeast. Yeast face the problem of taking in iron from the environment, a process similar to that of intestinal mucosal cells. Dancis et al. used genetic selection to isolate Sacchromyces cerevisiae mutants with defective iron transport. They constructed an expression plasmid in which an enzyme necessary for histidine biosynthesis was under the control of an iron-repressible promoter. The plasmid was introduced into a yeast histidine auxotroph (i.e. a strain of yeast that requires histidine to survive). Mutants were selected in the absence of histidine, in the presence of high levels of iron. Among the mutats they isolated, were cells with defective iron uptake. They discovered that membrane iron transport depends absolutely upon copper transport. In this model, ferric iron in yeast culture medium is reduced to its ferrous form by an externally oriented reductase (FRE1). The element is shuttled rapidly into the cell by a ferrous transporter, which appears to be coupled to an externally oriented copper-dependent oxidase (FET3) embedded in the cell membrane. FET3 is strikingly homologous to the mammalian copper oxidase ceruloplasmin. The re-oxidation of ferrous to ferric iron is apparently an obligatory step in the transport mechanism, although the coupling mechanism of oxidation and membrane transport is unclear. Although the genetic evidence for this scheme is compelling, the central component, the ferrous transporter itself, remains elusive. These investigators speculate that mammalian intestinal iron transport is analogous to the yeast iron uptake process. This assertion is supported by studies of copper-deficient swine, which show co-existing iron deficiency unresponsive to iron therapy.

Genetic Insights into Mammalian Iron Absorption

Mouse genetics provides a different perspective on mammalian intestinal iron transport. Mouse breeders readily recognize pale animals, and have developed anemic stocks with various mutations. Intestinal mucosal iron transport is defective in two mutant strains. Microcytic (mk) mice and sex-linked anemia (sla) mice have severe iron deficiency due apparently to defects in iron uptake and release, respectively, from the intestinal cell. Mice with the homozygous autosomal recessive mk mutation absorb iron poorly, have low serum iron levels, and lack stainable iron in intestinal mucosal cells. These findings are consistent with a defect in an apical iron transport molecule. Intriguingly, mk/mk mice are not rescued by parenteral iron replacement. Anemia develops in normal mice tranplanted with mk bone marrow, indicating that mk erythroid precursor cells also have a defect in red cell iron uptake. A common component to iron transport may therefore exist in intestinal cells and red cell precursors.

Mice that are homozygous or heterozygous for the sla mutation (sla/sla or sla/y) also have low serum iron levels. In contrast to mk mice, they have abnormal iron deposits within intestinal mucosal cells, suggesting that this X-linked defect impairs intracellular iron trafficking or basolateral export of iron to the plasma. The sla animals differ further from the mk mice by correction of anemia by parenteral iron. Based on studies of these mutants, distinct apical and basolateral iron transport systems possibly exist that function coordinately to transfer iron from intestinal lumen to plasma.

Whatever the mechanism of iron uptake, normally only about 10% of the elemental iron entering the duodenum is absorbed. However, this value increases markedly with iron deficiency (Finch, 1994). In contrast, iron overload reduces but does not eliminate absorption, reaffirming the fact that absorption is regulated by body iron stores. In addition, both anemia and hypoxia boost iron absorption. A portion of the iron that enters the mucosal cells is retained sequestered within ferritin. Intracellular intestinal iron is lost when epithelial cells are sloughed from the lining of the gastrointestinal tract. The remaining iron traverses the mucosal cells, to be coupled to transferrin for transport through the circulation.

> An average adult produces 2 × 1011 red cells daily, for a red cell renewal rate of 0.8 percent per day. Each red cell contains more than a billion atoms of iron, and each ml of red cells contains 1 mg of iron. To meet this daily need for 2 × 1020 atoms (or 20 mg) of elemental iron, the body has developed regulatory mechanisms whereby erythropoiesis profoundly influences iron absorption

Erythropoiesis and Iron AbsorptionApproximately 80% of total body iron is ultimately incorporated into red cell hemoglobin. Plasma iron turnover (PIT) represents the mass turnover of transferrin-bound iron in the circulation, expressed as mg/kg/day. Accelerated erythropoiesis increases plasma iron turnover, which is associated with enhanced iron uptake from the gastrointestinal tract. The mechanism by which PIT alters iron absorption is unknown. A circulating factor related to erythropoiesis that modulates iron absorption has been hypothesized, but not identified. Several candidate factors have been excluded, including transferrin and erythropoietin. Clinical manifestations of this apparent communication between the marrow and the intestine includes iron overload that develops in patients with severe thalassemia in the absence of transfusion. The accelerated (but ineffective) erythropoiesis in this condition substantially boosts iron absorption. In some cases, the coupling of increased PIT and increased gastrointestinal iron absorption is beneficial. In pregnancy, placental removal of iron raises the PIT. This process enhances gastrointestinal iron absorption thereby increasing the availability of the element to meet the needs of the growing and developing fetus. Competition studies suggest that several other heavy metals share the iron intestinal absorption pathway. These include lead, manganese, cobalt and zinc. Enhanced iron absorption induced by iron deficiency also augments the uptake of these elements. As iron deficiency often coexists with lead intoxication, this interaction can produce particularly serious medical complications in children. Interestingly, copper absorption and metabolism appear to be handled mechanisms different to those of iron.

Iron Transport

Only a small proportion of total body iron daily enters or leaves the body's stores on a daily basis (Figure 5.57). Consequently, intercellular iron transport, as a part of the iron reutilization process, is quantitatively more important that intestinal absorption. The greatest mass of iron is found in erythroid cells, which contain about 80% of the total body endowment. The reticuloendothelial system recycles a substantial amount of iron from effete red cells, approximating the amount used by the erythron for new hemoglobin production.

Transferrin

Of the approximate 3 grams of body iron in the adult male, approximately 3mg or 0.1% circulates in the plasma as an exchangeable pool (Table above). Essentially all circulating plasma iron normally is bound to transferrin. This chelation serves three purposes: it renders iron soluble under physiologic conditions, it prevents iron-mediated free radical toxicity, and it facilitates transport into cells. Transferrin is the most important physiological source of iron for red cells. The liver synthesizes transferrin and secretes it into the plasma. Transferrins are produced locally in the testes and CNS. These two sites are relatively inaccessible to proteins in the general circulation (blood: testis barrier, blood:brain barrier). The locally synthesized transferrin could play a role in iron metabolism in these tissues. Information on the function of transferrin produced in these localized sites is sparce, however. Plasma transferrin is an 80 kDa glycoprotein with homologous N-terminal and C-terminal iron-binding domains. The molecule is related to several other proteins, including ovotransferrin in bird and reptile eggs, lactoferrin in extracellular secretions and neutrophil granules and melanotransferrin, a protein produced by melanoma cells. Ovotransferrin may help protect the developing embryo in the semi-permeable egg by sequestering iron that microbes need to grow. Lactoferrin, in secretions such as milk and tears, might have a similar function. One recent report indicates that lactoferrin can act as a site-specific DNA binding protein, and could mediate transcriptional activation.

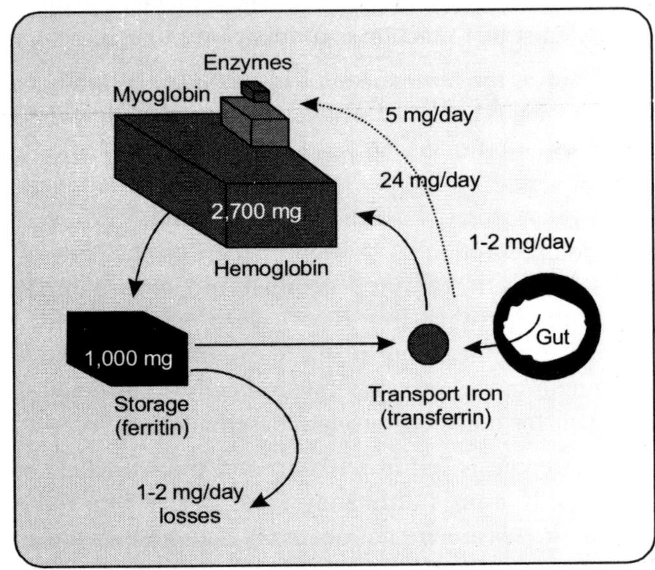

Fig. 5.57 : The iron is assiduously conserved and recycled for use in heme and non-heme enzymes. About 1 to 2 mg of iron are lost each day to sloughing of skin and mucosal cells of the gastrointestinal and genitouretal tracts. This obligate iron loss is balanced by iron absorption from the gastrointestinal tract. Only a small fraction of the 4 grams of body iron circulate as part of transferrin at any given time. Body iron is most prominently represented in hemoglobin and in ferritin

Such a function is, however, at odds with its existence as an extracellular protein. X-ray crystal structures exist for human lactoferrin and rabbit transferrin. All members of the transferrin protein superfamily have similar polypeptide folding patterns. N-terminal and C-terminal domains are globular moieties of about 330 amino acids; each of these is divided into two sub-domains, with the iron- and anion-binding sites in the intersubdomain cleft. The binding cleft opens with iron release, and closes with iron binding. N- and C-terminal binding sites are highly similar.

Iron binding by Transferrin

The precise mechanics of iron loading onto transferrin as it leaves intestinal epithelial cells or reticuloendothelial cells is unknown. The copper-dependent ferroxidase, ceruloplasmin, may play a role. Compelling evidence indicates that the protein is involved in mobilizing tissue iron stores to produce diferric transferrin. Transferrin binds iron avidly with a dissociation constant of approximately 1022 M-1. Ferric iron couples to transferrin only in the company of an anion (usually carbonate) that serves as a bridging ligand between metal and protein, excluding water from two coordination sites. Without the anion cofactor, iron binding to transferrin is negligible. With it, ferric transferrin is resistant to all but the most potent chelators. The remaining four coordination sites are provided by the transferrin protein - a histidine nitrogen, an aspartic acid carboxylate oxygen, and two tyrosine phenolate oxygens. Available evidence suggests that anion-binding takes place prior to iron-binding. Iron release from transferrin involves protonation of the carbonate anion, loosening the metal-protein bond.

	Iron	
Compartment	**Iron (grams)**	**Percent of Total**
Hemoglobin	2.7	66
Myoglobin	0.2	3
Heme Enzymes	0.008	0.1
Non-heme Enzymes	< 0.0001	—
Intracellular Storage (Ferritin)	1.0	30
Intracellular Labile Iron (Chelatable Iron)	0.07 (?)	1
Intercellular Transport (Transferrin)	0.003	0.1
Transferrin/Iron Physiology		

The sum of all iron binding sites on transferrin constitutes the total iron binding capacity (TIBC) of plasma. Under normal circumstances, about one-third of transferrin iron-binding pockets are filled. Consequently, with the exception of iron overload where all the transferrin binding sites are occupied, non-transferrin-bound iron in the circulation is virtually nonexistent. Distribution of plasma and tissue iron can be traced using 59Fe as a radioactive tag. The subject receives autologous transferrin loaded with radioactive iron that then can be monitored. Blood samples can be analyzed at timed intervals to determine the rate of loss of the radioactive label. Such ferrokinetic studies indicate that the normal half-life of iron in the circulation is about 75 minutes. The absolute amount of iron released from transferrin per unit time is the plasma iron turnover (PIT).

Such radioactive tracer studies indicate that at least eighty percent of the iron bound to circulating transferrin is delivered to the bone marrow and incorporated into newly formed erythrocytes. Other major sites of iron delivery include the liver, which is a primary depot for stored iron, and the spleen. Hepatic iron is found in both reticuloendothelial cells and hepatocytes. Reticuloendothelial cells acquire iron primarily by phagocytosis and breakdown of aging red cells These cells extract the iron from heme and return it to the circulation bound to transferrin. Hepatocytes take up iron by at least two different pathways. The first involves receptor-mediated endocytosis of transferrin. In addition, hepatocytes can take up ionic iron by a process independent of transferrin.

Ferrokinetics and the Bone Marrow

Given the preeminent role of the bone marrow in the clearance of labeled iron from the circulation, ferrokinetics provide a window on erythropoietic activity. Conditions that augment erythrocyte production increase the PIT. For example, hemolytic anemias such as hereditary spherocytosis and sickle cell disease induce rapid delivery of transferrin-bound iron to the marrow. In contrast, disorders that reduce red cell production prolong the PIT. This picture is seen, for example, with anemia due to Diamond Blackfan anemia. When erythrocytes are produced and released into the circulation in a normal fashion, the process of erythropoiesis is termed "effective". In patients with certain hemolytic anemias, however, the nascent red cells are so abnormal they are destroyed before leaving the marrow cavity. In this circumstance, the erythropoiesis is "ineffective", meaning simply that the erythropoietic precursors have failed to accomplish their primary task: the delivery of intact erythrocytes to the circulation. The ferrokinetic profiles such cases show rapid removal of iron from transferrin with a delayed entry of label into the pool of circulating red cell hemoglobin. ß+-thalassemia is an important example of this pattern of hemolytic anemia with ineffective erythropoiesis. In ß+-thalassemia, ineffective erythropoiesis is coupled with a markedly enhanced PIT.

Cellular Iron Uptake

Although transferrin was characterized fifty years ago, its receptor eluded investigators until the early 1980s. In a quest to better understand the behavior of neoplastic cells, investigators prepared monoclonal antibodies against tumor cells. The target of these monoclonal antibodies later was found to be the cell surface transferrin receptor glycoprotein. A broad body of literature now supports the concept that the iron-transferrin complex is internalized by receptor-mediated endocytosis. The general structure of the transferrin receptor is shown in Figure 180. This disulfide-linked homodimer has subunits containing 760 amino acids each. Oligosaccharides account for about 5% of the 90 kDa subunit molecular mass. Four glycosylation sites (three N-linked and one O-linked) line the protein. Glycosylation-defective mutants have fewer disulfide bridges, bind transferrin less efficiently and are expressed less prominently on the surface expression than are normal receptors. The transmembrane domain, between amino acids 62 and 89, functions as an internal signal peptide, as none exits at the N-terminal end. A molecule of fatty acid (usually palmitate) covalently links each subunit to the internal edge of the transmembrane domain and could play a role in membrane localization. Interestingly, non-acylated mutants mediate faster iron uptake than normal receptors. The transferrin binding regions of the protein are unidentified. Efforts to crystallize transferrin receptor protein are underway.

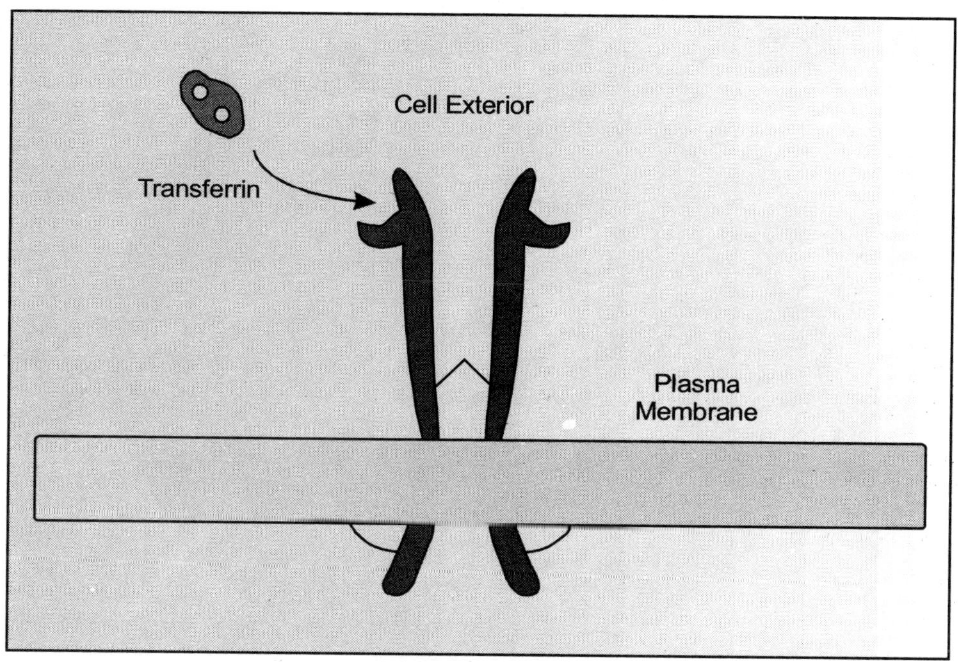

Fig. 5.58 : Schematic representation of the transferrin receptor. The molecule is a transmembrane homodimer linked by disulphide bonds. An acyl group attached to the cytoplasmic tail of the molecule anchors the assembly to the plasma membrane

Iron is taken into cells by receptor-mediated endocytosis of monoferric and diferric transferrin (Fig. 5.59). Receptors on the outer face of the plasma membrane bind iron-loaded transferrin with a very high affinity. The C-terminal domain of transferrin appears to mediate receptor binding. Diferric transferrin binds with higher affinity than monoferric transferrin or apotransferrin. The dissociation constant (Kd) for bound diferric transferrin ranges from 10-7 M to 10-9 M at physiologic pH, depending on the species and tissue assayed. The Kd of monoferric transferrin is approximately 10-6 M. The concentration of circulating transferrin is about 25 mM. Therefore, cellular transferrin receptors ordinarily are fully saturated.

After binding to its receptor on the cell surface, transferrin is rapidly internalized by invagination of clathrin-coated pits with formation of endocytic vesicles (Figure 5.59). This process requires the short, 61 amino acid intracellular tail of the transferrin receptor molecule (Rothenberger et al., 1987); (Alvarez et al., 1990); (McGraw and Maxfield, 1990); (Girones et al., 1991); (Miller et al., 1991). Receptors with truncated N-terminal cytoplasmic domains do not recycle (Rothenberger et al., 1987). This portion of the molecule contains a conserved tyrosine-threonine-arginine-phenylalanine (YTRF) sequence which functions as a signal for endocytotic internalization (Collawn et al., 1993). Genetically engineered addition of a second YTRF sequence enhances receptor endocytosis (Collawn et al., 1993).

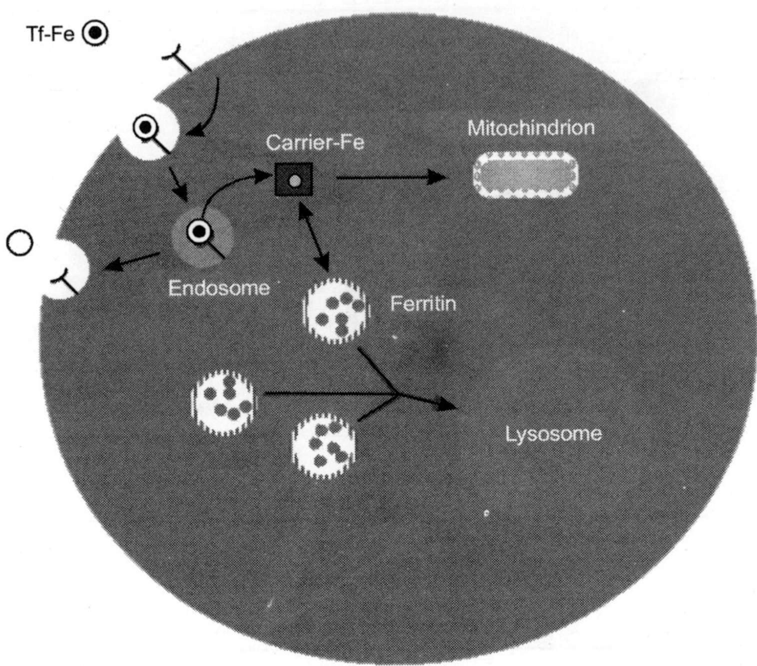

Fig. 5.59 : Receptor-mediated transferrin endocytosis. Ferro-transferrin binds to transferrin receptors on the external surface of the cell. The complex is internalized into an endosome, where the pH is lowered to about 5.5. Iron separates from the transferrin molecule, moving into the cell cytoplasm. Here, an iron transport molecule shuttles the iron to various points in the cell, including mitochondria and ferritin. Ferritin molecules accumulate excess iron. Lysosomes engulf aggregates of ferritin molecules in a process termed "autophagy"

A number of stimuli reversibly phosphorylate the serine residue adjacent to the YTRF sequence, at position 24 by the action of protein kinase C. The role of receptor phosphorylation is unclear. Despite removal of the phosphorylation site by site-directed mutagenesis, the transferrin receptor recycles normally. An ATP-dependent proton pump lowers the pH of the endosome to about 5.5. The acidification of the endosome weakens the association between iron and transferrin. Even at pH 5.5, Fe3+ would not normally dissociate from transferrin in the several minutes between its endocytosis and the return of transferrin apoprotein to the cell surface. A plasma membrane oxidoreductase reduces transferrin bound iron from the Fe3+ state to Fe2+, directly or indirectly facilitating the removal of iron from the protein. Conformational changes in the transferrin receptor also play a role in iron release.

Rather than entering lysosomes for degradation, as do ligands in other receptor-mediated endocytosis pathways, intact receptor-bound apotransferrin recycles to the cell surface, where neutral pH promotes detachment into the circulation. Thus the preservation and re-use of transferrin are accomplished by pH-dependent changes in the affinity of transferrin for its receptor. Exported apotransferrin binds additional iron and undergoes further rounds of iron delivery to cells. The average transferrin molecule, with a half-life of eight days, may be used up to one hundred times for iron delivery.

Topologically, the cell exterior and the endosome interior are equivalent compartments. The primary role of the transferrin-transferrin receptor interaction is to bring iron into the vicinity of the cell surface, thereby increasing the likelihood of iron uptake. Following its release from transferrin within the endosome, iron must traverse the plasma membrane to enter the cytosol proper. The molecules effecting this transport have not been identified, but the process may be carrier-mediated. Two anemic, mutant animals, the Belgrade rat (b/b) and the hemoglobin deficit mouse (hbd/hbd) appear to have lesions at or near this step. Their cells take up ferrotransferrin into endosomes, but fail to release iron into the cytoplasm. The molecular basis of the defects in these animals have not been elucidated.

The endosomal transporter may reside on the plasma membrane of the cell prior to endocytosis. If so, it should be oriented to transport iron directly into the cell, without the assistance of transferrin. Such non-transferrin-bound iron uptake activities have been characterized in tissue culture. This uptake system could function constitutively but inefficiently. Coupling the transferrin cycle to transport across the plasma membrane might augment iron uptake by creating an iron-rich environment for the transporter within the endosome. This same elusive transport molecule could also be involved in intestinal iron uptake. The phenotype of the mk/mk mouse suggests that red cell iron uptake and intestinal iron uptake share a common component which could be the 'endosomal' transporter.

Once inside the cell cytoplasm, iron appears to be bound by a low molecular weight carrier molecule, which may assist in delivery to various intracellular locations including mitochondria (for heme biosynthesis) and ferritin (for storage). The identity of the intracellular iron carrier molecule(s) remains unknown. The amount of iron in transit within the cell at any given time is minuscule and defies precise measurement. This minute pool of transit iron, which is believed to be in the Fe^{2+} oxidation state, is the biologically active form of the element. Metabolically inactive iron, stored in ferritin and hemosiderin, is in equilibrium with exchangeable iron bound to the low molecular weight carrier molecule.

Both prokaryotes and eukaryotes produce ferritin molecules for iron storage. Ferritins are complex twenty-four subunit heteropolymers of H (for heavy or heart) and L (for light or liver) protein subunits. L subunits are 19.7 kDa in mass, with isoelectric points of 4.5-5.0; H subunits are 21 kDa with isoelectric points of 5.0-5.7. The subunits of the ferritin molecule form a sphere with a central cavity in which up to 4500 atoms of crystalline iron is stored in the form of poly-iron-phosphate oxide. Eight channels through the sphere are lined by hydrophilic amino acid residues (along the three-fold axes of symmetry) and six more are lined by hydrophobic residues (along the four-fold axes). Strong interspecies amino acid conservation exists in the residues that line the hydrophilic channels, while marked variation exists in those along the hydrophobic passages. Hydrophilic channels terminate with aspartic acid and glutamic acid residues , and are lined by serine, histidine and cysteine residues (all of which potentially bind metal ligands). The evolutionary conservation of the hydrophilic channels suggests that they provide the route for iron entry and exit from the ferritin shell, but this contention remains unproved. Little is known about how iron is released from ferritin for use.

Although the two ferritin chains are highly homologous, only H ferritin has ferroxidase activity. A mechanism involving dioxygen converts ferrous to ferric iron, promoting incorporation into ferritin. The composition of ferritin shells varies from H-subunit homopolymers to L-subunit homopolymers, and includes all possible combinations between the two. Isoelectric focusing of ferritin from a particular tissue reveals multiple bands representing shells with different subunit compositions. These isoferritins, as they are called, show tissue specific variation. Ferritin from liver, for instance, is rich in L-subunits, as is that from the spleen. In contrast, the heart has ferritin rich in H-subunits. Increased H subunit content correlates with increased iron utilization, while increased L subunit content correlates with increased iron storage. The H:L ratio rises with activation of heme synthesis or cell proliferation. Ferritin thus provides a flexible reserve of iron. Ferritin molecules aggregate over time to form clusters, which are engulfed by lysosomes and degraded. The end-product of this process, hemosiderin, is an amorphous agglomerate of denatured protein and lipid interspersed with iron oxide molecules. In cells overloaded with iron, lysosomes accumulate large amounts of hemosiderin which can be visualized by Prussian blue staining. Although the iron enmeshed in this insoluble compound constitutes an endstage product of cellular iron storage, it remains in equilibrium with soluble ferritin. Ferritin iron, in turn, is in equilibrium with iron complexed to low molecular weight carrier molecules. Therefore the introduction into the cell of an effective chelator captures iron from the low molecular weight "toxic iron" pool, draws iron out of ferritin, and eventually depletes iron from hemosiderin as well, though only very slowly. As might be expected, the bioavailability of hemosiderin iron is much lower than that of iron stored in ferritin.

Non-Transferrin-Bound Iron Uptake

Alhough compelling evidence exists that the transferrin cycle is important for iron acquisition by the erythron, other tissues can import iron by alternative mechanisms. Some patients and mutant mice that have little or no circulating transferrin. Despite severe hypochromic, microcytic anemia, non-erythroid tissues are grossly normal. While the red cells suffer from iron deficiency, serum iron levels (iron not bound to transferrin) are elevated, and excess iron is deposited in the liver. The iron-deprived bone marrow likely signals the gut to increase absorption, exacerbating tissue iron excess. Ponka and Schulman speculate that non-erythroid cells depend less on transferrin because their modest iron needs can be met by turnover of endogenous ferritin and heme iron. Red cells are more vulnerable because of greater iron use to form hemoglobin. The transferrin cycle could serve primarily to enhance iron uptake by tissues with a great demand for the element. Iron overload produces fully saturated transferrin and non-transferrin bound iron circulating in a chelatable, low molecular weight form. This iron is weakly complexed to albumin, citrate, amino acids and sugars, and behaves differently from iron associated with transferrin. Non-hematopoietic tissues, particularly the liver, endocrine organs, kidneys and heart preferentially take up this iron.

Radiolabeled iron administered to mice with and without available transferrin binding capacity has quite different patterns of distribution. In normal animals, hematopoietic tissues are the prime sites of uptake. When free transferrin sites are absent, however, most iron is deposited in the liver and pancreas, indicating that these organs serve as iron reservoirs in the situation of iron overload. Notably, this pattern of distribution is similar to that seen in idiopathic hemochromatosis. These data support the idea that, while the transferrin pathway is important for meeting the needs of the erythron, it is not essential for iron uptake by all tissues. Kaplan and coworkers have studied iron incorporation from $FeNH_4$ citrate. Intriguingly, they find that transferrin-independent uptake increases in direct proportion to the concentration of this compound, similar to hepatic uptake of non-transferrin-bound iron in patients with saturated transferrin. They speculate that this is a protective alternative pathway that removes the toxic metal from the circulation. Other investigators have described similar uptake in HepG2 cells, and shown that it is reversible by addition of chelating compounds. A non-transferrin iron uptake mechanism with different properties has been described in K562 erythroleukemia cells. In the absence of ferric transferrin, iron uptake into K562 cells is sensitive to treatment with trypsin, suggesting that it requires a protein carrier. Higher ambient iron concentrations do not increase cellular iron uptake. As discussed above, this transport may be accomplished by the same machinery responsible for passage of iron out of transferrin cycle endosomes into the cytoplasm. These two processes accomplish essentially the same task. The putative endosomal iron transporter must be oriented to transport iron from an endocytosed extracellular compartment into the cytoplasm. This transporter may exist on the cell surface prior to receptor-mediated endocytosis, with the capacity to transport iron to a modest extent. This activity is not restricted to erythroid cells. PHA-stimulated human peripheral lymphocytes have a similar transferrin-independent iron uptake mechanism.

IODINE

MOST ESSENTIAL TRACE ELEMENT

DAILY REQUIREMENT: MALE: 0.05 g; FEMALE 0.05 g

Infants 0-6 months	15 mg
Infants 7-12 months	15 mg
Children 1-6 years	6 mg
School children 7-12 years	4 mg
Adolescents and adults (12+ years)	2 mg

Dietary Source of Iodine

This element is an essential constituent of the body. It is estimated that a total 25 mg found in the human organism, 15 mg is in the thyroid. In this organ, a large percentage (but not all) of the iodine is in the combination in the organic molecule called thyroxine a hormone.

An average person needs 0.05 mg per day. Since the ocean is rich in iodine, see foods, fish and oysters, are the food sources of this element. The table below shows the different sources of diet enriched in iodine.The iodine content of food depends on the iodine content of the soil in which it is grown. The iodine present in the upper crust of earth is leached by glaciation and repeated flooding and is carried to the sea. Sea water is, therefore, a rich source of iodine. The seaweed located near coral reefs has an inherent biologic capacity to concentrate iodine from the sea. The reef fish which thrive on seaweed are rich in iodine. Thus, a population consuming seaweed and reef fish has a high intake of iodine, as the case in Japan. The amount of iodine intake by the Japanese is in the range of 2-3 mg/day. In several areas of Asia, Africa, Latin America, and parts of Europe, iodine intake varies from 20 to 80 mg/day. In the United States and Canada and some parts of Europe, the intake is around 500 mg/ day. The average iodine content of foods (fresh and dry basis) as reported by Koutras is given in Tables.

Average Iodine Content of Foods (in μg/g)

Food	Fresh basis		Dry basis	
	Mean	Range	Mean	Range
Fish (fresh water)	30	17-40	116	68-194
Fish (marine)	832	163-3180	3715	471-4591
Shellfish	798	308-1300	3866	1292-4987
Meat	50	27-97	-	-
Milk	47	35-56	-	-
Eggs	93	-	-	-
Cereal grains	47	22-72	65	34-92
Fruits	18	10-29	154	62-277
Legumes	30	23-36	234	223-245
Vegetables	29	12-201	385	204-1636

Iodine Rich Foods List	μg	Portion
Haddock	up to 300	100
Cod	90*	100
Condensed milk	up to 70	100
Trifle	60	100
Eggs	50*	100
Mayonnaise	35	100
Jaffa cakes	32*	100
Cheddar cheese	40	100
Malt bread	29	100
Naan Bread	28	100
Pudding	26	100
Cheese cake	24	100
Sea kelp	High **	100
Seaweed	High **	100
Sea foods	High **	100

** high for the amounts consumed!

The RNI* for iodine is about 140 μg daily for adults.

Iodine Absorption

At present, the only physiologic role known for iodine in the human body is in the synthesis of thyroid hormones by the thyroid gland. Therefore, the dietary requirement of iodine is determined by normal thyroxine (T4) production by the thyroid gland without stressing the thyroid iodide trapping mechanism or raising thyroid stimulating hormone (TSH) levels.

Iodine from the diet is absorbed throughout the gastrointestinal tract. Dietary iodine is converted into the iodide ion before it is absorbed. The iodide ion is bio-available and absorbed totally from food and water. This is not true for iodine within thyroid hormones ingested for therapeutic purposes. Iodine enters the circulation as plasma inorganic iodide, which is cleared from circulation by the thyroid and kidney. The iodide is used by the thyroid gland for synthesis of thyroid hormones, and the kidney excretes iodine with urine. The excretion of iodine in the urine is a good measure of iodine intake. In a normal population with no evidence of clinical iodine deficiency either in the form of endemic goitre or endemic cretinism, urinary iodine excretion reflects the average daily iodine requirement. Therefore, for determining the iodine requirements, the important indexes are serum T4 and TSH levels (indicating normal thyroid status) and urinary iodine excretion. The simplified diagram of metabolic circuit of iodine is given in

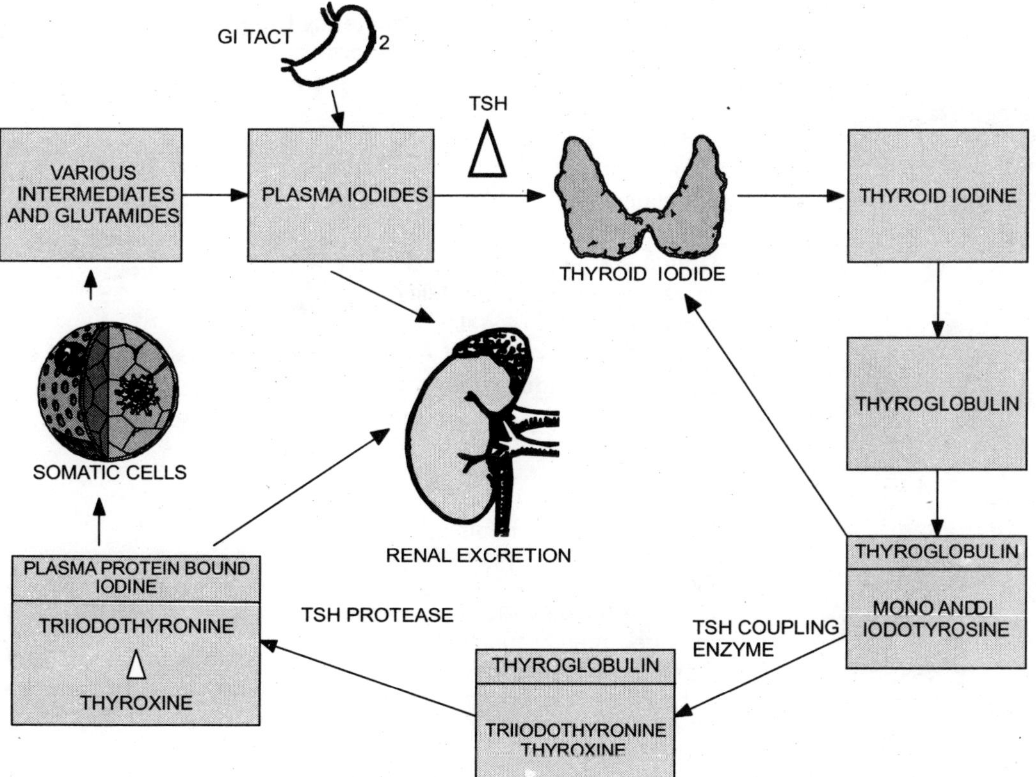

Fig. 5.60 : Simplified Diagram of the Metabolic Circuit of Iodine

All biologic actions of iodide are attributed to the thyroid hormones. The major thyroid hormone secreted by the thyroid gland is T_4 (tetra-iodo-thyronine). T_4 in circulation is taken up by the cells and is de-iodinated by the enzyme 5' prime-mono-de-iodinase in the cytoplasm to convert it into tri-iodo-thyronine (T_3), the active form of thyroid hormone. T_3 traverses to the nucleus and binds to the nuclear receptor. All the biologic actions of T_3 are mediated through the binding to the nuclear receptor, which controls the transcription of a particular gene to bring about the synthesis of a specific protein.

The physiologic actions of thyroid hormones can be categorised as growth and development and control of metabolic processes in the body. Thyroid hormones play a major role in the growth and development of brain and central nervous systems in humans from the 15th week of gestation to age 3 years. If iodine deficiency exists during this period and results in thyroid hormone deficiency, the consequence is derangement in the development of brain and central nervous system. These derangements are irreversible, the most serious form being that of cretinism. The effect of iodine deficiency at different stages of life is given in Table below

The other physiologic role of thyroid hormone is to control several metabolic processes in the body. These include carbohydrate, fat, protein, vitamin, and mineral metabolism. For example, thyroid hormone increases energy production, increases lipolysis, and regulates neoglucogenesis, and glycolysis.

Table showing the Spectrum of Iodine Deficiency Disorders
Spectrum of Iodine Deficiency Disorders

Life stage Stage in life	Effects
Foetus	Abortions Stillbirths Congenital anomalies Increased perinatal mortality Increased infant mortality Neurological cretinism: mental deficiency, deaf mutism, spastic diplegia, and squint Myxedematous cretinism: mental deficiency and dwarfism Psychomotor defects
Neonate	Neonatal goitre Neonatal hypothyroidism
Child and Adolescent	Goitre Juvenile hypothyroidism Impaired mental function Retarded physical development
Adult	Goitre with its complications Hypothyroidism Impaired mental function Population at risk

Iodine deficiency affects all stages of human life, from the intra-uterine stage to old age, as shown in the table above. However, pregnant women, lactating women, women of reproductive age, and children younger than 3 years are considered to be at high risk. During foetal and neonatal growth and development, iodine deficiency leads to irreversible damage to the brain and central nervous system.

The iodine content of food varies with geographic location because there is a large variation in the iodine content of the inorganic world. Thus, the average iodine content of foods shown in the table cannot be used universally for estimating iodine intake.

Recommended Intake

The daily intake of iodine recommended by the National Research Council of the US National Academy of Sciences in 1989 was 40 µg/day for young infants (0-6 months), 50 µg/day for older infants (6-12 months), 60-100 µg/day for children (1-10 years), and 150 µg/day for adolescents and adults. These values approximate 7.5 µg/kg/day for age 0-12 months, 5.4 µg/kg/day for age 1-10 years, and 2 µg/kg/day for adolescents and adults. These amounts are proposed to allow normal T4 production without stressing the thyroid iodide trapping mechanism or raising TSH levels.

Iodine Content of the Inorganic World

Location	Iodine content
Terrestrial air	1.0 µg/l
Marine air	100.0 µg/l
Terrestrial water	5.0 µg/l
Sea water	50.0 µg/l
Igneous rocks	500.0 µg/kg
Soils from igneous rocks	9000.0 µg/kg
Sedimentary rocks	1500.0 µg/kg
Soils from sedimentary rocks	4000.0 µg/kg
Metamorphic rocks	1600.0 µg/kg
Soils from the metamorphic rocks	5000.0 µg/kg

Iodine Requirements in Infancy

The US recommendation of 40 µg/day for infants aged 0-6 months (or 8 µg/kg/day, 7 µg/100 kcal, or 50 µg/l milk) is probably derived from the observation that until the late 1960s the iodine content of human milk was approximately 50 µg/l and from the concept that nutrition of the human-milk-fed infant growing at a satisfactory rate has been the standard against which nutrition requirements have been set. However, more recent data indicate that the iodine content of human milk varies markedly as a function of the iodine intake of the population. For example, it ranges from 20 to 330 µg/l in Europe and from 30 to 490 µg/l in the United States. It is as low as 12 µg/l under conditions of severe iodine deficiency. An average human-milk intake of 750 ml/day would give an intake of iodine of about 60 µg/day in Europe and 120 µg/day in the United States. The upper US value (490 µg/l) would provide 368 µg/day or 68 µg/kg/day for a 5-kg infant. Positive iodine balance in the young infant, which is required for the increasing iodine stores of the thyroid, is achieved only when the iodine intake is at least 15 µg/kg/day in full-term infants and 30 µg/kg/day in pre-term infants. The iodine requirement of pre-term infants is twice that of term infants because of a 50 percent lower retention of iodine by pre-term infants. This corresponds approximately to an iodine intake of 90 µg/day. (This is probably based on the assumption of average body weight of 6 kg for a child of 6 months, the mid-age of an infant.) This value is twofold higher than the US recommendations.

On the basis of these considerations, a revision is proposed for the earlier World Health Organization (WHO), United Nations Children's Fund (UNICEF), and International Council for the Control of Iodine Deficiency Disorders (ICCIDD) recommendations : an iodine intake of 90 μg/day from birth onwards is suggested. To reach this objective, and based on an intake of milk of about 150 ml/kg/day, the iodine content of formula milk should be increased from 50 to 100 μg/l for full-term infants and to 200 μg/l for pre-term infants.

For a urine volume of about 4-6 dl/day from 0 to 3 years, the urinary concentration of iodine indicating iodine repletion should be in the range of 150-220 μg/l (1.18-1.73mmol/l) in infants aged 0-36 months. Such values have been observed in iodine-replete infants in Europe, Canada , and the United States. Under conditions of moderate iodine deficiency, as seen in Belgium, the average urinary iodine concentration is only 50-100 μg/l (0.39-0.79mmol/l) in this age group. It reaches a stable normal value of 180-220 μg/l (1.41-1.73mmol/l) only after several months of daily iodine supplementation with a physiologic dose of 90 μg/day .

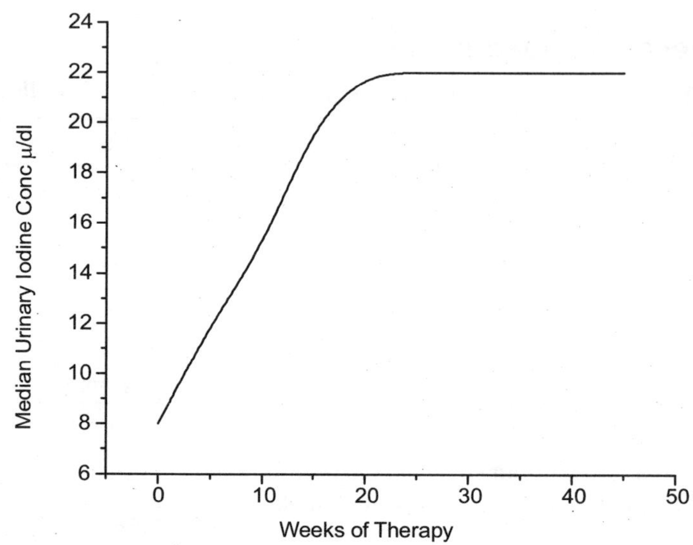

Fig. 5.61 : Changes over time for the median urinary concentration of iodine in healthy Belgian infants aged 6-36 months and supplemented with iodine at 90 μg/kg/day for 44 weeks

Each point represents 32-176 iodine determinations.

When the urinary iodine concentration in neonates and young infants is below a threshold of 50-60 μg/l (0.39-0.47mmol/l), corresponding to an intake of 25-35 μg/day, there is a sudden increase in the prevalence of neonatal serum TSH values in excess of 50 mU/ml, indicating sub-clinical hypothyroidism and eventually complicated by transient neonatal hypothyroidism. When the urinary iodine concentration is in the range of 10-20 μg/l (0.08-0.16mmol/l), as observed in severe endemic goitre regions, up to 10 percent of the neonates have overt severe hypothyroidism, with serum TSH levels above 100 mU/mL and serum T3 values below 30 μg/l (39 nmol/L) . Untreated, these infants progress to myxedematous endemic cretinism.

Thus, the iodine requirement of the young infant approximates 15 μg/kg/day (30 μg/kg/day in pre-term infants). Hyperthyrotropinemia (high levels of serum TSH), indicating sub-clinical hypothyroidism with the risk of brain damage, occurs when the iodine intake is about one-third of this value, and dramatic neonatal hypothyroidism resulting in endemic cretinism occurs when the intake is about one-tenth of this value.

Iodine Requirements in Children

The daily iodine need on a body weight basis decreases progressively with age. A study by Tovar and colleagues correlating 24-hour thyroid radioiodine uptake and urinary iodine excretion in 9-13-year-old schoolchildren in rural Mexico suggested that an iodine intake in excess of 60 μg/day is associated with a 24-hour thyroidal radioiodine uptake below 30 percent. Lower excretion values are associated with higher uptake values. This would approximate 3 μg/kg/day in an average size 10-year-old child (approximate body weight of 20 kg), so that an intake of 60-100 μg/day for child of 1-10 years seems appropriate. These requirements are based on the body weight of Mexican children who participated in this study. The average body weight of a 10-year-old child, as per the Food and Agriculture Organization references, is 25 kg. Thus, the iodine requirement for a 1-10-year-old child would be 90-120 μg/day.

Iodine Requirements in Adults

Iodine at 150 μg/day for adolescents and adults is justified by the fact that it corresponds to the daily urinary excretion of iodine and to the iodine content of food in non-endemic areas (areas where iodine intake is adequate. It also provides the iodine intake necessary to maintain the plasma iodide level above the critical limit of 0.10 μg/dl, which is the average level likely to be associated with the onset of goitre. Moreover, this level of iodine intake is required to maintain the iodine stores of the thyroid above the critical threshold of 10 mg, below which an insufficient level of iodisation of thyroglobulin leads to disorders in thyroid hormone synthesis.

Data reflecting either iodine balance or its effect on thyroid physiology can help to define optimal iodine intake. In adults and adolescents in equilibrium with their nutritional environment, most dietary iodine eventually appears in the urine, so the urinary iodine concentration is a useful measure for assessing iodine intake. For this, casual samples are sufficient if enough are collected and if they accurately represent a community. A urinary iodine concentration of 100 μg/L corresponds to an intake of about 150 μg/day in the adult. Median urinary iodine concentrations below 100 μg/l in a population are associated with increases in median thyroid size and in serum TSH and thyroglobulin values. Correction of the iodine deficiency will bring all these measures back into the normal range. Recent data from the Thyro-Mobil project in Europe have confirmed these relations by showing that the largest thyroid sizes are associated with the lowest urinary iodine concentrations. Once a median urinary iodine excretion of about 100 μg/L is reached, the ratio of thyroid size to body size remains fairly constant. Moulopoulos et al reported that a urinary iodine excretion between 151 and 200 μg/g creatinine (1.18-1.57 mmol/g creatinine), corresponding to a concentration of about 200 μg/l (1.57 mmol/l), gave the lowest values for serum TSH in a non-goitrous population. Similar recent data from Australia show that the lowest serum TSH and thyroglobulin values were associated with urine containing 200-300 μg iodine/g creatinine (1.57-2.36mmol/g creatinine).

Other investigations followed serum TSH levels in subjects without thyroid glands who were given graded doses of T4 and found that euthyroidism established in adults with an average daily dose of 100 μg T4 would require at least 65 μg of iodine with maximal efficiency of iodine use by the thyroid. In practice such maximal efficiency is never obtained and therefore considerably more iodine is necessary.

Data from controlled observations associated a low urinary iodine concentration with a high goitre prevalence, high radioiodine uptake, and low thyroidal organic iodine content. Each of these measures reached a steady state once the urinary iodine excretion was 100 μg/l (0.78 mmol/l) or greater.

Iodine Requirements in Pregnancy

The iodine requircment during pregnancy is increased to provide for the needs of the foetus and to compensate for the increased loss of iodine in the urine resulting from an increased renal clearance of iodine during pregnancy. These requirements have been derived from studies of thyroid function during pregnancy and in the neonate under conditions of moderate iodine deficiency. For example, in Belgium, where the iodine intake is estimated to be 50-70 μg/day, thyroid function during pregnancy is characterised by a progressive decrease of the serum concentrations of thyroid hormones and an increase in serum TSH and thyroglobulin. Thyroid volume progressively increases and is above the upper limit of normal in 10 percent of the women by the end of pregnancy. Serum TSH and thyroglobulin are still higher in the neonates than in the mothers. These abnormalities are prevented only when the mother receives a daily iodide supplementation of 161 μg/day during pregnancy (derived from 131 μg potassium iodide and 100 μg T4 given daily). T4 with iodine was probably administered to the pregnant women to rapidly correct sub-clinical hypothyroidism, which would not have occurred if iodine had been administered alone. These data indicate that the iodine intake required to prevent the onset of sub-clinical hypothyroidism of mother and foetus during pregnancy, and thus to prevent the possible risk of brain damage of the foetus, is approximately 200 μg/day.

On the basis of the considerations reviewed above for the respective population groups to meet the daily iodine requirements, revisions of the current recommendations for daily iodine intake by WHO, UNICEF, and ICCIDD are proposed; these proposed revisions are presented in the table below:

Proposed revision for daily iodine intake recommendations of 1996 by the World Health Organization, United Nations Children's Fund, and International Council for the Control of Iodine Deficiency Disorders

Population sub-groups	Total iodine intake μg/day	Iodine ìg/kg/day
Infants (first 12 months)	90[a]	15.0
Children (1-6 years)	90	6.0
Schoolchildren (7-12 years)	120	4.0
Adults (12+ years)	150	2.0
Pregnant and lactating women	200	3.5

*Revised to 90 μg from the earlier recommendation of 50 μg.

Upper Limit of Iodine intake for Different Age Groups

An iodine excess also can be harmful to the thyroid of infants by inhibiting the process of synthesis and release of thyroid hormones (Wolff-Chaikoff effect). The threshold upper limit of iodine intake (the intake beyond which thyroid function is inhibited) is not easy to define because it is affected by the level of iodine intake before exposure to iodine excess. Indeed, long-standing moderate iodine deficiency is accompanied by an accelerated trapping of iodide and by a decrease in the iodine stores within the thyroid. Under these conditions, the critical ratio between iodide and total iodine within the thyroid, which is the starting point of the Wolff-Chaikoff effect, is more easily reached during iodine depletion than under normal conditions. In addition, the neonatal thyroid is particularly sensitive to the Wolff-Chaikoff effect because the immature thyroid gland is unable to reduce the uptake of iodine from the plasma to compensate for increased iodine ingestion. For these reasons transient neonatal hypothyroidism or transient hyperTSHemia after iodine overload of the mother, especially after the use of povidone iodine, has been reported more frequently in European countries such as in Belgium, France, and Germany, which have prevailing moderate iodine deficiency.

Iodine Intake in Areas of Moderate Iodine Deficiency

In a study in Belgium, iodine overload of mothers (cutaneous povidone iodine) increased the milk iodine concentration and increased iodine excretion in the term newborns (mean weight about 3 kg). Mean milk iodine concentrations of 18 and 128 μg/dl were associated with average infant urinary iodine excretion levels of 280 and 1840 μg/l (2.20-14.48 mmol/l), respectively. Estimated average iodine intakes would be 112 and 736 μg/day, or 37 and 245 μg/kg/day, respectively. The lower dose significantly increased the peak TSH response to exogenous thyroid releasing hormone but did not increase the (secretory) area under the TSH response curve. The larger dose increased both the peak response and secretory area as well as the baseline TSH concentration. Serum T4 concentrations were not altered, however. Thus, these infants had a mild and transient, compensated hypothyroid state. Non-contaminated mothers secreted milk containing 9.5 μg iodine/dl, and the mean urinary iodine concentration of their infants was 144 μg/l (1.13 mmol/l). These data indicate that modest iodine overloading of term infants in the neonatal period in an area of relative dietary iodine deficiency (Belgium) also can impair thyroid hormone formation.

Similarly, studies in France indicated that premature infants exposed to cutaneous povidone iodine or fluorescinated alcohol-iodine solutions and excreting iodine in urine in excess of 100 μg/day manifested decreased T4 and increased TSH concentrations in serum. The extent of these changes was more marked in premature infants with less than 34 weeks gestation than in those with 35-37 weeks gestation. The full-term infants were not affected. These studies suggest that in Europe the upper limit of iodine intake, which predisposes to blockage of thyroid secretion in premature infants (about 200 μg/day) is 2 to 3 times the average intake from human milk and about equivalent to the upper range of intake.

Iodine Intake in Areas of Iodine Sufficiency

Similar studies have not been conducted in the United States, where transient hypothyroidism is rarely seen perhaps because iodine intake is much higher. For example, urinary concentrations of 50 μg/dl and above in neonates, which can correspond to a Wolff-Chaikoff effect in Europe, are frequently seen in healthy neonates in North America.

The average iodine intake of infants in the United States in 1978, including infants fed whole cow milk, was estimated by the market-basket approach to be 576 μg/day (standard deviation [SD] 196);

that of toddlers was 728 μg/day (SD 315) and of adults was 952 μg/day (SD 589). The upper range for infants (968 μg/day) would provide a daily intake of 138 μg/kg for a 7-kg infant, and the upper range for toddlers (1358 μg/day) would provide a daily intake of 90 μg/kg for a 15-kg toddler.

Table below summarises the recommended dietary intake of iodine for age and approximate level of intake which appear not to impair thyroid function in the European studies of Delange in infants, in the loading studies of adults in the United States, or during ingestion of the highest estimates of dietary intake (just reviewed) in the United States.

Except for the values for premature infants, these probably safe limits are 15-20 times more than the recommended intakes. These data refer to all sources of iodine intake. The average iodine content of infant formulas is approximately 5 μg/dl. The upper limit probably should be one that provides a daily iodine intake of no more than 100 μg/kg. For this limit and with the assumption that the total intake is from infant formula, with a daily intake of 150 ml/kg (100 kcal/kg), the upper limit of the iodine content of infant formula would be about 65 μg/dl. The current suggested upper limit of iodine in infant formulas of 75 μg/100 kcal (89μg/500 kJ or 50 μg/dl), therefore, seems reasonable.

Recommended dietary intakes of iodine and probable safe upper limits

Group	Recommended μg/kg/day	Upper limit[a] μg/kg/day
Premature infants	30	100
Infants 0-6 months	15	150
Infants 7-12 months	15	140
Children 1-6 years	6	50
School children 7-12 years	4	50
Adolescents and adults (12+ years)	2	30
Pregnancy and lactation	3.5	40

[a] Probably safe.

Excess iodine intake

Excess iodine intake is more difficult to define. Many people regularly ingest huge amounts of iodine - in the range 10-200 mg/day - without apparent adverse effects. Common sources are medicines (e.g., amiodarone contains 75 mg iodine per 200-mg capsule), foods (particularly dairy products), kelp (eaten in large amounts in Japan), and iodine-containing dyes (for radiologic procedures). Excess consumption of salt has never been documented to be responsible for excess iodine intake. Occasionally each of these may have significant thyroid effects, but generally they are tolerated without difficulty. Researchers showed that people without evidence of underlying thyroid disease almost always remain euthyroid in the face of large amounts of excess iodine and escape the acute inhibitory effects of excess intra-thyroidal iodide on the organification (i.e., attachment of 'oxidized iodine' species to throsyl residues in the thyroid gland for the synthesis of thyroid hormones) of iodide and on subsequent hormone synthesis (escape from or adaptation to the acute Wolff-Chaikoff effect). This adaptation most likely involves a decrease in thyroid iodide trapping, perhaps corresponding to a decrease in the thyroid sodium-iodide transporter recently cloned. Some people, especially those with long-standing nodular goitre who live in iodine-deficient regions and are generally ages 40 years or older, may develop iodine-induced hyperthyroidism after ingestion of excess iodine in a short period of time.

Iodine Fortification

Iodine deficiency is present in almost all parts of the developed and developing world, and environmental iodine deficiency is the main cause of iodine deficiency disorders. Iodine is irregularly distributed over the earth's crust, resulting in acute deficiencies in areas such as mountainous regions and flood plains. The problem is aggravated by accelerated deforestation and soil erosion. Thus, the food grown in iodine-deficient regions can never provide enough iodine for the people and livestock living there. The iodine deficiency results from geologic rather than social and economic conditions. It cannot be eliminated by changing dietary habits or by eating specific kinds of foods but must be corrected by supplying iodine from external sources. It has, therefore, been a common practice to use common salt as a vehicle for iodine fortification for the past 75 years. Salt is consumed at approximately the same level throughout the year by the entire population of a region. Universal salt iodisation is now a widely accepted strategy for preventing and correcting iodine deficiency disorders.

There are areas where consumption of goitrogens in the staple diet (e.g., cassava) affects the proper utilisation of iodine by the thyroid gland. For example, in Congo, Africa, as a result of cassava diets there is an overload of thiocyanate. To overcome this problem, appropriate increases in salt iodisation are required to ensure the recommended dietary intake. The iodisation of salt is done either by spraying potassium iodate or potassium iodide in amounts that ensure a minimum of 150 μg iodine/day. Both of these forms of iodine are absorbed as iodide ions and are completely bio-available. Other methods of iodine prophylaxis are also used: iodised oil (capsule and injections), iodised water, iodised bread, iodised soya sauce, iodoform compounds used in dairy and poultry, and certain food additives.

Iodine loss occurs as a result of improper packaging, Humidity and moisture, and transport in open trucks and railway wagons exposed to sunlight. To compensate for these losses, higher levels of iodine are used during the production of iodised salt. Losses during the cooking process vary from 20 percent to 40 per cent depending on the type of cooking used.

To ensure the consumption of recommended levels of iodine, the iodine content of salt at the production level should be monitored with proper quality assurance programmes. Regular evaluation of the urinary iodine excretion pattern in the population consuming iodised salt or exposed to other iodine prophylactic measures would help the adjusting of iodine intake.

Recommendations

Recommendations for Future Research

- elaborate the role of T_4 in brain development at the molecular level;
- investigate the relation between selenium and iodine deficiency, which has been reported in certain areas of Africa; and

- Investigate the possible interference of infections and other systemic illnesses with iodine or thyroid hormone use (such interference has not been reported on a population basis).

Recommendations for Future Actions

- establish quality assurance procedures at iodised salt production sites;
- track the progress of iodine deficiency disease elimination through the implementation of cyclic monitoring, which involves division of the country into five zones and carrying out the assessment in one zone each year; and
- Develop and validate quantitative testing kits for iodised salt.

COPPER

Essential Trace Element — Daily requirement; Male 2 mg; Female 2 mg

Dietary Sources

Dietary sources of copper is shown in the table below:

Nuts	11.6mg
Legumes	9.0mg
Cereals	4.7 mg
Fruits	4.2 mg
Poultry	3.0 mg
Fish	2.5 mg
Green legumes	1.7 mg
Leafy veg	1.2 mg

Copper after Absorption

The claim has been made that copper plays some role in skin and hair pigmentation. It is said that the element accelerates the oxidation of dopa by dopa oxidase an enzyme present in the skin.

Some experiment have shown that anemia in the rat occurs due to copper deficiency, it is further accompanied by cytochrome oxidase activity of bone marrow it should be emphasized is one of the chief centres of hematopoiesis it would be seen that as if copper plays an important part in this reaction.

ZINC

Essential Trace Element — Daily requirement; male 15 -20 mg; female 15 – 20 mg

Dietary Source

Although we need zinc in only tiny amounts, the body makes use of it in many important ways. As with other trace elements, it is essential for the action of enzymes - the proteins that initiate vital chemical reactions in the body. It is present in the skin, eyes and bones, and in high concentrations in the liver

and pancreas. Together with vitamin B6 (pyridoxine), it aids in cell formation. It has two other vital uses: aiding growth and sexual maturation, and keeping skin healthy. In the latter role, it has been found to have important healing properties in the treatment of wounds, burns and acne.

Zinc Rich Foods List	Milligrams	Portion
Oysters	25 +	100g
Shellfish	20	100g
Brewers Yeast	17	100g
Wheat Germ	17	100g
Wheat Bran	16	100g
All Bran cereal	6.8	100g
Pine Nuts	6.5	100g
Pecan Nuts	6.4	100g

Ok Sources of Zinc	Milligrams	Portion
Liver	6	100g
Cashew Nuts	5.7	100g
Parmesan Cheese	5.2	100g
Fish	3	100g
Eggs	2	100g

Pregnant and breast-feeding women must ensure they are getting adequate supplies of zinc from their diet and/or from supplements, as a lack of this mineral could lead to foetal abnormalities and stunted growth in their babies.

People with skin complaints, and particularly adolescents with acne, should consider supplementation, together with vitamins A, B2, B6 and E. Heavy drinkers, the elderly, convalescents, anyone with a diet high in processed foods and women on the pill MAY require extra zinc intake.

Zinc after Absorption

Only a small percentage of dietary zinc is absorbed ans the absorption occurs mainly from duodenum and ileum. It acts a binding factor which is secreated from pancreas, which forms complex with zinc and helps in its absorption.

Functions in the Body

1. Role in the Enzyme action
2. Role in the Vitamin a Absorption

3. Role in the insulin Secretion
4. Role in the Growth and Reproduction
5. Role in Would healing
6. Roile in Biosynthesis of Mononucleotides

SODIUM

Essential Primary Element

Daily requirement: Male 1 - 3.5 g; Female - 1 - 3.5 g; Infants 0.1 gm - 0.5 g; Children 0.5 - 2.5 g

This is the most important and falls in the category of the principal elements. Sodium acts as an electrolyte which sis found in large concentration in extracellular fluid compartment.

Approximate body distribution of sodium is as follows:

	Total m/Mol	Conc m/Mol
Total body	3150	—
Intracellular	250	10
Extracellular	2900	140
Plama	400	140

The sodium is found in the body mainly associated with chloride or carbonate, NaCl or NaHCO3

Dietary Source

Sodium is widely distributed in food material the table below shows the sodium content in foodsGenerally all natural foods such as fruits and vegetables, are low in sodium. Whereas, all processed food products are often high in sodium, this is mainly due to the addition of salt as a preservative. Products like soy sauce, other sauces, powdered soups, marmite, gravy, mustard, oxo cubes and many pickled products are very high in sodium, however, they rarely contribute too much because they are usually eaten in small quantities!

Sodium Rich Foods List	Milligrams	Portion
Anchovies (canned)	4000mg	100g
Cockles (boiled)	3500mg	100g
Olives (brine)	2200mg	100g
Bacon Rashers	2000mg	100g
Salami	1800mg	100g

Sodium Rich Foods List	Milligrams	Portion
Smoked Salmon	1800mg	100g
Prawns (boiled)	1600mg	100g
Savoury Rice	1400mg	100g
Feta Cheese	1400mg	100g
Danish Blue	1200mg	100g
Black Pudding (fried)	1200mg	100g
Some Smoked Fish	up to 1200mg	100g
Most processed foods / canned products	up to 1000+	100g
Processed Meats	up to 1000+	100g
Some Breakfast Cereals	up to 1000mg	100g
Most other Cheeses	up to 1000mg	100g
Bread	up to 800mg	100g
Pastries	up to 600mg	100g

Sodium Absorption

Sodium is widely absorbed via sodium pump situated in basal and lateral plams membrane of the intestinal and renal cells, Na pump actively transports sodium into extracelllar fluid.

Sodium Pump

Na^+/K^+-ATPase was discovered by Jens Christian Skou in 1957 while working as assistant professor at the Department of Physiology, University of Aarhus, Denmark. He published his work in 1957.

In 1997, he received one-half of the Nobel Prize in Chemistry "for the first discovery of an ion-transporting enzyme, Na^+, K^+-ATPase."

Na^+/K^+-ATPase (also known as the Na^+/K^+ pump, sodium-potassium pump, or simply NAKA, for short) is an enzyme

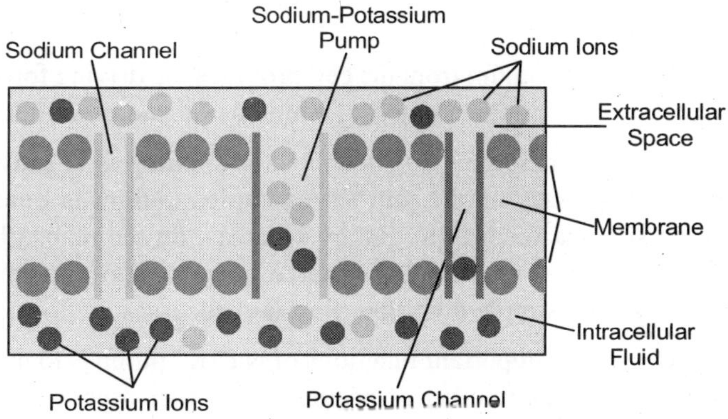

Fig. 5.62 : The Sodium-Potassium Pump

(EC 3.6.3.9) located in the plasma membrane (specifically an electrogenic transmembrane ATPase). It is found in the human cell and is found in all metazoa (animals).

Active transport is responsible for the well-established observation that cells contain relatively high concentrations of potassium ions but low concentrations of sodium ions. The mechanism responsible for this is the sodium-potassium pump which moves these two ions in opposite directions across the plasma membrane. This was investigated by following the passage of radioactively labeled ions across the plasma membrane of certain ones. It was found that the concentrations of sodium and potassium ions on the two other sides of the membrane are interdependent, suggesting that the same carrier transports both ions. It is now known that the carrier is an ATP-ase and that it pumps three sodium ions out of the cell for every two potassium ions pumped in.

The sodium-potassium pump was discovered in the 1950s by a Danish scientist, Jens Skou, who was awarded a Nobel Prize in 1997. It marked an important step forward in our understanding of how ions get into and out of cells, and it has a particular significance for excitable cells such as nervous cells, which depend on it for responding to stimuli and transmitting impulses.

The Na^+/K^+-ATPase helps maintain resting potential, avail transport and regulate cellular volume. It also functions as signal transducer/integrator to regulate MAPK pathway, ROS as well as intracellular calcium.

In order to maintain the cell membrane potential, cells must keep a low concentration of sodium ions and high levels of potassium ions within the cell (intracellular). After the formation of an action potential, when the cell is repolarizing, that is within the cell it is becoming more and more negative as K^+ ions flood out, there is a stage where the membrane potential undershoots its resting membrane potential as K^+ channels take too long to close. This is called hyperpolarization. In order to restore the appropriate concentrations, the sodium-potassium pump pumps 3 sodium ions out by hydrolysing ATP and allows 2 potassium ions in through active transport. The carrier (pump) undergoes a conformational change in order to do this. The end result is that the resting membrane potential is restored and another action potential can occur. However, one must not be mistaken as the Na^+/K^+-ATPase has no direct action in the formation of an action potential. The resting potential avails action potentials of nerves and muscles.

Export of sodium from the cell provides the driving force for several facilitated membrane transport proteins, which import glucose, amino acids and other nutrients into the cell.

Another important task of the Na^+-K^+ pump is to provide an Na^+ gradient that is used by certain carrier processes. In the gut, for example, sodium is transported out of the reabsorbing cell on the blood side via the Na^+-K^+ pump, whereas, on the reabsorbing side, the Na^+-Glucose symporter uses the created Na^+ gradient as a source of energy to import both Na^+ and Glucose, which is far more efficient than simple diffusion. Similar processes are located in the renal tubular system.

One of the important functions of Na^+-K^+ pump is to maintain the volume of the cell. Inside the cell there are a large number of proteins and other organic compounds that cannot escape from the cell. Most, being negatively charged, collect around them a large number of positive ions. All these substances tend to cause the osmosis of water into the cell which, unless checked, can cause the cell to swell up and lyse. The Na^+-K^+ pump is a mechanism to prevent this. The pump transports 3 Na^+ ions out of the cell and in exchange takes 2 K^+ ions into the cell. As the membrane is far less permeable to Na^+ ions than K^+ ions the sodium ions have a tendency to stay there. This represents a continual net loss of ions out of the cell. The opposing osmotic tendency that results operates to drive the water molecules out of the cells. Furthermore, when the cell begins to swell, this automatically activates the Na^+-K^+ pump, which moves still more ions to the exterior.

Within the last decade, many independent labs have demonstrated that in addition to the classical ion transporting, this membrane protein can also relay extracellular ouabain binding signalling into the cell through regulation of protein tyrosine phosphorylation. The downstream signals through ouabain-triggered protein phosphorylation events include the activation of mitogen-activated protein kinase (MAPK) signal cascades, mitochondrial reactive oxygen species (ROS) production as well as activation of phospholipase C (PLC) and inositol triphosphate (IP3) receptor (IP3R) in different intracellular compartments.

Protein-protein interactions play very important role in Na^+-K^+ pump - mediated signal transduction. For example, Na^+-K^+ pump interacts directly with Src, a non-receptor tyrosine kinase, to form receptor complex. Na^+-K^+ pump also interacts with ankyrin, IP3R, PI3K, PLC-gamma and cofilin.

Mechanism

The pump, with bound ATP, binds 3 intracellular Na^+ ions.

- ATP is hydrolyzed, leading to phosphorylation of the pump at a highly conserved aspartate residue and subsequent release of ADP.
- A conformational change in the pump exposes the Na^+ ions to the outside. The phosphorylated form of the pump has a low affinity for Na^+ ions, so they are released.
- The pump binds 2 extracellular K^+ ions. This causes the dephosphorylation of the pump, reverting it to its previous conformational state, transporting the K^+ ions into the cell.

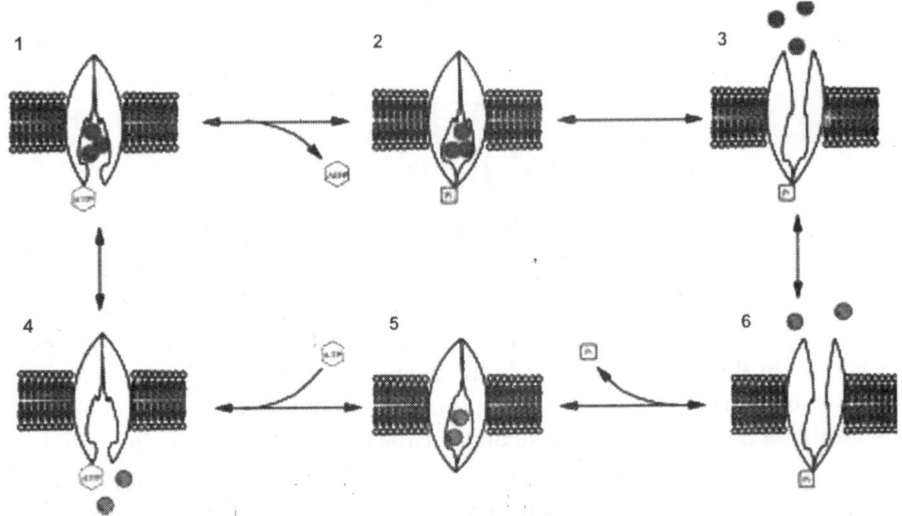

Fig. 5.63 : The Mechanism of the Sodium Pump action

- The unphosphorylated form of the pump has a higher affinity for Na^+ ions than K^+ ions, so the two bound K+ ions are released. ATP binds, and the process starts again.

The Na^+/K^+-ATPase is upregulated by cAMP. Thus, substances causing an increase in cAMP upregulate the Na^+/K^+-ATPase. These include the ligands of the Gs-coupled GPCRs. In contrast, substances causing a decrease in cAMP downregulate the Na^+/K^+-ATPase. These include the ligands of the Gi-coupled GPCRs.

The Na^+-K^+-ATPase can be pharmacologically modified by administrating drugs exogenously.

For instance, Na^+-K^+-ATPase found in the membrane of heart cells is an important target of cardiac glycosides (for example digoxin and ouabain), inotropic drugs used to improve heart performance by increasing its force of contraction. Contraction of any muscle is dependent on a 100- to 10,000-times higher-than-resting intracellular Ca^{2+} concentration, which, as soon as it is put back again on its normal level by a carrier enzyme in the plasma membrane, and a calcium pump in sarcoplasmic reticulum, muscle relaxes.

Since this carrier enzyme (Na^+-Ca^{2+} translocator) uses the Na gradient generated by the Na^+-K^+ pump to remove Ca^{2+} from the intracellular space, slowing down the Na^+-K^+ pump results in a permanently-higher Ca^{2+} level in the muscle, which will eventually lead to stronger contractions.

Genes involed are stated below:

- Alpha: ATP1A1, ATP1A2, ATP1A3, ATP1A4. #1 predominates in kidney. #2 is also known as "alpha(+)"
- Beta: ATP1B1, ATP1B2, ATP1B3, ATP1B4

Functions of Soidum

1. *Fluid Balance*: Sodium maintains crystalloid osmotic pressure of extra cellular fluids and helps in retaining water in ECF (extra cellular fluid)
2. *Neuro Molecular Excitability*: Although other cations Na^+ is also involved in neuromolecular irritability which is given.

$$\text{Neuromolecular Irritability } \alpha = \frac{(K^+)(Na^+)}{(Ca^{++})+(Mg^{++})+(H^+)}$$

3. *Acid-base Balance*: Na^+ - H^+ exchange in renal tubule to acidify the urine.
4. *Maintanance of Viscosity of the Blood:* The salts of Na with globulins are soluble and further Na^+ and K^+ both regulate in maintaining the degree of hydration of the plasma proteins.
5. *Role in Resting Membrane Potential:* Plasma membrane has a poor Na^+ permeability and passive Na^+ in flow through it. Na^+ pump keeps Na^+ conc far higher outside than inside. This separation of charges is called polarization pf the membrane. It creates potential difference of 70-95 milivolt across the membrane and is called resting membrane potential.
6. *Role in Action Potential*: A local depolarization of nerve or muscle fibre is observed in stimulation. This rapidly increases its pemiability to Na^+ causing considerable transmembrane influx of Na^+ down its inward con-gradient.

Hypernatremia or **hypernatraemia** is an electrolyte disturbance that is defined by an elevated sodium level in the blood. Hypernatremia is generally not caused by an excess of sodium, but rather by a relative deficit of free water in the body. For this reason, hypernatremia is often synonymous with the less precise term, dehydration.Water is lost from the body in a variety of ways, including perspiration, insensible losses from breathing, and in the feces and urine. If the amount of water ingested consistently falls below the amount of water lost, the serum sodium level will begin to rise, leading to hypernatremia. Rarely, hypernatremia can result from massive salt ingestion, such as may occur from drinking seawater.Ordinarily, even a small rise in the serum sodium concentration above the normal range results in a strong sensation of thirst, an increase in free water intake, and correction of the abnormality. Therefore, hypernatremia most often occurs in people such as infants, those with impaired mental status, or the elderly, who may have an intact thirst mechanism but are unable to ask for or obtain water.

Common causes of hypernatremia include:

- Inadequate intake of water, typically in elderly or otherwise disabled patients who are unable to take in water as their thirst dictates. This is the most common cause of hypernatremia.
- Inappropriate excretion of water, often in the urine, which can be due to medications like diuretics or lithium or can be due to a medical condition called diabetes insipidus
- Intake of a hypertonic fluid (a fluid with a higher concentration of solutes than the remainder of the body). This is relatively uncommon, though it can occur after a vigorous resuscitation where a patient receives a large volume of a concentrated sodium bicarbonate solution. Ingesting seawater also causes hypernatremia because seawater is hypertonic.
- Mineralcorticoid excess due to a disease state such as Conn's syndrome or Cushing's Syndrome

Symptoms

Clinical manifestations of hypernatremia can be subtle, consisting of lethargy, weakness, irritability, and edema. With more severe elevations of the sodium level, seizures and coma may occur.

Severe symptoms are usually due to acute elevation of the plasma sodium concentration to above 158 mEq/L (normal is typically about 135-145 mEq/L). Values above 180 mEq/L are associated with a high mortality rate, particularly in adults. However such high levels of sodium rarely occur without severe coexisting medical conditions.

Treatment

The cornerstone of treatment is administration of free water to correct the relative water deficit. Water can be replaced orally or intravenously. However, overly rapid correction of hypernatremia is potentially very dangerous. The body (in particular the brain) adapts to the higher sodium concentration. Rapidly lowering the sodium concentration with free water, once this adaptation has occurred, causes water to flow into brain cells and causes them to swell. This can lead to cerebral edema, potentially resulting in seizures, permanent brain damage, or death. Therefore, significant hypernatremia should be treated carefully by a physician or other medical professional with experience in treatment of electrolyte imbalances.

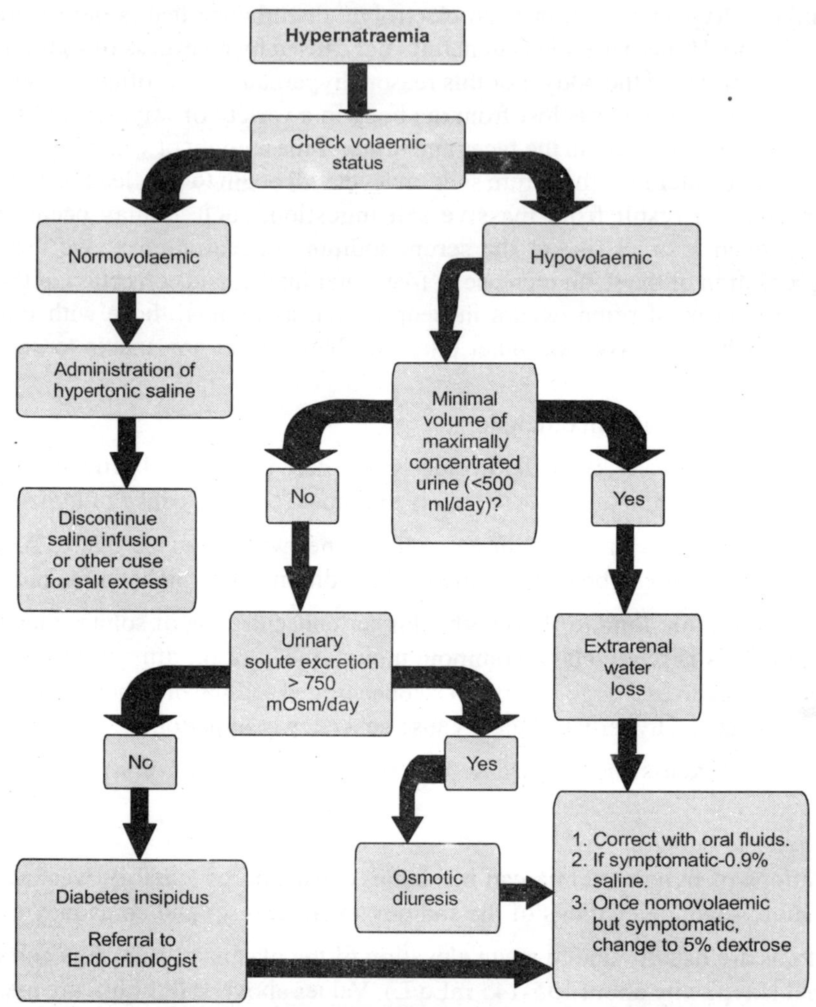

Fig. 5.64 : The Flow Chart for the Management of Hypernatremia

POTASSIUM

Essential Primary Element   Daily Requirement: Male; Female; Children

Potassium is the major intracellular cation. It is widely distributed in the body fluids and tissues as follows:

Whole blood	200mg/dl
Plasma	20mg/dl
Cells	440mg/dl
Muscles tissue	250 - 400mg/dl

Dietary Source

Potassium rich foods	Potassium Content	Sodium content	RDA %*	Calories
Soya flour	1650mg	9mg	47%	450
Black treacle	1500mg	97mg	43%	260
Apricots ready-to-eat	1380mg	15mg	39%	160
Bran Wheat	1160mg	28mg	33%	200
Tomato Puree	1150mg	240mg	33%	70
Sultanas	1050mg	20mg	30%	275
Raisins	1020mg	60mg	30%	270
Potato chips (crisps UK)	1000mg	1000mg	29%	450
All Bran	1000mg	900mg	29%	260
Wheatgerm	950mg	5mg	27%	300
Figs	900mg	60mg	26%	100
Dried mixed fruit	880mg	48mg	25%	230
Bombay Mix	790mg	800mg	23%	500
Papadums	750mg	2400mg	22%	370
Currants	720mg	14mg	22%	270
Sultana Bran	660mg	700mg	19%	300
Seeds average	650mg	20mg	18%	500
Nuts average (unsalted)	600mg	300mg	17%	600
Baked Potato + skin	600mg	12mg	17%	130
Roast Potato	550mg	9mg	16%	160
Oven chips	530mg	50mg	15%	170
Bran Flakes	530mg	1000mg	15%	320
Gammon lean	520mg	2200mg	15%	170
Soya beans boiled	510mg	2mg	15%	140
Plantain boiled	500mg	4mg	14%	112
Raisin Splitz	500mg	10mg	14%	340
Weetos	500mg	300mg	14%	370
Crispbread	500mg	220mg	14%	320
Muesli low salt	450mg	390mg	13%	360
Sardines	430mg	650mg	12%	200
Pilchards	420mg	370mg	12%	125
Veal	420mg	110mg	12%	230
Wholemeal Pasta	400mg	130mg	11%	320
Banana	400mg	1mg	11%	96

Absorption of Potassium

As soon as it is absorbed potassium enters the cells. It is excreted in the urine. The amount of potassium excretion increases when there isan excessive dietary intake of sodium. Average nornmal human body contains 3.6 mols.

Many functions of potassium and sodium are carried out in co-ordination with each other and are common. These function have already been described under sodium. Briefly

1. Influence of the muscular activity.
2. Involved in acid-base balance.
3. It has an important role in cardiac function.
4. Certain enzymes such as pyruvate kinase require K^+ as a cofactor.
5. Involved in neuromolecular irritability and nerve condition process.

Functions of Potassium

Potassium, the most abundant cation in the human body, regulates intracellular enzyme function and neuromuscular tissue excitability. Serum potassium is normally maintained within the narrow range of 3.5 to 5.5 mEq/L. The intracellular-extracellular potassium ratio (K_i/K_e) largely determines neuromuscular tissue excitability. Because only a small portion of potassium is extracellular, neuromuscular tissue excitability is markedly affected by small changes in extracellular potassium. Thus, the body has developed elaborate regulatory mechanisms to maintain potassium homeostasis. Because dietary potassium intake is sporadic and it cannot be rapidly excreted renally, short-term potassium homeostasis occurs via transcellular potassium shifts. Ultimately, long-term maintenance of potassium balance depends on renal excretion of ingested potassium. The illustrations in this chapter review normal transcellular potassium homeostasis as well as mechanisms of renal potassium excretion.

With an understanding of normal potassium balance, disorders of potassium metabolism can be grouped into those that are due to altered intake, altered excretion, and abnormal transcellular distribution.

The diagnostic algorithms that follow allow the reader to limit the potential causes of hyperkalemia and hypokalemia and to reach a diagnosis as efficiently as possible. Finally, clinical manifestations of disorders of potassium metabolism are reviewed, and treatment algorithms for hypokalemia and hyperkalemia are offered. Recently, the molecular defects responsible for a variety of diseases associated with disordered potassium metabolism have been discovered. Hypokalemia and Liddle's syndrome and hyperkalemia and pseudohypoaldosteronism type I result from mutations at different sites on the epithelial sodium channel in the distal tubules. The hypokalemia of Bartter's syndrome can be accounted for by two separate ion transporter defects in the thick ascending limb of Henle's loop. Gitelman's syndrome, a clinical variant of Bartter's syndrome, is caused by a mutation in an ion cotransporter in a completely different segment of the renal tubule. The genetic mutations responsible for hypokalemia in the syndrome of apparent mineralocorticoid excess and glucocorticoidremediable aldosteronism have recently been elucidated and are illustrated below:

Physiology of Potassium Balance : Distribution of Potassium

ECF 350 mEq (10%)	ICF 3150 mEq (90%)
Plasma 15 mEq (0.4%)	Musde 2650 mEq (76%)
Interstitial fluid 35 mEq (1%)	Liver 250 mEq (7%)
Bone 300 mEq (8.6%)	Erythrocytes 250 mEq (7%)
$[K^+]$ = 3.5 – 5.0 mEq/L	[K+] = 140 -150 mEq/L
Urine 90-95 mEq/d	Urine 90-95 mEq/d
Stool 5-10 mEq/d	Stool 5-10mEq/d
Sweat < 5 mEq/d	Sweat < 5 mEq/d

External balance and distribution of potassium. The usual Western diet contains approximately 100 mEq of potassium per day. Under normal circumstances, renal excretion accounts for approximately 90% of daily potassium elimination, the remainder being excreted in stool and (a negligible amount) in sweat. About 90% of total body potassium is located in the intracellular fluid (ICF), the majority in muscle. Although the extracellular fluid (ECF) contains about 10% of total body potassium, less than 1% is located in the plasma. Thus, disorders of potassium metabolism can be classified as those that are due to (1) altered intake, (2) altered elimination, or (3) deranged transcellular potassium shifts.

Factors causing Transcellular Potassium Shifts

Factor	Δ Plasma K+
Acid-base status	
Metabolic acidosis	
Hyperchloremic acidosis	↑↑
Organic acidosis	↔
Respiratory acidosis	↓
Metabolic alkalosis	↑
Respiratory alkalosis	↑
Pancreatic hormones	
Insulin	↓↓
Glucagon	↑
Catecholamines	
β-Adrenergic	↓
α-Adrenergic	↑
Hyperosmolarity	↑
Aldosterone	↓, ↔
Exercise	↑

Extrarenal Potassium Homeostasis: Insulin and Catecholamines

Schematic representation of the cellular mechanisms by which insulin and β-adrenergic stimulation promote potassium uptake by extrarenal tissues. Insulin binding to its receptor results in hyperpolarization of cell membranes, which facilitates potassium uptake. After binding to its receptor, insulin also activates Na^+-K^+-ATPase pumps, resulting in cellular uptake of potassium. The second messenger that mediates this effect has not yet been identified. Catecholamines stimulate cellular potassium uptake via the β_2 adrenergic receptor (β_2R). The generation of cyclic adenosine monophosphate (3′, 5′ cAMP) activates Na^+-K^+-ATPase pumps, causing an influx of potassium in exchange for sodium [10]. By inhibiting the degradation of cyclic AMP, theophylline potentiates catecholaminestimulated potassium uptake, resulting in hypokalemia.

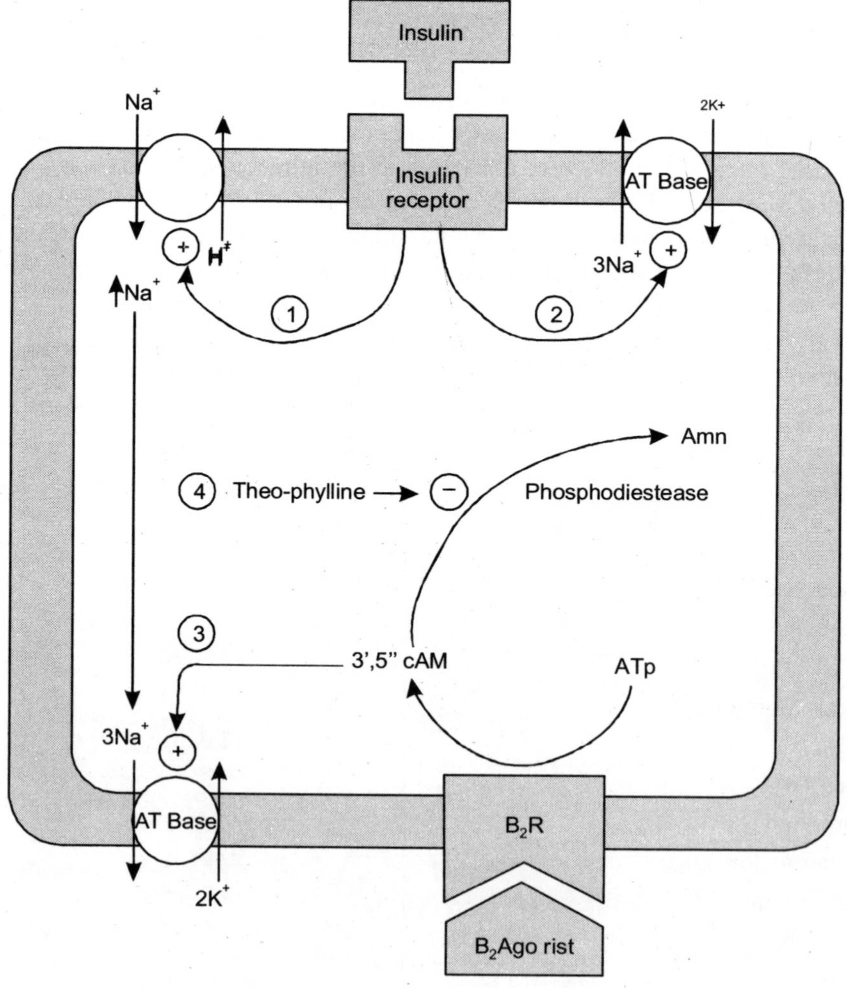

Fig. 5.65 : The Disorder of Potassium Metabolism

Renal Potassium Handling

More than half of filtered potassium is passively reabsorbed by the end of the proximal convolted tubule (PCT). Potassium is then added to tubular fluid in the descending limb of Henle's loop. The major site of active potassium reabsorption is the thick ascending limb of the loop of Henle (TAL), so that, by the end of the distal convoluted tubule (DCT), only 10% to 15% of filtered potassium remains in the tubule lumen. Potassium is secreted mainly by the principal cells of the cortical collecting duct (CCD) and outer medullary collecting duct (OMCD). Potassium reabsorption occurs via the intercalated cells of the medullary collecting duct (MCD). Urinary potassium represents the difference between potassium secreted and potassium reabsorbed. During states of total body potassium depletion, potassium reabsorption is enhanced. Reabsorbed potassium initially enters the medullary interstitium, but then it is secreted into the pars recta (PR) and descending limb of the loop of Henle (TDL). The physiologic role of medullary potassium recycling may be to minimize potassium "backleak" out of the collecting tubule lumen or to enhance renal potassium secretion during states of excess total body potassium. The percentage of filtered potassium remaining in the tubule lumen is indicated in the corresponding nephron segment.

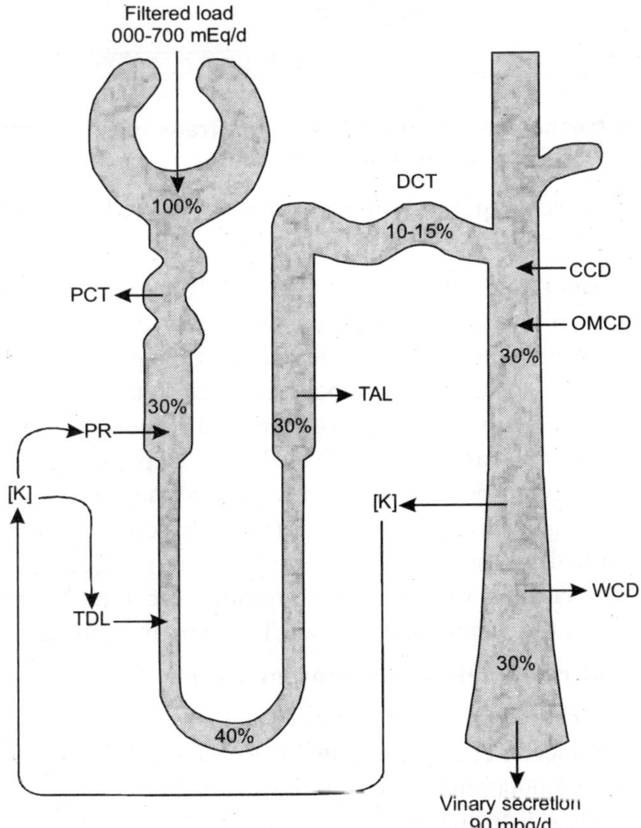

Fig. 5.66 : The Renal Potassium Handling

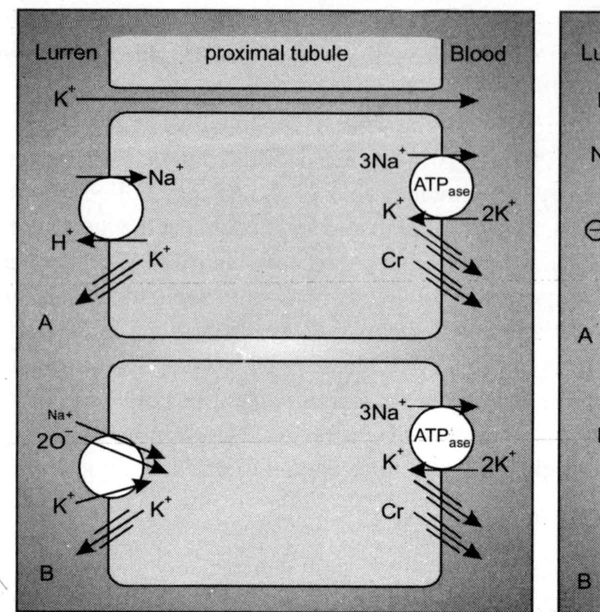

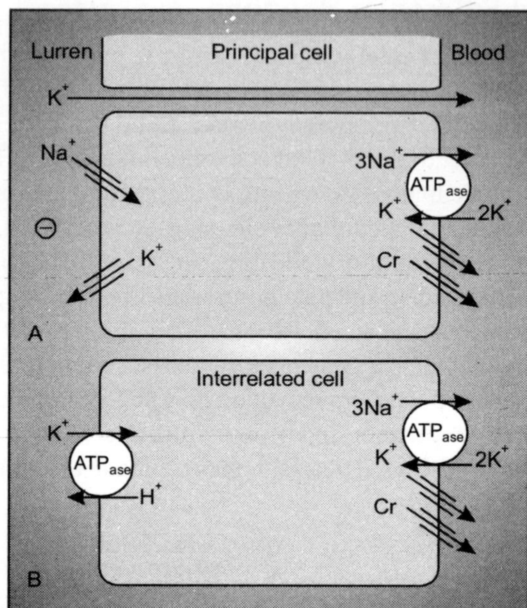

Fig. 5.67 : Cellular Mechanisms of Renal Potassium Transport: Proximal Tubule and Thick Ascending Limb.

Cellular mechanisms of renal potassium transport: proximal tubule and thick ascending limb (Fig. 5.67).

A, Proximal tubule potassium reabsorption is closely coupled to proximal sodium and water transport. Potassium is reabsorbed through both paracellular and cellular pathways. Proximal apical potassium channels are normally almost completely closed. The lumen of the proximal tubule is negative in the early proximal tubule and positive in late proximal tubule segments. Potassium transport is not specifically regulated in this portion of the nephron, but net potassium reabsorption is closely coupled to sodium and water reabsorption. B, In the thick ascending limb of Henle's loop, potassium reabsorption proceeds by electroneutral Na^+-K^+-$2Cl^-$ cotransport in the thick ascending limb, the low intracellular sodium and chloride concentrations providing the driving force for transport. In addition, the positive lumen potential allows some portion of luminal potassium to be reabsorbed via paracellular pathways. The apical potassium channel allows potassium recycling and provides substrate to the apical Na^+-K^+-$2Cl^-$ cotransporter. Loop diuretics act by competing for the Cl^- site on this carrier.

Cellular mechanisms of renal potassium transport: cortical collecting tubule (Fig. 5.68).

A, Principal cells of the cortical collecting duct: apical sodium channels play a key role in potassium secretion by increasing the intracellular sodium available to Na^+-K^+-ATPase pumps and by creating a favorable electrical potential for potassium secretion. Basolateral Na^+-K^+-ATPase creates a favorable concentration gradient for passive diffusion of potassium from cell to lumen through potassium-selective channels. B, Intercalated cells. Under conditions of potassium depletion, the cortical collecting duct becomes a site for net potassium reabsorption. The H^+-K^+-ATPase pump is regulated by potassium intake. Decreases in total body potassium increase pump activity, resulting in enhanced potassium reabsorption. This pump may be partly responsible for the maintenance of metabolic alkalosis in conditions of potassium depletion.

Hypokalemia : Diagnostic Approach

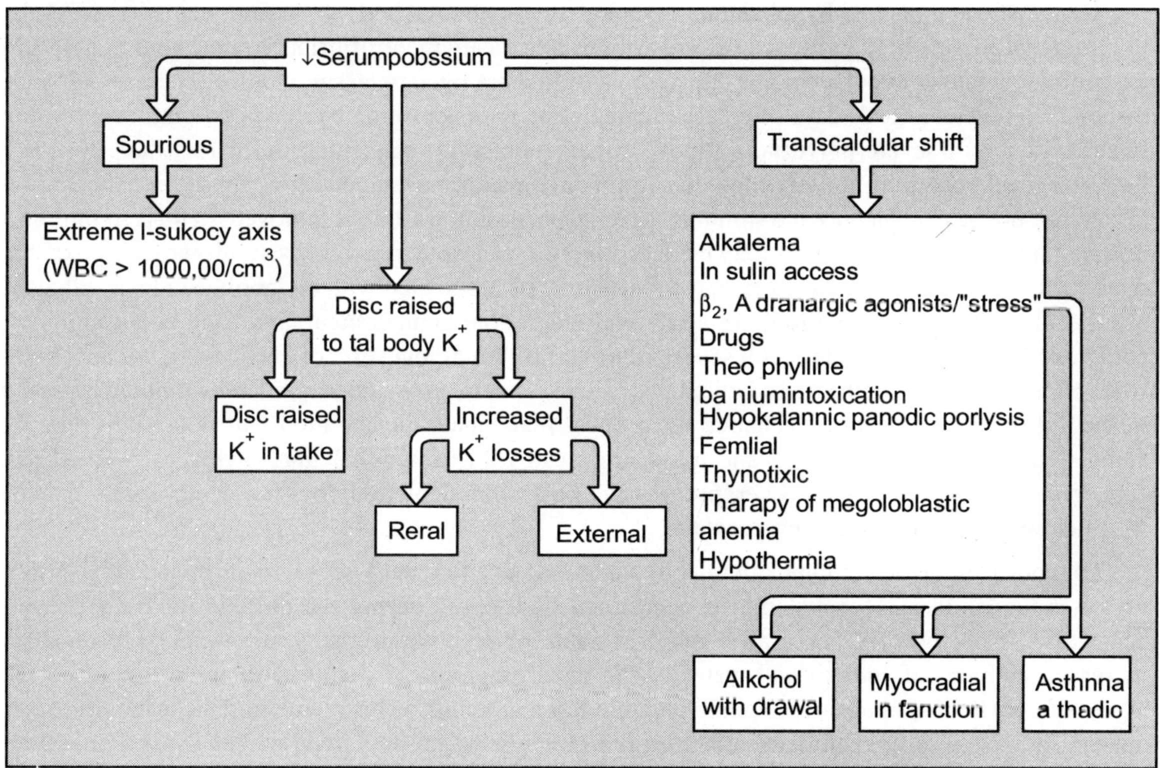

Fig. 5.68 : The Diagnostic Approach of the Hyperkalemia

Overview of diagnostic approach to hypokalemia: Hypokalemia without total body potassium depletion. Hypokalemia can result from transcellular shifts of potassium into cells without total body potassium depletion or from decreases in total body potassium. Perhaps the most dramatic examples occur in catecholamine excess states, as after administration of β_2 adreneric receptor (β_2AR) agonists or during "stress." It is important to note that, during some conditions (e.g., ketoacidosis), transcellular shifts and potassium depletion exist simultaneously. Spurious hypokalemia results when blood specimens from leukemia patients are allowed to stand at room temperature; this results in leukocyte uptake of potassium from serum and artifactual hypokalemia. Patients with spurious hypokalemia do not have clinical manifestations of hypokalemia, as their in vivo serum potassium values are normal. Theophylline poisoning prevents cAMP breakdown (see Fig.3.65). Barium poisoning from the ingestion of soluble barium salts results in severe hypokalemia by blocking channels for exit of potassium from cells. Episodes of hypokalemic periodic paralysis can be precipitated by rest after exercise, carbohydrate meal, stress, or administration of insulin. Hypokalemic periodic paralysis can be inherited as an autosomal-dominant disease or acquired by patients with thyrotoxicosis, especially Chinese males. Therapy of megaloblastic anemia is associated with potassium uptake by newly formed cells, which is occasionally of sufficient magnitude to cause hypokalemia.

Diagnostic approach to hypokalemia: hypokalemia with total body potassium depletion secondary to extrarenal losses. In the absence of redistribution, measurement of urinary potassium is helpful in determining whether hypokalemia is due to renal or to extra renal potassium losses. The normal kidney responds to several (3 to 5) days of potassium depletion with appropriate renal potassium conservation. In the absence of severe polyuria, a "spot" urinary potassium concentration of less than 20 mEq/L indicates renal potassium conservation. In certain circumstances (e.g., diuretics abuse), renal potassium losses may not be evident once the stimulus for renal potassium wasting is removed. In this circumstance, urinary potassium concentrations may be deceptively low despite renal potassiumlosses. Hypokalemia due to colonic villous adenoma or laxative abuse may be associated with metabolic acidosis, alkalosis, or no acid-base disturbance. Stool has a relatively high potassium content, and fecal potassium losses could exceed 100 mEq per day with severe diarrhea. Habitual ingestion of clay (pica), encountered in some parts of the rural southeastern United States, can result in potassium depletion by binding potassium in the gut, much as a cation exchange resin does. Inadequate dietary intake of potassium, like that associated ith anorexia or a "tea and toast" diet, can lead to hypokalemia, owing to delayed renal conservation of potassium; however, progressive potassium depletion does not occur unless intake is well below 15 mEq of potassium per day.

Diagnostic approach to hypokalemia: hypokalemia due to renal losses with normal acidbase status or metabolic acidosis. Hypokalemia is occasionally observed during the diuretic recovery phase of acute tubular necrosis (ATN) or after relief of acute obstructive uropathy, presumably secondary to increased delivery of sodium and water to the distal nephrons. Patients with acute monocytic and myelomonocytic leukemias occasionally excrete large amounts of lysozyme in their urine. Lysozyme appears to have a direct kaliuretic effect on the kidneys (by an undefined mechanism). Penicillin in large doses acts as a poorly reabsorbable anion, resulting in obligate renal potassium wasting. Mechanisms for renal potassium wasting associated with aminoglycosides and cisplatin are illdefined. Hypokalemia in type I renal tubular acidosis is due in part to secondary hyperaldosteronism, whereas type II renal tubular acidosis can result in a defect in potassium reabsorption in the proximal nephrons. Carbonic anhydrase inhibitors result in an acquired form of renal tubular acidosis. Ureterosigmoidostomy results in hypokalemia in 10% to 35% of patients, owing to the sigmoid colon's capacity for net potassium secretion. The osmotic dieresis associated with diabetic ketoacidosis. results in potassium depletion, although patients may initially present with a normal serum potassium value, owing to altered transcellular potassium distribution.

Hypokalemia and magnesium depletion. Hypokalemia and magnesium depletion can occur concurrently in a variety of clinical settings, including diuretic therapy, ketoacidosis, aminoglycoside therapy, and prolonged osmotic diuresis (as with poorly controlled diabetes mellitus). Hypokalemia is also a common finding in patients with congenital magnesium-losing kidney disease. The patient depicted was treated with cisplatin 2 months before presentation. Attempts at oral and intravenous potassium replacement of up to 80 mEq/day were unsuccessful in correcting the hypokalemia. Once serum magnesium was corrected, however, serum potassium quickly normalized.

CHLORINE

Essential Primary Element Daily Intake: 100 – 200mMol

Chlorine is taken in diet as sodium chloride. Many vegetables and meats have small amount of chlorine in the form of chlorides. It is also available in the 'chloronated water which is normally supplied as a process of putrification of water for drinking purpose

Daily Requirement and Distribution

About 100-200mMol is taken in diet as sodiunm chloride

Distribution

Whole Blood	250 mg/dl
Plasma	375 mg/dl
CSF	440 mg/dl
Cells	190 mg/dl
Muscle	40 mg/dl

Absorption

It takes place in small intestines, the mechanism of chloride uptake is not clear, but it appears to depend upon the exchange process with the HCO^{-}_{3} whilst the accompanying sodium exchange for hydroxyl ion.

Extraction

1. Sweat 5mMol/day depends upon whether
2. Through Faces 5mMol/day
3. Renal 100 - 200mMol/day, 99% of the chlorine in the glomarular filtrate in reabsorbed by renal tubes mainly in proximal tubule (60-70%) and then in ascending loop of Hencle (20-25%) followed by the distatt tubule collecting duct (10-15%)

Regulation

Control of absorption and excretion of chloride appears to be similar that of sodium. Increase in clood volume decrease reabsorption of chloride and vice - versa. Plasma level of chloride carry with abd to a great extent depends upon the plasma concentration of HCO_3^-

↓ Na associate with Cl^- ↓

↑ Na associate with Cl^- ↑

↑ HCO_3^- associate with Cl^- ↓

↓ HCO_3^- associate with Cl^{-1} ↑

Functions

1. It is important in the production of HCL in gastric juice
2. It is important in chloride shift CO_2 that is derived as a end product from cellular metabolism, diffuses from tissues through the plasma and into RBC where the CO_2 conc is relatively low, within the erythrocyte the CO_2 is combined with the water to form H_2CO_3 by carbonic anhydrous. The acid then dissociates into a bicarbonate and H^+. The H^+ is buffered by Hb and the HCO_3^- diffuses from the red cell into the plams in exchange of Cl^-

The reverse reactions occur when the erythrocytes reach lungs where CO_2 content exceeds that of the alveoli. Thus arterial and venus plams will differ slightly (2-3 mMol/L) in their constitution.

Clinical Importance

Spot urinary chloride is useful in classifying metabolic alterations into the saline responsive and saline non-responsive type.

Urine $Cl^- < 10$mMol/L-Saline responsive, metabolic alkalosis - vomiting, previous diuretic therapy chloride diarrhea igestion of alkali.

Urine Cl^-. 20m/Mol/L-Saline unresponsive metabolic alkalosis mineralocorticoids. Excess Barter's Syndrome⁻ severe K^+ deficiency, current diuretic therapy*. In certain conditions the urinary chloride concentration is useful tool in assessment of volume deplement, Hyperchloremis - Respiratory alkalosis and metabolic acidisis. Hyperchloremia may be associated with chloride loss, vomiting.

CALCIUM

Most Essential Primary Element	Daily req: Male: - 1000mg; Female: - 1000 - 1500mg; Children : - 210 - 800 mg

Calcium is the most essential primary element mainly utilized for bone and teeth.

[1] Bartter syndrome is a rare inherited defect in the thick ascending limb of the loop of Henle. It is characterized by low potassium levels (hypokalemia), decreased acidity of blood (alkalosis), and normal to low blood pressure. There are two types of Bartter syndrome: neonatal and classic. A closely associated disorder, Gitelman syndrome, is milder than both subtypes of Bartter syndrome.

* A diuretic therapy is the application of any drug that elevates the rate of urination and thus provides a means of forced diuresis.

Dietary Source

It is available in almost all food stuffs, the table below shows the foods containing calcium:

Calcium intakc Chart

	Milligrams Per Day
Infants	
0-5 months	210 mg
6-11 months	270 mg
Children	
1-3 years	500 mg
4-8 years	800 mg
Males/Females	
9-18 years	1,300 mg
19-50 years	1,000 mg
51-70+ years	1,200 mg
50 + years (women not on HRT)	1,500 mg
Pregnant and Lactating	
< 18 years	1,300 mg
19 +	1,000 mg

Calcium-Rich Foods List	Milligrams	Portion
Hard Cheese	300-800	100g
Whitebait	800mg	100g (3.5oz)
Sardines	500mg	100g (3.5oz)
Tofu	500mg+	100g
Milk	300mg	1 glass (8 oz)
Anchovies	300mg	100g
Almonds	245mg	100g (3.5oz)
Milk Chocolate	220mg	100g
Fish Paste	200mg	100g
Most stuff made with milk	up to 200g	100g
Spinach	150mg	1 cupful
Yogurt	150mg	100g
Breads	up to 150	100g
Broccoli	70mg	1 cupful

After Absorption

The total calcium of the body is 25-35 mol (100-170 g). About 99% of its is found in the bones. It exists as carbonates or phosphate of calcium, about 0.5% is in soft tissue and 1% is in ECF (Eosinophil Chemotactic Factor). The normal level of claicum is 9-11mg/dl. The calcium in plasma is of 3 types namely, ionized calcium (diffusion) protein bound calcium and complexed calcium; it is probably complexed with organic acids.

About 40% of total calcium is in ionized form, Albumin is the major protein with which calcouim is bound. All the three forms if calcium plams remain in equilibrium with each other ionized calcium is physiologically active form of calcium.

About 40% of average daily intake of calcium is absorbed by the gut. The actual amount taken upon depends on:

1. Amount of ionized calcium Ca^{2+} available. Levels of Ca^{2+} in the intestinal lumen are reduced by dietary substance that form calcium Acid pH increases the ionization of calcium. Alkalinity promotes complex formation and diminished absorption.
2. Presence of the active metabolites of Vitamin D – 1 $25(OH)_2D_3$

Calcium is secreted into gut* as a normal constituent of bile and intestinal fluids. Fæcal output of calcium couldexceed intestinal absorption of situations, whre the dietr contains high levels of phylates or other sequestrating substances. Under normal circumstances the fæces are not an important excretion rout for calcium.

Action of Calcium in Kidney

Kindney filter about 250mMol of Ca^{++} every day, some 95% of which is reabsorbed by the tubules. The major portion of this filtered Ca^{+2} is taken up by proximal tubule without hormonal regulation. A fine adjustment to that amount reabsorbed occurs in distal tubules under the influence of PTH (PTH – uptake). Plasma level of ionized calcium conc is the principal regulator of PTH secretion by a simple negative feedback mechanism. A threshold level of magnesium is required for the PTH release. Hyper magnesemia inhibits PTH seretion. PTH secretion is also subject to negative feedback by the vitamin D metabolite 1, 25 $(OH)_2D_3$ PTH rapidly stimulates osteoclast activity, the increased bone resorption causing an increase in plams Ca^{+2} and PO_4 Vitamin D_3 plays a permissive role for thid effect.

PTH stimulates more slowly osteoblast activity PTH via c-AMP increases the distal nephron reap reabsorption of calcium and decreased that of PO_4 in the proximal tubule. In bonding so, PTH increases the tubular synthesis and excretion of c-AMP. PTH also stimulates the enzyme complex that converts 25 OHD_3 to $125(OH)_2D_3$ there by increasing calcium update form the gut.

Hypercalcemis stimulates calcitonin katacalin release while hypercalemia has inhibitory effect. Cacitonin strongly inhibits osteoblastic bone resorption. However the role of calcinations in calcium regulation is controvercial.

* The gut flora consists of the microorganisms that normally live in the digestive tract of animals. The gut flora includes much of the human flora. The term "gut flora" is interchangeable with intestinal microflora and intestinal microbiota.

Thyroid, ACTH and prostaglandins have some effect on calevel of plasma (see Fig. 5.65 and 5.70).

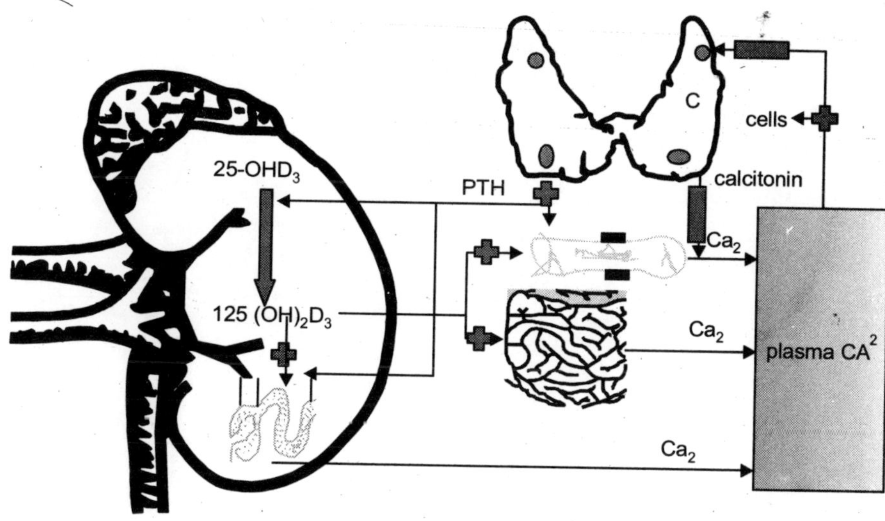

Fig. 5.69 : The Calcium Metabolim in Vital Organs

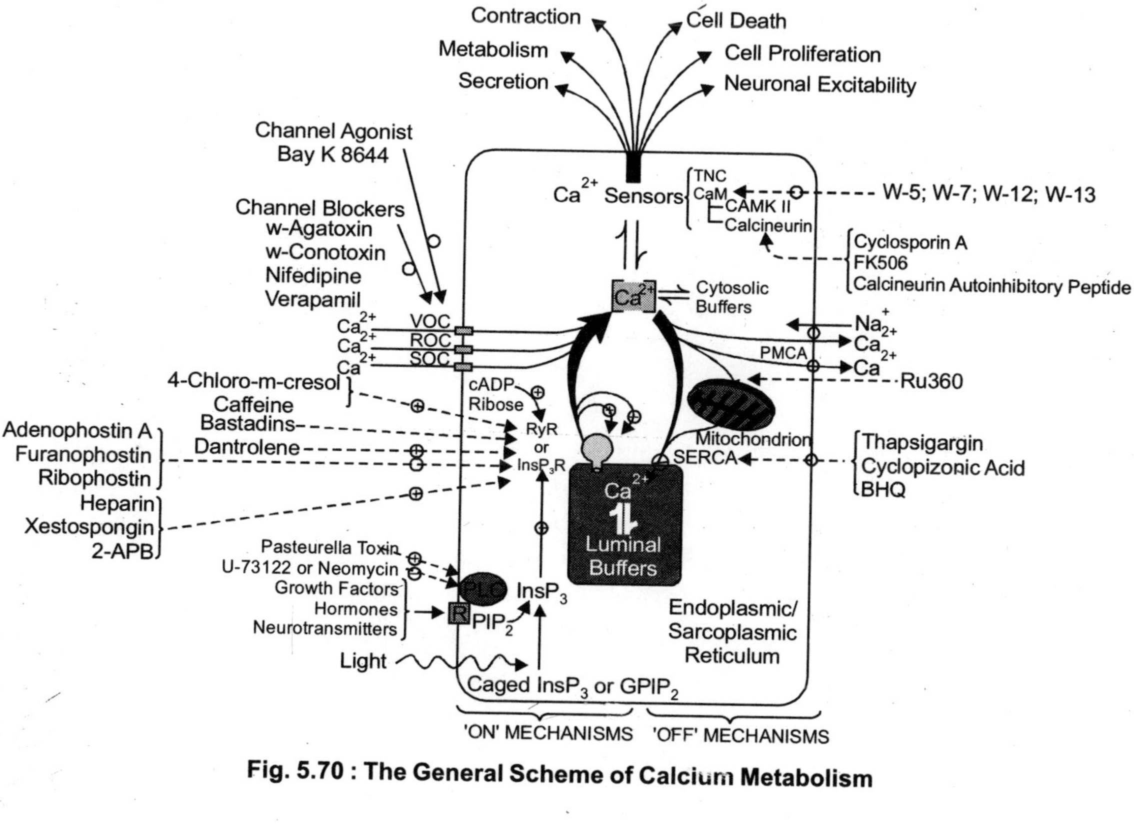

Contraction
Cell Death
Metabolism
Cell Proliferation
Secretion
Neuronal Excitability
Channel Agonist
Bay K 8644
Channel Blockers
w-Agatoxin
w-Conotoxin
Nifedipine
Verapamil
Ca^{2+} Sensors
TNC
CaM
CAMK II
Calcineurin
W-5; W-7; W-12; W-13
Cyclosporin A
FK506
Calcineurin Autoinhibitory Peptide
Cytosolic Buffers
Ca^{2+} VOC
Ca^{2+} ROC
Ca^{2+} SOC
Na^{+}
Ca^{2+}
PMCA
Ca^{2+}
Ru360
4-Chloro-m-cresol
Caffeine
cADP Ribose
Adenophostin A
Furanophostin
Ribophostin
Bastadins
Dantrolene
RyR or $InsP_3R$
Mitochondrion
SERCA
Thapsigargin
Cyclopizonic Acid
BHQ
Heparin
Xestospongin
2-APB
Ca^{2+}
Luminal Buffers
Pasteurella Toxin
U-73122 or Neomycin
PLC
Growth Factors
Hormones
Neurotransmitters
R
PIP_2
$InsP_3$
Endoplasmic/ Sarcoplasmic Reticulum
Light
Caged $InsP_3$ or $GPIP_2$
'ON' MECHANISMS
'OFF' MECHANISMS

Fig. 5.70 : The General Scheme of Calcium Metabolism